Large Eddy Simulation for Compressible Flows

Scientific Computation

For other titles published in this series, go to
www.springer.com/series/718

E. Garnier • N. Adams • P. Sagaut

Large Eddy Simulation for Compressible Flows

Springer

Dr. E. Garnier
Office National d'Études
et de Recherches Aérospatiales
8 rue des Vertugadins
92190 Meudon
France
e-mail: eric.garnier@onera.fr

Prof. Dr. N. Adams
Institute of Aerodynamics
Technische Universität München
85747 Garching
Germany
e-mail: nikolaus.adams@tum.de

Prof. P. Sagaut
Institut Jean Le Rond d'Alembert
Université Pierre et Marie Curie
Boite 162
4, place Jussieu
75252 Paris Cedex 5
France
e-mail: pierre.sagaut@upmc.fr

ISBN 978-94-007-3670-2
e-ISBN 978-90-481-2819-8
DOI 10.1007/978-90-481-2819-8

Softcover reprint of the hardcover 1st edition 2009

Printed on acid-free paper

Springer is part of Springer Science+Business Media (www.springer.com)

Contents

E. Garnier et al., *Large Eddy Simulation for Compressible Flows*,
Scientific Computation,

1

Introduction

Turbulent flows are ubiquitous in most application fields, ranging from engineering to earth sciences and even life sciences. Therefore, simulation of turbulent flows has become a key tool in both fundamental and applied research. The complexity of Navier–Stokes turbulence, which is illustrated by the fact that the number of degrees of freedom of turbulence grows faster than $O(Re^{11/4})$, where Re denotes the Reynolds number, renders the Direct Numerical Simulation (DNS) of turbulence inapplicable to most flows of interest. To alleviate this problem, truncated solutions in both frequency and wavenumber may be sought, whose computational cost is much lower and may ideally be arbitrarily adjusted. The most suitable approach to obtain such a low-cost three-dimensional unsteady simulation of a turbulent flow is Large-Eddy Simulation (LES), which was pioneered to compute meteorological flows in the late 1950s and the early 1960s.

One of the main issues raised by LES is a closure problem: because of the non-linearity of the Navier–Stokes equations, the effect of unresolved scales must be taken into account to recover a reliable description of resolved scales of motion (Chap. 2). This need to close the governing equations of LES has certainly been the main area of investigation since the 1960s, and numerous closures, also referred to as subgrid models, have been proposed. Most existing subgrid models have been built using simplified views of turbulence dynamics, the main physical phenomenon taken into account being the direct kinetic energy cascade from large to small scales that is observed in isotropic turbulence and high-Reynolds fully developed turbulent flows. The most popular paradigm for interscale energy transfer modeling is subgrid viscosity (Chap. 4), which is an easy way to account for the net pumping of resolved kinetic energy by unresolved scales. Other models have been based on mathematical manipulations of governing equations, such as approximate deconvolution, and are, at least theoretically, more general since they are not based on a priori assumptions on turbulence dynamics (Chap. 5). An important observation is that the vast majority of existing works dealing with subgrid modeling is devoted to incompressible flows, the main extension being for variable-density

E. Garnier et al., *Large Eddy Simulation for Compressible Flows*,
Scientific Computation,

flows (e.g. for meteorological flows) and low-speed reacting flows. In most cases, we are faced with ad hoc modifications of subgrid models developed for incompressible flows rather than new models developed *ab initio*.

The case of high-speed compressible flows (Chap. 3), in which compressibility effects are associated with high values of the Mach number, and which may exhibit typical compressible phenomena such as compression shocks is even more problematic, since the issue of accounting for true compressibility effects has been hardly addressed up to now. One of the main objectives of the present book is to analyze existing work and to provide a critical survey of existing closures. The main reasons for the complexity of compressible turbulent flows are:

- The governing equations are more complex: we have five conservation equations supplemented by an equation of state, instead of three momentum equations and a divergence free condition in the incompressible case; there are more nonlinear mechanisms and more unknowns in the compressible case.
- While the subgrid closure issue for the incompressible case is a pure interscale energy transfer modeling problem, the complexity is dramatically increased for the compressible case: one must account for both interscale and intermodal[1] energy transfer. Energy can be transferred from one scale to another, leading to an energy cascade phenomenon, but also from one mode to another (e.g. energy of the vortical modes can be transformed in to acoustic energy or heat). By their very nature, subgrid models developed within the incompressible flow framework do not account for intermodal transfer. It is also worth noting that intermodal energy transfer, but also self-interaction of acoustic and entropy fluctuations is not governed by the same mechanisms as the kinetic energy cascade. Therefore, modeling paradigms such as subgrid viscosity are irrelevant to parametrize them.

Recent works dealing with LES theory have emphasized new important issues. A first one is that the governing equations for LES, which are usually obtained by applying a scale separation operator to the original Navier–Stokes equations, are nothing but a model of what is really done in practical simulations. A real LES simulation is carried out on a given computational grid with a given numerical method. Therefore, the removal of some small scales of the full solution of the exact Navier–Stokes equations originates in a complex combination of truncation in the space-time resolution[2] and numerical errors which is still not well understood. A direct consequence is that governing equations found in the LES literature must be interpreted as ad hoc

[1] We anticipate that compressible turbulent fluctuations can be viewed as combination of three fundamental physical modes: vortical modes, acoustic modes and entropy modes.

[2] This truncation is intuitively understood considering the Nyquist theorem, which states that there exists an upper limit in the spectral content of a finite set of samples.

tools which mimic true LES solutions rather than exact mathematical models (Chap. 6). Pioneering work in this research area has emphasized several open problems:

- The definition of a formal scale separation operator which mimics real LES on bounded domains and which accounts for features of the computational grid and those of the numerical method.
- Discretization errors cannot be neglected, since they can overwhelm subgrid-model effects if dissipative/stabilized numerical methods are used without care.
- Discretization errors and subgrid models can interact, sometime leading to an unexpected increase in result accuracy, due to partial cancellation between discretization and modeling errors.

Once again, the problem is much more complex when compressible flows are addressed. The main reason is that the number of mathematical and physical symmetries of the continuous equations to be preserved by the numerical method is larger than in the incompressible case, and that additional constraints, such as preservation of fundamental thermodynamic laws, arise. Here, both the numerical method and the subgrid models, or at least their sum, should satisfy these new requirements. These new aspects have been hardly considered up to now. Another point is that many popular stabilized methods designed to compute flows with shocks within the RANS framework have been observed to be badly suited for LES purposes, since they are too dissipative.

Another important issue is the proper formulation of boundary conditions for LES (Chap. 7), the main problems being the definition of unsteady turbulent inflow conditions and wall models.[3] Wall models have been investigated in the early 1970s, and since that time several different models have been proposed and assessed. However no genuine extension for compressible flows is available. The main strategy used so far was to assume that subgrid compressibility effects are negligible in the near-wall region if the computational grid is not too coarse. Research on turbulent inflow conditions is much more recent since it has been identified as a key issue only in the late 1990s. Existing work mostly addresses the incompressible flow case, and often is applied directly to compressible flow simulations. Nothing is done to reconstruct acoustic and entropy fluctuations at the turbulent inlet.

In order to illustrate the state-of-the art of modeling applied to flow simulation, three chapters summarize significant applications in the field of LES for compressible flows. Chapter 8 gives an overview of contributions dedicated to subsonic flows, Chap. 9 focuses on applications dedicated to supersonic flows without shock, and Chap. 10 reviews applications with shock turbulence interactions.

[3] A wall model is a specific subgrid model used to prescribe boundary conditions on solid surfaces when the LES grid is too coarse to allow for the use of the usual no-slip boundary condition.

LES of compressible flows remains unexplored, and based on variable-density extensions of models, methods and paradigms developed within the incompressible-flow framework. Limitations of the available compressible LES theory are evident, and may prohibit improvement of the results in many cases. The objective of this book is to provide the reader with a comprehensive state-of-the-art presentation of compressible LES, but also to point out gaps in the theoretical framework, with the hope to help both the fluid engineer in an educated application of compressible LES, and the specialist in further model development.

2

LES Governing Equations

This chapter is divided in five main parts. The first one is devoted to the presentation of the chosen set of equations. The second part deals with the filtering paradigm and its peculiarities in the framework of compressible flows. In particular the question of discontinuities is addressed and the Favre filtering is introduced. Since the formulation of the energy equation is not unique, the third part first presents different popular formulations. Physical assumptions which permit a simplification of the system of equations are discussed. Furthermore, additional relationships relevant to LES modeling are introduced. Finally, in the last part, fundamentals of LES modeling are established and the distinction of the models according to functional and structural approaches is introduced.

2.1 Preliminary Discussion

Large-eddy simulation relies on the idea that some scales of the full turbulent solutions are discarded to obtain a desired reduction in the range of scales required for numerical simulation. More precisely, small scales of the flow are supposed to be more universal (according to the celebrated local isotropy hypothesis by Kolmogorov) and less determined by boundary conditions than the large ones in most engineering applications. Very large scales are sometimes also not directly represented during the computation, their effect must also be modeled. This mesoscale modeling is popular in the field of meteorology and oceanography. Let us first note here that small and large scales are not well defined concepts, which are flow dependent and not accurately determined by the actual theory of LES.

In practice, as all simulation techniques, LES consists of solving the set of governing equations for fluid mechanics (usually the Navier–Stokes equations, possibly supplemented by additional equations) on a discrete grid, i.e. using a finite number of degrees of freedom. The essential idea is that the spatial distribution of the grid nodes implicitly generates a scale separation, since

E. Garnier et al., *Large Eddy Simulation for Compressible Flows*,
Scientific Computation,

scales smaller than a typical scale associated to the grid spacing cannot be captured. It is also worthy noting that numerical schemes used to discretize continuous operators, because they induce a scale-dependent error, introduce an additional scale separation between well resolved scales and poorly resolved ones.

As a consequence, the LES problem make several subranges of scales appearing:

- *represented resolved scales*, which are scales large enough to be accurately captured on the grid with a given numerical method.
- *represented non-resolved scales*, which are scales larger than the mesh size, but which are corrupted by numerical errors. These scales are the smallest represented scales.
- *non-represented scales*, i.e. scales which are too small to be represented on the computational grid.

One of the open problem in the field of LES is to understand and model the existence of these three scale subranges and to write governing equations for them. To address the modeling problem, several mathematical models for the derivation of LES governing equations have been proposed since Leonard in 1973, who introduced the filtering concept for removing small scales to LES.

The filtering concept makes it possible to address some problems analytically, including the closure problem and the definition of boundary conditions. One the other hand the filtering concept introduces some artefacts, i.e. conceptual problems which are not present in the original formulation. An example is the commutation error between the convolution filter and a discretization scheme.

The most popular filter concept found in the literature for LES of compressible flows is the convolution filter approach, which will be extensively used hereafter. Several other concepts have been proposed for incompressible flow simulation, the vast majority of which having not been extended to compressible LES.

2.2 Governing Equations

2.2.1 Fundamental Assumptions

The framework is restricted to compressible gas flows where the continuum hypothesis is valid. This implies that the chosen set of equations will be derived in control volumes that will be large enough to encompass a sufficient number of molecules so that the concept of statistical average hold. The behavior of the fluid can then be described by its macroscopic properties such as its pressure, its density and its velocity. Even if one can expect that the Knudsen number (ratio of the mean free path of the molecules over a characteristic dimension of the flow) be of the order of 1 in shocks, Smits and Dussauge [266] notice that for shocks of reasonable intensity (where the shock thickness is of the order of few mean free paths) the continuum equations for the gas give shock structure in agreement with experiments.

For sake of simplicity, we consider only gaseous fluid: multi-phase flows are not considered. Furthermore, we restrict our discussion to non-reactive mono-species gases. With respect to issues related to combustion the reader may consult Ref. [220]. Moreover, the scope of this monograph is restricted to non-hypersonic flows (Mach < 6 in air) for which dissociation and ionization effects occurring at the molecular level can be neglected. Temperature differences are supposed to be sufficiently weak so that radiative heat transfer can be neglected. Furthermore, a local thermodynamic equilibrium is assumed to hold everywhere in the flow. With the aforementioned assumptions a perfect gas equation of state can be employed. We restrict ourselves to Newtonian fluids for which the dynamic viscosity varies only with temperature. Since we consider non-uniform density fields, gravity effects could appear. Nevertheless, the Froude number which describes the significance of gravity effects as computed to inertial effects is assumed to be negligible regarding the high velocity of the considered flows (Mach > 0.2).

Finally, the compressible Navier-Stokes equations which express the conservation of mass, momentum, and energy are selected as a mathematical model for the fluids considered in this textbook. These differential equations are supplemented by an algebraic equation, the perfect gas equation of state.

2.2.2 Conservative Formulation

The way the energy conservation is expressed in the Navier-Stokes equations is not unique. Formulations exist for the temperature, pressure, enthalpy, internal energy, total energy, and entropy. Nevertheless, the only way to formulate this equation in conservative form is to chose the total energy. The conservative formulation is necessary for capturing possible discontinuities of the flow at the correct velocity in numerical simulations [155].

Using this form, the Navier-Stokes equations can be written as:

$$\frac{\partial \rho}{\partial t} + \frac{\partial \rho u_j}{\partial x_j} = 0, \tag{2.1}$$

$$\frac{\partial \rho u_i}{\partial t} + \frac{\partial \rho u_i u_j}{\partial x_j} + \frac{\partial p}{\partial x_i} = \frac{\partial \sigma_{ij}}{\partial x_j}, \tag{2.2}$$

$$\frac{\partial \rho E}{\partial t} + \frac{\partial (\rho E + p) u_j}{\partial x_j} = \frac{\partial \sigma_{ij} u_i}{\partial x_j} - \frac{\partial}{\partial x_j} q_j, \tag{2.3}$$

where t and x_i are independent variables representing time and spatial coordinates of a Cartesian coordinate system $\mathbf{x}$, respectively. The three components of the velocity vector $\mathbf{u}$ are denoted u_i ($i = 1, 2, 3$). The summation convention over repeated indices applies. The total energy per mass unit E is given by:

$$\rho E = \frac{p}{\gamma - 1} + \frac{1}{2}\rho u_i u_i \tag{2.4}$$

and the density ρ, the pressure p and the static temperature T are linked by the equation of state:

$$p = \rho RT. \tag{2.5}$$

The gas constant is $R = C_p - C_v$ where C_p and C_v are the specific heats at constant pressure and constant volume, respectively. The variation of these specific heats with respect to temperature is very weak and should not be taken into account according to the framework defined in the previous section. For air, R is equal to 287.03 $\mathrm{m}^2\,\mathrm{s}^2\,\mathrm{K}$.

According to the Stokes's hypothesis which assumes that the bulk viscosity can be neglected, the shear-stress tensor for a Newtonian fluid is given by:

$$\sigma_{ij} = 2\mu(T)S_{ij} - \frac{2}{3}\mu(T)\delta_{ij}S_{kk}, \tag{2.6}$$

where S_{ij}, the components of rate-of-strain tensor $S(\mathbf{u})$ are written as:

$$S_{ij} = \frac{1}{2}\left(\frac{\partial u_i}{\partial x_j} + \frac{\partial u_j}{\partial x_i}\right). \tag{2.7}$$

The variation of the dynamic viscosity μ with temperature can be accounted for by the Sutherland's law

$$\frac{\mu(T)}{\mu(T_0)} = \left(\frac{T}{T_0}\right)^{\frac{3}{2}} \frac{T_0 + S_1}{T + S_1}, \quad \text{with } S_1 = 110.4 \text{ K} \tag{2.8}$$

which is valid from 100 K to 1900 K [266]. It is often approximated by the power law

$$\frac{\mu(T)}{\mu(T_0)} = \left(\frac{T}{T_0}\right)^{0.76}, \tag{2.9}$$

which is valid between 150 K and 500 K. The use of these laws introduces an additional non-linearity in the momentum and energy equations.

The heat flux q_j is given by

$$q_j = -\kappa \frac{\partial T}{\partial x_j}, \tag{2.10}$$

where κ is the thermal conductivity which can be expressed as $\kappa = \mu C_p / Pr$. The Prandtl number is the ratio of the kinematic viscosity $\nu = \mu/\rho$ and thermal diffusivity $\kappa/(\rho C_p)$ and, is assumed to be constant equal to 0.72 for air.

2.2.3 Alternative Formulations

Four alternative formulations have been employed in the literature for LES of compressible flows.

The enthalpy form has been used by Erlebacher et al. [75]

$$\frac{\partial \rho h}{\partial t} + \frac{\partial \rho h u_j}{\partial x_j} = \frac{\partial p}{\partial t} + u_j \frac{\partial p}{\partial x_j} + \frac{\partial}{\partial x_j}\left(\kappa \frac{\partial}{\partial x_j} T\right) + \Phi, \tag{2.11}$$

where the enthalpy $h = C_p T$, and the viscous dissipation Φ is defined as:

$$\Phi = \sigma_{ij} \frac{\partial u_i}{\partial x_j}. \tag{2.12}$$

It can be noted that this form includes the temporal derivative of the pressure on the right hand side.

The temperature form has been used by Moin et al. [201]. As the following pressure form, it corresponds to an equation for the internal energy $\epsilon = C_v T$

$$\frac{\partial}{\partial t}(\rho C_v T) + \frac{\partial}{\partial x_j}(\rho u_j C_v T) = -p \frac{\partial u_j}{\partial x_j} + \Phi + \frac{\partial}{\partial x_j}\left(\kappa \frac{\partial T}{\partial x_j}\right). \tag{2.13}$$

The pressure form can be found in the work by Zang et al. [323]

$$\frac{\partial p}{\partial t} + u_j \frac{\partial p}{\partial x_j} + \gamma p \frac{\partial u_j}{\partial x_j} = (\gamma - 1)\Phi + (\gamma - 1)\frac{\partial}{\partial x_j}\left(\kappa \frac{\partial T}{\partial x_j}\right), \tag{2.14}$$

where $\gamma = C_p / C_v$. Its value is fixed to 1.4 for air according to the aforementioned framework.

The entropy form has been employed by Mathew et al. [194].

$$\frac{\partial \rho s}{\partial t} + \frac{\partial \rho s u_j}{\partial x_j} = \frac{1}{T}\left[\Phi + \frac{\partial}{\partial x_j}\left(\kappa \frac{\partial T}{\partial x_j}\right)\right], \tag{2.15}$$

with $s = C_v \ln(p \rho^{-\gamma})$.

2.3 Filtering Operator

Large-eddy simulation is based on the idea of scale separation or filtering with a mathematically well-established formalism. We restrict our presentation here to the fundamental definitions. The entire formalism can be found in Ref. [244]. The specifics for compressible flows such as the notion of filtering of discontinuous flows and the Favre variables are detailed.

2.3.1 Definition

The framework of this section is restricted to the ideal case of homogeneous turbulence. This implies that the filter should respect the physical properties of isotropy and homogeneity. Subsequently, the filter properties are independent of the position and of the orientation of the frame of reference in space. As a result, its cut-off scale is constant and identical in all spatial directions. To address the issue of discontinuous flows, non-centered filters which are not isotropic may be defined. We put the emphasis on isotropic filters on which LES is grounded. Reference [244] provides an extension to inhomogeneous filters.

Scales are separated using a scale high-pass filter which is also a low-pass filter in frequency. Filtering is represented mathematically in physical space as a convolution product. The resolved part $\bar{\phi}(\mathbf{x},t)$ of a space-time variable $\phi(\mathbf{x},t)$ is defined formally by the relation

$$\bar{\phi}(\mathbf{x},t) = \frac{1}{\Delta}\int_{-\infty}^{\infty}\int_{-\infty}^{\infty} G\left(\frac{\mathbf{x}-\boldsymbol{\xi}}{\Delta}, t-t'\right)\phi(\xi,t')dt'd^3\boldsymbol{\xi}, \tag{2.16}$$

where the convolution kernel G is characteristic of the filter used, and is associated with the cut-off scale in space and time, Δ and τ_c, respectively. This relation is denoted symbolically by

$$\bar{\phi} = G \star \phi. \tag{2.17}$$

The dual definition in Fourier space is obtained by multiplying the spectrum $\hat{\phi}(\mathbf{k},\omega)$ of $\phi(\mathbf{x},t)$ by the transfer function $\hat{G}(\mathbf{k},\omega)$ of the kernel $G(\mathbf{x},t)$:

$$\bar{\hat{\phi}}(\mathbf{k},\omega) = \hat{G}(\mathbf{k},\omega)\hat{\phi}(\mathbf{k},\omega), \tag{2.18}$$

or in symbolic form

$$\bar{\hat{\phi}} = \hat{G}\hat{\phi}, \tag{2.19}$$

where $\mathbf{k}$ and ω are wave number and frequency, respectively. The spatial cutoff length Δ is associated to the cutoff wave number k_c and time τ_c with the cutoff frequency ω_c.

The non-resolved part of $\phi(\mathbf{x},t)$, denoted $\phi'(\mathbf{x},t)$ is defined as:

$$\phi'(\mathbf{x},t) = \phi(\mathbf{x},t) - \bar{\phi}(\mathbf{x},t), \tag{2.20}$$

or

$$\phi' = (1-G)\star\phi. \tag{2.21}$$

Fundamental Properties

For further manipulating of the Navier-Stokes equations after filter application, we require the following three properties:

- Consistency

$$\bar{a} = a \longleftrightarrow \int_{-\infty}^{+\infty} \int_{-\infty}^{+\infty} G(\boldsymbol{\xi}, t') d^3\boldsymbol{\xi} dt' = 1. \quad (2.22)$$

- Linearity

$$\overline{\phi + \psi} = \bar{\phi} + \bar{\psi} \quad (2.23)$$

which is satisfied by the convolution form of filtering.
- Commutation with differentiation[1]

$$\overline{\frac{\partial \phi}{\partial s}} = \frac{\partial \bar{\phi}}{\partial s}, \quad s = \mathbf{x}, t \quad (2.24)$$

Introducing the commutator $[f, g]$ of two operators f and g applied to a dummy variable ϕ

$$[f, g](\phi) = f \circ g(\phi) - g \circ f(\phi) = f\left(g(\phi)\right) - g\left(f(\phi)\right) \quad (2.25)$$

the relation (2.24) can be rewritten as

$$\left[G\star, \frac{\partial}{\partial s}\right] = 0. \quad (2.26)$$

This commutator satisfies the Leibniz identity

$$[f \circ g, h] = [f, h] \circ g + f \circ [g, h]. \quad (2.27)$$

The filter G is not *a priori* a Reynolds operator, since the following property of this kind of operator is not satisfied in general

$$\overline{\phi \bar{\psi}} = \bar{\phi}\bar{\psi}. \quad (2.28)$$

The filter $\overline{(.)}$ is not necessarily idempotent (i.e. G is not a projector), and the large scale component of a fluctuating quantity does not vanish

$$\bar{\bar{\phi}} = G \star G \star \phi = G^2 \phi \neq \bar{\phi}, \quad (2.29)$$

$$\overline{\phi'} = G \star (1 - G) \star \phi \neq 0. \quad (2.30)$$

Let us note that some filter can be inverted, leading to the preservation of the information present in the full exact solution. Conversely, if the filter is a Reynolds operator, the inversion is no longer possible since its kernel $ker(G) = \phi'$ is no longer reduced to the zero element. In this case, the filtering induces an irremediable loss of information.

[1] The space commutation property is satisfied only if the domain is unbounded and if the convolution kernel is homogeneous (Δ constant and independent of space). It is however necessary to vary the cut-off length in order to adapt it to the structure of the solution. For example, this adaptation is mandatory for wall-bounded flows for which the filter length scale must diminish close to the wall in order to capture the smallest dynamically active scales. Ghosal and Moin [96] have shown for the case of homogeneous filters that the commutation error is not bounded. They propose a method to guarantee a second-order commutation error. More recently, Vasilyev et al. [299] have defined filters commuting at an arbitrarily high order.

Additional Hypothesis

The framework presented above is very general. In practice, additional constraints are needed. We now assume that the space-time convolution kernel $G(\mathbf{x}-\boldsymbol{\xi}, t-t')$ is obtained by tensorial extension of one-dimensional kernels

$$G\left(\frac{\mathbf{x}-\boldsymbol{\xi}}{\Delta}, t-t'\right) = G_t(t-t')G\left(\frac{\mathbf{x}-\boldsymbol{\xi}}{\Delta}\right) = G_t(t-t') \prod_{i=1,3} G_i\left(\frac{x_i-\xi_i}{\Delta}\right). \tag{2.31}$$

Since up to now there is no example of LES of compressible flow based on temporal filtering, we restrict our discussion to spatial filtering. Mathematically, this additional restriction is expressed by

$$G_t(t-t') = \delta(t-t'). \tag{2.32}$$

Nevertheless, one has to keep in mind that the spatial filtering implies a temporal filtering since the dynamics of the Navier-Stokes equations make it possible to associate a characteristic time scale with a length scale. The latter one, denoted t_c following dimensional argument can be computed as

$$t_c = \Delta/\sqrt{k_c E(k_c)}, \tag{2.33}$$

where $k_c E(k_c)$ is the kinetic energy associated to the cutoff wave number $k_c = \pi/\Delta$. Suppressing the spatial scales corresponding to wave numbers higher than k_c implies the suppression of frequencies higher than the cutoff frequency $\omega_c = 2\pi/t_c$.

Three Classical Filters for Large Eddy Simulation

Three particular convolution filters are commonly used for performing the spatial scale separation, the Box or top hat filter, the Gaussian filter, and the spectral or sharp cutoff filter. Their kernel functions are given in Table 2.1 both in physical and spectral space for one spatial dimension. The parameter ς of the Gaussian filter is generally taken equal to 6. Both Gaussian and Box filters have a compact support in physical space.

Table 2.1. Kernels of three classical filters

Filter	Kernel in physical space G	Kernel in spectral space $\widehat{G}$
Box filter	$G(x-\xi) = \begin{cases} \frac{1}{\Delta} & \text{if} \lvert x-\xi\rvert \le \frac{\Delta}{2} \\ 0 & \text{otherwise} \end{cases}$	$\hat{G}(k) = \frac{\sin(k\Delta/2)}{k\Delta/2}$
Gaussian	$G(x-\xi) = (\frac{\varsigma}{\pi\Delta^2})^{1/2}\exp(\frac{-\varsigma(x-\xi)^2}{\Delta^2})$	$\hat{G}(k) = e^{-(\Delta^2 k^2)/4\varsigma}$
Sharp cutoff	$G(x-\xi) = \frac{\sin(k_c(x-\xi))}{k_c(x-\xi)}$ with $k_c = \frac{\pi}{\Delta}$	$\hat{G}(k) = \begin{cases} 1 & \text{if } \lvert k\rvert < k_c \\ 0 & \text{otherwise} \end{cases}$

Differential Interpretation of the Filters

For filters defined on the compact support $[\Delta\alpha, \Delta\beta]$ (with $\alpha \neq \beta$), the following definition of filtering can be adopted in the one-dimensional case:

$$\bar{\phi}(x,t) = \frac{1}{\Delta}\int_{x-\Delta\beta}^{x-\Delta\alpha} G\left(\frac{x-\xi}{\Delta}\right)\phi(\xi,t)d\xi \tag{2.34}$$

$$= \int_{\alpha}^{\beta} G(z)\phi(x-\Delta z,t)dz \tag{2.35}$$

In order to facilitate the following developments, the change of variable $z = (x-\xi)/\Delta$ was employed to derive (2.35) from (2.34).

To go toward a differential interpretation of the filter, we perform a Taylor series expansion of the $\phi(\xi,t)$ term at (x,t):[2]

$$\phi(\xi,t) = \phi(x,t) + \sum_{l=1}^{\infty} \frac{(\xi-x)^l}{l!}\frac{\partial^l \phi(x,t)}{\partial x^l}. \tag{2.36}$$

With the aforementioned change of variable, equation (2.36) can be recast as

$$\phi(x-\Delta z,t) = \phi(x,t) + \sum_{l=1}^{\infty} \frac{(-1)^l(\Delta z)^l}{l!}\frac{\partial^l \phi(x,t)}{\partial x^l}, \tag{2.37}$$

Introducing this expansion into (2.35) and considering the symmetry and the conservation properties of the constants of the kernel G, we get

$$\bar{\phi}(x,t) = \int_{\alpha}^{\beta} G(z)\phi(x,t)dz + \int_{\alpha}^{\beta}\sum_{l=1}^{\infty}\frac{(-1)^l}{l!}\Delta^l z^l G(z)\frac{\partial^l \phi(x,t)}{\partial x^l}dz$$

$$= \phi(x,t) + \sum_{l=1}^{\infty}\frac{(-1)^l}{l!}\Delta^l M_l\frac{\partial^l \phi(x,t)}{\partial x^l} \tag{2.38}$$

where M_l is the lth-order moment of the convolution kernel

$$M_l = \int_{\alpha}^{\beta} z^l G(z)dz \tag{2.39}$$

Odd moments vanish for a centered kernel. The differential form (2.38) is well posed if and only if $\forall l\ |M_l| < \infty$ meaning that the kernel G decays rapidly in space. The first five non zero moments for both box and Gaussian filters are given in Table 2.2.

---

[2] This implies that the turbulent field is smooth enough so that a Taylor series expansion exists.

Table 2.2. Values of the first five non-zero moments for the Box and Gaussian filters

M_l	$l=0$	$l=2$	$l=4$	$l=6$	$l=8$
Box	1	1/12	1/80	1/448	1/2304
Gaussian	1	1/12	1/48	5/576	35/6912

2.3.2 Discrete Representation of Filters

From practical considerations, the filter must be expressed in a discrete form. The filtered field at the ith grid point $\bar{\phi}_i$ obtained by applying a discrete filter with a $(M+N+1)$ points stencil to the variable ϕ, is formally defined on a uniform grid with mesh size Δx as

$$\bar{\phi}_i = G\phi_i = \sum_{n=-M}^{N} a_n \phi_{i+n}. \tag{2.40}$$

where the real coefficients a_n specify the filter. The preservation of a constant variable is ensured under the condition

$$\sum_{n=-M}^{N} a_n = 1 \tag{2.41}$$

The transfer function of this filter kernel can be expressed as

$$\hat{G}(k) = \sum_{n=-M}^{N} a_n e^{jkn\Delta x}, \quad \text{with } j^2 = -1. \tag{2.42}$$

Introducing the Taylor series for each n

$$\phi_{i\pm n} = \sum_{l=0}^{\infty} \frac{(\pm n\Delta x)^l}{l!} \frac{\partial^l \phi}{\partial x^l}, \tag{2.43}$$

gives on substitution into (2.40)

$$\bar{\phi}_i = \sum_{n=-M}^{N} a_n \sum_{l=0}^{\infty} \frac{n^l \Delta x^l}{l!} \frac{\partial^l \phi}{\partial x^l}, \tag{2.44}$$

which can be recast as

$$\bar{\phi}_i = \left(1 + \sum_{l=1}^{\infty} a_n^* \Delta x^l \frac{\partial^l}{\partial x^l}\right) \phi_i. \tag{2.45}$$

where we have introduced the abbreviation

$$a_n^* = \frac{1}{l!} \sum_{n=-M}^{N} a_n n^l. \tag{2.46}$$

Additionally, (2.38) can be recast by virtue of the parameter $\epsilon = \Delta/\Delta x$ which represents the ratio of the mesh size Δx to the cut-off length scale Δ, as

$$\bar{\phi}(x,t) = \phi(x,t) + \sum_{l=1}^{\infty} \frac{(-1)^l}{l!} (\epsilon \Delta x)^l M_l \frac{\partial^l \phi(x,t)}{\partial x^l}. \tag{2.47}$$

The a_n are obtained as solution of a linear system which results from computing terms of like order in (2.47) and (2.45).

As an example, using the values of the moments M_l provided in Table 2.2 up to $l = 4$, the fourth-order approximation of the Gaussian filter is

$$\bar{\phi}_i = \frac{\epsilon^4 - 4\epsilon^2}{1152} (\phi_{i+2} + \phi_{i-2}) + \frac{16\epsilon^2 - \epsilon^4}{288} (\phi_{i+1} + \phi_{i-1}) + \frac{\epsilon^4 - 20\epsilon^2 + 192}{192} \phi_i. \tag{2.48}$$

Up to second order, the approximation of Box and Gaussian filters is identical, and for $\epsilon = \sqrt{6}$, one can derive the very popular three-point symmetric filter

$$\bar{\phi}_i = \frac{1}{4}(\phi_{i-1} + 2\phi_i + \phi_{i+1}). \tag{2.49}$$

The Simpson integration rule can be applied with $\epsilon = 2$ to obtain

$$\bar{\phi}_i = \frac{1}{6}(\phi_{i-1} + 4\phi_i + \phi_{i+1}). \tag{2.50}$$

In order to ensure that derivation commutes with filtering, one can define high-order commuting filters that have vanishing moments up to an arbitrary order (for boundary conditions, non symmetric filters are considered) [299]. An additional constrain is added to ensure that the transfer function of the filter is null at the cut-off wave number $k_c = \pi/\Delta x$. In discrete form, this leads to

$$\hat{G}(k_c) = \sum_{n=-M}^{N} (-1)^n a_n = 0. \tag{2.51}$$

This kind of filters belongs to the category of linearly constrained filters [299]. Increasing the number of vanishing moments also allows to find a better approximation of a sharp cutoff filter.

Considering the particular case $N = M$, the complex transfer function (2.42) can be decomposed in real and imaginary parts:

$$\Re\{\hat{G}(k)\} = a_0 + \sum_{n=1}^{N} (a_n + a_{-n}) \cos(kn\Delta x), \tag{2.52}$$

$$\Im\{(\hat{G}(k)\} = \sum_{n=1}^{N} (a_n - a_{-n}) \sin(kn\Delta x), \tag{2.53}$$

The latter of which vanishes in the particular case on a symmetric filter ($a_n = a_{-n}$).

For optimal filters, these coefficients are computed as to minimize the functional

$$\int_0^{\pi/\Delta x} (\Re\{\hat{G}(k) - \hat{G}_{target}(k)\})^2 dk + \int_0^{\pi/\Delta x} (\Im\{\hat{G}(k) - \hat{G}_{target}(k)\})^2 dk \quad (2.54)$$

where $G_{target}(k)$ is the target transfer function. Such filters have been proposed by [240, 299].

Finally, implicit filters defined as

$$\sum_{n=-P}^{P} b_n \bar{\phi}_{i+n} = a_n \phi_{i+n}. \quad (2.55)$$

These are an important part of the Approximate Deconvolution Method [277] (see Chap. 5).

2.3.3 Filtering of Discontinuities

As remarked by Lele [165], for a shock wave occurring in a turbulent flow the classical jump conditions hold for the instantaneous flow. Sagaut and Germano [243] have noticed that the usual filtering procedures, based on a central spatial filter that provides information from both sides, when applied around the discontinuity, produce parasitic contributions that affect the filtered quantities. This issue is developed in the following. Let us consider an unsteady fluctuating variable ϕ defined in a region Ω. We consider the case where this variable oscillates around a mean value U_0 in the subdomain Ω_0 and around the mean value U_1 in the subdomain Ω_1, where $\Omega_0 \bigcup \Omega_1 = \Omega$. The two subdomains do not overlapp and have an interface Γ. Using this domain decomposition, we obtain

$$\phi(x,t) = \begin{cases} \phi_0(x,t) = U_0(t) + \varrho_0(x,t) & x \in \Omega_0, \\ \phi_1(x,t) = U_1(t) + \varrho_1(x,t) & x \in \Omega_1 \end{cases} \quad (2.56)$$

where ϱ_p, $p = 0, 1$ represent the "turbulent" contribution around the mean value U_p. We assume in the following for the sake of simplicity that the function G has a compact support denoted as $S(x)$ at point x, i.e.

$$G\left(\frac{x-\xi}{\Delta}\right) = 0 \quad \text{if } \xi \notin S(x). \quad (2.57)$$

Filter applied to $\phi(x,t)$ gives

$$\bar{\phi}(x,t) = \frac{1}{\Delta} \int_\Omega G\left(\frac{x-\xi}{\Delta}\right) \phi(\xi) d\xi \quad (2.58)$$

We assume a central filter, i.e. its kernel is isotropic. By combination of (2.56) and (2.58), we get the following expression for $\bar{\phi}(x,t)$

$$\bar{\phi}(x,t) = \frac{1}{\Delta}\int_{\Omega_0 \cap S(x)} G\left(\frac{x-\xi}{\Delta}\right)(U_0 + \varrho_0(\xi,t))d\xi + \frac{1}{\Delta}\int_{\Omega_1 \cap S(x)} G\left(\frac{x-\xi}{\Delta}\right)(U_1 + \varrho_1(\xi,t))d\xi \tag{2.59}$$

$$= U_0\frac{1}{\Delta}\int_{\Omega_0 \cap S(x)} G\left(\frac{x-\xi}{\Delta}\right)d\xi + \frac{1}{\Delta}\int_{\Omega_0 \cap S(x)} G\left(\frac{x-\xi}{\Delta}\right)\varrho_0(\xi,t)d\xi + U_1\frac{1}{\Delta}\int_{\Omega_1 \cap S(x)} G\left(\frac{x-\xi}{\Delta}\right)d\xi + \frac{1}{\Delta}\int_{\Omega_1 \cap S(x)} G\left(\frac{x-\xi}{\Delta}\right)\varrho_1(\xi,t)d\xi \tag{2.60}$$

$$= \underbrace{[[U]]\frac{1}{\Delta}\int_{\Omega_1 \cap S(x)} G\left(\frac{x-\xi}{\Delta}\right)d\xi + U_0}_{I} + \underbrace{\frac{1}{\Delta}\int_{\Omega_0 \cap S(x)} G\left(\frac{x-\xi}{\Delta}\right)\varrho_0(\xi,t)d\xi + \frac{1}{\Delta}\int_{\Omega_1 \cap S(x)} G\left(\frac{x-\xi}{\Delta}\right)\varrho_1(\xi,t)d\xi}_{II} \tag{2.61}$$

$$= \underbrace{-[[U]]\frac{1}{\Delta}\int_{\Omega_0 \cap S(x)} G\left(\frac{x-\xi}{\Delta}\right)d\xi + U_1}_{III} + \underbrace{\frac{1}{\Delta}\int_{\Omega_0 \cap S(x)} G\left(\frac{x-\xi}{\Delta}\right)\varrho_0(\xi,t)d\xi + \frac{1}{\Delta}\int_{\Omega_1 \cap S(x)} G\left(\frac{x-\xi}{\Delta}\right)\varrho_1(\xi,t)d\xi}_{II} \tag{2.62}$$

where the jump operator is defined as

$$[[U]] = U_1 - U_0. \tag{2.63}$$

It can be seen that terms *I* and *III* in relations (2.61) and (2.62) are not related to "turbulent" fluctuations ϱ_p, $p = 0, 1$, but only to the discontinuity in the mean field. A first look at these terms shows that the filtered variable $\bar{\phi}(x,t)$ is not discontinuous, the sharp interface having been smoothed to become a graded solution over a region of thickness $2R$, where R is the radius of the kernel support S. The subgrid fluctuation $\phi'(x,t) \equiv \phi(x,t) - \bar{\phi}(x,t)$ is therefore equal to

$$\phi'(x,t) = \begin{cases} \varrho_0(x,t) - (II) - [[U]]\frac{1}{\Delta}\int_{\Omega_1 \cap S(x)} G(\frac{x-\xi}{\Delta})d\xi & x \in \Omega_0, \\ \varrho_1(x,t) - (II) + [[U]]\frac{1}{\Delta}\int_{\Omega_1 \cap S(x)} G(\frac{x-\xi}{\Delta})d\xi & x \in \Omega_1. \end{cases} \tag{2.64}$$

It is observed that the jump in the mean field appears as a contribution in the definition of the subgrid fluctuation, which is an artefact of the filtering procedure. In the case of shocks, the contribution of the jump in the fluctuation will dominate the turbulent part of the subgrid fluctuations in most practical applications, rendering subgrid models which rely explicitly on the assumption that the subgrid fluctuations are of turbulent nature as inadequate. Sagaut and Germano [243] have defined non-centered filters which should be used to avoid this unphysical effect.

2.3.4 Filter Associated to the Numerical Method

The accuracy of a numerical scheme is traditionally associated to the order of its truncation error. However, in the framework of LES where the kinetic energy spectrum spreads over a wide range of scales, it seems more appropriate to compute the spectral distribution of the truncation error which can be associated to the filter transfer function in the wavenumber space. The notion of effective (or modified) wave number can be introduced [301]. To this end, the effect of the discretization on a periodic function $e^{j\alpha x}$ for which the exact derivative is $j\alpha e^{j\alpha x}$ is studied.

Consider the approximation of the first derivative $\frac{\partial f}{\partial x}$ at the ith node of a uniform grid

$$\frac{\partial f(x)}{\partial x} \simeq \frac{1}{\Delta x} \sum_{l=-N}^{M} a_l f_{i+l}. \tag{2.65}$$

The Fourier transform of f can be defined as

$$\tilde{f}(\alpha) = \frac{1}{2\pi} \int_{-\infty}^{\infty} f(x) e^{-j\alpha x} dx. \tag{2.66}$$

This transform is applied to both sides of (2.65)

$$j\alpha \tilde{f} \simeq \left(\frac{1}{\Delta x} \sum_{l=-N}^{M} a_l e^{j\alpha l \Delta x} \right) \tilde{f}. \tag{2.67}$$

The quantity

$$\bar{\alpha} = \frac{-j}{\Delta x} \sum_{l=-N}^{M} a_l e^{j\alpha l \Delta x} \tag{2.68}$$

is the modified wave number of the Fourier transform of the finite difference scheme (2.65).

For example, for a second-order accurate centered scheme ($a_1 = a_{-1} = 1/2$), one obtains

$$\bar{\alpha} = \frac{-j}{\Delta x} \left(\frac{1}{2} e^{j\alpha \Delta x} - \frac{1}{2} e^{-j\alpha \Delta x} \right) = \frac{\sin(\alpha \Delta x)}{\Delta x} \tag{2.69}$$

which is second order accurate for small values of α. More generally, for a centered scheme where N=M and $a_l = a_{-l}$, the modified wave number is real. Conversely, if $N \neq M$ or $a_l \neq a_{-l}$, $\bar{\alpha}$ is complex, its imaginary part being associated to the dissipative character of the scheme. This character is shared by every schemes that are able to capture the discontinuities susceptible to occur in compressible flows.

As noted in Ref. [246], the filter transfer function can be written as

$$\hat{G} = \frac{\bar{\alpha}}{\alpha}. \tag{2.70}$$

The convolution kernel of few classical centered schemes are represented in Fig. 2.1 using the general formulae given by Lele [166] for both explicit and implicit (compact) centered schemes up to $M = N = 3$. The spectral scheme is optimal since its kernel is equal to 1. For low order schemes, the errors are large even for small wave numbers (large scales).

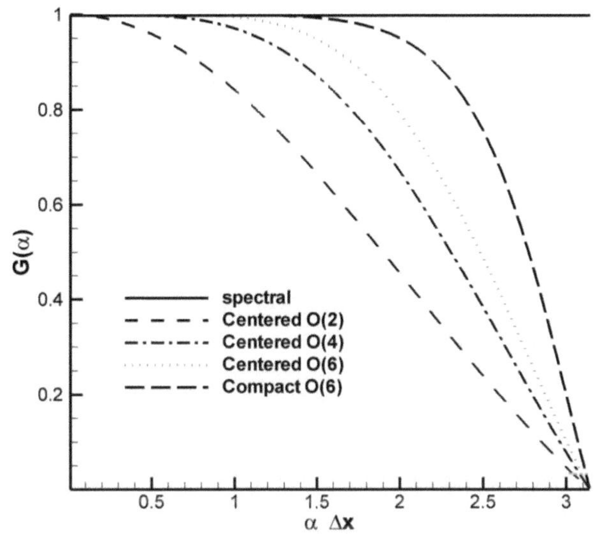

Fig. 2.1. Equivalent convolution kernel for some first order derivative schemes in Fourier space

Nevertheless, since the numerical schemes used are consistent, the numerical error cancels out as Δx tends towards zero. It is then possible to minimize the numerical error by employing a large $\Delta/\Delta x$ ratio. This technique, based on the decoupling of two length scales, is called pre-filtering and aims at ensuring the convergence of the solution regardless of the grid.[3]

[3] This technique is rarely used in practice since it leads to a simulation cost increased by a factor $(\Delta/\Delta x)^4$.

2.3.5 Commutation Error

Every product of two variables ϕ and ψ occurring in the Navier-Stokes equations give rise to a term $\overline{\phi\psi}$ whereas the computable variables are $\bar{\phi}$ and $\bar{\psi}$. Replacing $\overline{\phi\psi}$ by the product $\bar{\phi}\bar{\psi}$ introduces an error which is the commutation error between the filtering operator and the multiplication operator B defined by the bilinear form

$$B(a, b) = ab. \tag{2.71}$$

Using the commutator operator of (2.25), this error can be expressed as

$$\overline{\phi\psi} = \bar{\phi}\bar{\psi} + [G\star, B](\phi, \psi). \tag{2.72}$$

$[G\star, B](\phi, \psi)$ is a subgrid term since it takes account of information contained at subfilter scales. For incompressible flows, the commutation error implies the presence in the momentum equations of subgrid stress scale tensor defined as

$$[G\star, B](u_i, u_j) = \overline{u_i u_j} - \bar{u}_i \bar{u}_j. \tag{2.73}$$

The main modeling effort of the LES community has been concentrated on this term which is the only one arising for the incompressible equations for single phase flows.

2.3.6 Favre Filtering

Most authors dealing with LES of compressible flows have used a change of variable in which filtered variables are weighted by the density.[4] Mathematically, this change of variables is written as

$$\overline{\rho\phi} = \bar{\rho}\tilde{\phi}. \tag{2.74}$$

Any scalar or vector variable can be decomposed into a low frequency part $\tilde{\phi}$ and a high frequency part ϕ''

$$\phi = \tilde{\phi} + \phi''. \tag{2.75}$$

The $(\tilde{\cdot})$ operator is linear but does not commute with the derivative operators in space and time

[4] To our knowledge, the classical LES filtering in compressible flows has been employed by Yoshizawa [320] and Bodony and Lele [23]. In Ref. [320], the analysis of compressible shear flows, realized with the aid of a multiscale Direct-Interaction Approximation, has been limited to weakly compressible flows. In Ref. [23] a simplified set of equations has been employed to compute cold and heated jets at Mach numbers ranging from 0.5 to 1.5. The merit of using the classical LES filtering is not discussed.

$$\widetilde{\frac{\partial \phi}{\partial x_j}} \neq \frac{\partial \tilde{\phi}}{\partial x_j}, \tag{2.76}$$

$$\widetilde{\frac{\partial \phi}{\partial t}} \neq \frac{\partial \tilde{\phi}}{\partial t}. \tag{2.77}$$

If the $\overline{(.)}$ operator is a Reynolds operator, the following relations can be established

$$\overline{\rho \phi''} = 0, \tag{2.78}$$

$$\bar{\phi} - \tilde{\phi} = \overline{\phi''} = -\frac{\overline{\rho' \phi'}}{\bar{\rho}} = -\frac{\overline{\rho' \phi''}}{\bar{\rho}}. \tag{2.79}$$

One can note the similarity of this change of variable with Favre [81] averaging. It is called "Favre filtering", keeping in mind that it is in fact a filtering expressed in terms of Favre variables by a change of variable.

The motivation of using such an operator is twofold:

- The term $\overline{\rho u_i}$ present in the filtered continuity equation can be decomposed following equation (2.72)

$$\overline{\rho u_i} = \bar{\rho}\,\overline{u_i} + [G\star, B](\rho, u_i) = \bar{\rho}\tilde{u}_i. \tag{2.80}$$

  With the filtering defined as in Sect. 2.3.1, the necessary transformation from $\overline{\rho u_i}$ to $\bar{\rho}\,\overline{u_i}$ would lead to another subgrid term which can be avoided by the change of variable (2.74) transforming $\overline{\rho u_i}$ to $\bar{\rho}\tilde{u}_i$.
- The "Favre-filtered" equations are structurally similar to their corresponding non filtered equations (with the exception of the subgrid terms). Moreover, in the framework of a RANS/LES coupling, the similarity with RANS equations can be beneficial. Generally, introducing the operator

$$H(a, b, c) = bc/a \tag{2.81}$$

  it is possible to recast the terms formally written as $\overline{\rho \phi \psi}$ in the following way

$$\overline{\rho \phi \psi} = \bar{\rho}\widetilde{\phi \psi} = \bar{\rho}\tilde{\phi}\tilde{\psi} + [G\star, H](\rho, \rho\phi, \rho\psi). \tag{2.82}$$

  For compressible flows, the subgrid scale (SGS) stress tensor results as

$$\tau_{ij} = [G\star, H](\rho, \rho u_i, \rho u_j) = \bar{\rho}(\widetilde{u_i u_j} - \tilde{u}_i \tilde{u}_j). \tag{2.83}$$

One should notice that this decomposition is not applied to the pressure and density fields. The filtered equation of state can then be written as

$$\bar{p} = \bar{\rho} R \tilde{T} \tag{2.84}$$

Quantities depending only on the temperature such as the enthalpy or the internal energy can also be Favre-filtered.

Theoretically, the use of this change of variable has important consequences concerning interpretation of results. HaMinh and Vandromme [107] remark that density weighted variables are well adapted to the comparison with experimental measurements carried out with hot wire anenometry. Conversely, they are less suitable for a comparison with data obtained by Laser Doppler Anenometry[5] for which the classical filtering operator is appropriate. Smits and Dussauge [266] evaluate the difference between mean velocity profiles of $\tilde{u}$ and $\bar{u}$ to about 1.5% in a $Ma = 3$ turbulent boundary layer. The comparison of LES results with DNS data must also be done with care, and, for proper comparisons DNS data should also be Favre-filtered.

2.3.7 Summary of the Different Type of Filters

As a summary, 3 different classes of filters have been identified in LES.

- The analytical filter represented by a convolution product is used for expressing the filtered Navier–Stokes equations.
- The filter associated to the computational grid. No frequency higher than the Nyquist frequency associated to this grid can be represented in the simulation.
- The filter induced by the numerical scheme. The error committed by approximating the partial derivative operators by discrete operators modifies the computed solution. This kind of error can be computed using the modified wave number formalism [283].

Additionally, it is possible to associate a filter to the model used to approximate the subgrid scale tensor.

The computed solution is the result of these filtering processes constituting the effective filter. When performing a computation the question arises as to what the effective filter is, that governs the dynamics of the numerical solution.

2.4 Formulation of the Filtered Governing Equations

In this section the different ways of filtering the momentum and energy equations are reviewed. Non conservative and conservative forms are presented. Due to the use of the aforementioned "Favre filtering", the continuity equation becomes

$$\frac{\partial \bar{\rho}}{\partial t} + \frac{\partial \bar{\rho}\tilde{u}_j}{\partial x_j} = 0. \tag{2.85}$$

In the particular case of an energy equation based on the total energy, the filtered momentum equation depends on the choice of the filtered pressure which may be different from the quantity obtained by applying a filter to the pressure.

[5] The same is also true for Particule Image Velocimetry.

2.4.1 Enthalpy Formulation

The enthalpy formulation is

$$\frac{\partial \bar{\rho}\tilde{h}}{\partial t} + \frac{\partial \bar{\rho}\tilde{h}\tilde{u}_j}{\partial x_j} - \frac{\partial \bar{p}}{\partial t} - \tilde{u}_j \frac{\partial \bar{p}}{\partial x_j} + \frac{\partial \check{q}_j}{\partial x_j} - \check{\Phi}$$
$$= -\left[\frac{\partial \bar{\rho} C_p(\widetilde{Tu_j} - \tilde{T}\tilde{u}_j)}{\partial x_j} - \left(\overline{u_j \frac{\partial p}{\partial x_j}} - \tilde{u}_j \frac{\partial \bar{p}}{\partial x_j} \right) - (\bar{\Phi} - \check{\Phi}) + \frac{\partial (\overline{q_j} - \check{q}_j)}{\partial x_j} \right] \tag{2.86}$$

where the filtered enthalpy $\tilde{h}$ is equal to $C_p\tilde{T}$ and the filtered computable viscous dissipation $\check{\Phi}$ is defined as

$$\check{\Phi} = \check{\sigma}_{ij} \frac{\partial \tilde{u}_i}{\partial x_j}. \tag{2.87}$$

where

$$\check{\sigma}_{ij} = \mu(\tilde{T}) \left(2\tilde{S}_{ij} - \frac{2}{3}\delta_{ij}\tilde{S}_{kk} \right) \tag{2.88}$$

which depends on the computable rate-of-strain tensor

$$\tilde{S}_{ij} = \frac{1}{2} \left(\frac{\partial \tilde{u}_i}{\partial x_j} + \frac{\partial \tilde{u}_j}{\partial x_i} \right). \tag{2.89}$$

The computable heat flux is

$$\check{q}_j = -\kappa(\tilde{T}) \frac{\partial \tilde{T}}{\partial x_j}. \tag{2.90}$$

The non linearity introduced by Sutherland's law to viscosity and conductivity gives rise to the additional term $\overline{q_j} - \check{q}_j$ in the energy equation.

Using the following decomposition of the filtered pressure-gradient velocity correlation

$$\overline{u_j \frac{\partial p}{\partial x_j}} = \frac{\partial \overline{pu_j}}{\partial x_j} - \overline{p \frac{\partial u_j}{\partial x_j}} \tag{2.91}$$
$$= \frac{\partial \overline{\rho R T u_j}}{\partial x_j} - \overline{p \frac{\partial u_j}{\partial x_j}} \tag{2.92}$$
$$= \frac{\partial \bar{\rho} R \tilde{T} \tilde{u}_j}{\partial x_j} + \frac{\partial \bar{\rho} R(\widetilde{Tu_j} - \tilde{T}\tilde{u}_j)}{\partial x_j} - \overline{p \frac{\partial u_j}{\partial x_j}} \tag{2.93}$$
$$= \bar{p}\frac{\partial \tilde{u}_j}{\partial x_j} + \tilde{u}_j \frac{\partial \bar{p}}{\partial x_j} + \frac{\partial \bar{\rho} R(\widetilde{Tu_j} - \tilde{T}\tilde{u}_j)}{\partial x_j} - \overline{p \frac{\partial u_j}{\partial x_j}} \tag{2.94}$$

in (2.86) leads to the following form for the enthalpy equation

$$\frac{\partial \bar{\rho}\tilde{h}}{\partial t} + \frac{\partial \bar{\rho}\tilde{h}\tilde{u}_j}{\partial x_j} - \frac{\partial \bar{p}}{\partial t} - \tilde{u}_j \frac{\partial \bar{p}}{\partial x_j} + \frac{\partial \check{q}_j}{\partial x_j} - \check{\Phi}$$
$$= -\left[\frac{\partial C_v Q_j}{\partial x_j} + \Pi_{dil} - \epsilon_v + \frac{\partial(\overline{q_j} - \check{q}_j)}{\partial x_j}\right]. \quad (2.95)$$

The SGS temperature flux is defined as

$$Q_j = \bar{\rho}(\widetilde{u_j T} - \tilde{u}_j \tilde{T}). \quad (2.96)$$

The SGS pressure-dilatation can be written as

$$\Pi_{dil} = \overline{p\frac{\partial u_j}{\partial x_j}} - \bar{p}\frac{\partial \tilde{u}_j}{\partial x_j}. \quad (2.97)$$

The SGS viscous dissipation is expressed as

$$\epsilon_v = \bar{\Phi} - \check{\Phi}. \quad (2.98)$$

2.4.2 Temperature Formulation

Applying the filtering operation to (2.13) gives

$$\frac{\partial \bar{\rho} C_v \tilde{T}}{\partial t} + \frac{\partial \bar{\rho}\tilde{u}_j C_v \tilde{T}}{\partial x_j} + \bar{p}\frac{\partial \tilde{u}_j}{\partial x_j} - \check{\Phi} + \frac{\partial \check{q}_j}{\partial x_j}$$
$$= -\left[\frac{\partial \overline{\rho u_j C_v T}}{\partial x_j} - \frac{\partial \bar{\rho}\tilde{u}_j C_v \tilde{T}}{\partial x_j} + \overline{p\frac{\partial u_j}{\partial x_j}} - \bar{p}\frac{\partial \tilde{u}_j}{\partial x_j} - (\bar{\Phi} - \check{\Phi}) + \frac{\partial(\overline{q_j} - \check{q}_j)}{\partial x_j}\right], \quad (2.99)$$

which can be recast as

$$\frac{\partial \bar{\rho} C_v \tilde{T}}{\partial t} + \frac{\partial \bar{\rho}\tilde{u}_j C_v \tilde{T}}{\partial x_j} + \bar{p}\frac{\partial \tilde{u}_j}{\partial x_j} - \check{\Phi} + \frac{\partial \check{q}_j}{\partial x_j}$$
$$= -\left[\frac{\partial C_v Q_j}{\partial x_j} + \Pi_{dil} - \epsilon_v + \frac{\partial(\overline{q_j} - \check{q}_j)}{\partial x_j}\right]. \quad (2.100)$$

The temperature formulation has been used with additional simplifications by Moin et al. [201]. It can also be found in its internal energy form in Ref. [192] by replacing $C_v\tilde{T}$ by $\tilde{\epsilon}$ in the first two terms.

2.4.3 Pressure Formulation

Applying the filtering operator to (2.14) gives

$$\frac{\partial \bar{p}}{\partial t} + \tilde{u}_j \frac{\partial \bar{p}}{\partial x_j} + \gamma \bar{p} \frac{\partial \tilde{u}_j}{\partial x_j} - (\gamma - 1)\breve{\Phi} + (\gamma - 1)\frac{\partial \breve{q}_j}{\partial x_j}$$
$$= -\left[\overline{u_j \frac{\partial p}{\partial x_j}} - \tilde{u}_j \frac{\partial \bar{p}}{\partial x_j} + \gamma \overline{p \frac{\partial u_j}{\partial x_j}} - \gamma \bar{p} \frac{\partial \tilde{u}_j}{\partial x_j} - (\gamma - 1)(\bar{\Phi} - \breve{\Phi}) \right.$$
$$\left. + (\gamma - 1)\frac{\partial (\overline{q_j} - \breve{q}_j)}{\partial x_j} \right]. \tag{2.101}$$

Using the following decomposition:

$$u_j \frac{\partial p}{\partial x_j} + \gamma p \frac{\partial u_j}{\partial x_j} = u_j \frac{\partial p}{\partial x_j} + \gamma \left(\frac{\partial p u_j}{\partial x_j} - u_j \frac{\partial p}{\partial x_j} \right) \tag{2.102}$$
$$= \gamma \frac{\partial p u_j}{\partial x_j} - (\gamma - 1) u_j \frac{\partial p}{\partial x_j} \tag{2.103}$$

which can be employed both globally filtered or only with computable variables, the filtered pressure equation can be written as

$$\frac{\partial \bar{p}}{\partial t} + \tilde{u}_j \frac{\partial \bar{p}}{\partial x_j} + \gamma \bar{p} \frac{\partial \tilde{u}_j}{\partial x_j} - (\gamma - 1)\breve{\Phi} + (\gamma - 1)\frac{\partial \breve{q}_j}{\partial x_j}$$
$$= -\left[\gamma R \frac{\partial Q_j}{\partial x_j} - (\gamma - 1)\left(\overline{u_j \frac{\partial p}{\partial x_j}} - \tilde{u}_j \frac{\partial \bar{p}}{\partial x_j} \right) - (\gamma - 1)(\bar{\Phi} - \breve{\Phi}) \right.$$
$$\left. + (\gamma - 1)\frac{\partial (\overline{q_j} - \breve{q}_j)}{\partial x_j} \right]. \tag{2.104}$$

It is possible to introduce a stronger separation between computable terms and terms to be modeled using (2.94). In this case (2.101) becomes

$$\frac{\partial \bar{p}}{\partial t} + \tilde{u}_j \frac{\partial \bar{p}}{\partial x_j} + \gamma \bar{p} \frac{\partial \tilde{u}_j}{\partial x_j} - (\gamma - 1)\breve{\Phi} + (\gamma - 1)\frac{\partial \breve{q}_j}{\partial x_j}$$
$$= -R \frac{\partial Q_j}{\partial x_j} - (\gamma - 1)\left[\Pi_{dil} - \epsilon_v + \frac{\partial (\overline{q_j} - \breve{q}_j)}{\partial x_j} \right]. \tag{2.105}$$

2.4.4 Entropy Formulation

An equation for the Favre-filtered entropy can be written as

$$\frac{\partial \bar{\rho}\tilde{s}}{\partial t} + \frac{\partial \bar{\rho}\tilde{s}\tilde{u}_j}{\partial x_j} - \frac{1}{\tilde{T}}\left[\breve{\Phi} + \frac{\partial}{\partial x_j}\left(\kappa(\tilde{T}) \frac{\partial \tilde{T}}{\partial x_j} \right) \right]$$
$$= -\frac{\partial \bar{\rho}(\widetilde{s u_j} - \tilde{s}\tilde{u}_j)}{\partial x_j} + \left(\overline{\frac{\Phi}{T}} - \frac{\breve{\Phi}}{\tilde{T}} \right) + \overline{\frac{1}{T}\frac{\partial}{\partial x_j}\left(\kappa(T) \frac{\partial T}{\partial x_j} \right)}$$
$$- \frac{1}{\tilde{T}}\frac{\partial}{\partial x_j}\left(\kappa(\tilde{T}) \frac{\partial \tilde{T}}{\partial x_j} \right), \tag{2.106}$$

with $\tilde{s} = \frac{1}{\bar{\rho}}\overline{\rho C_v \ln(p\rho^{-\gamma})}$. Nevertheless, $\tilde{s}$ can not be easily linked to the computable entropy $\breve{s} = C_v \ln(\bar{p}\bar{\rho}^{-\gamma})$. This reason may explain the fact that the equation for the filtered entropy has not yet been used in the literature in a form similar to (2.106). However, Mathew et al. [194] have used the entropy formulation in the ADM framework (see Chap. 5) which does not require any explicit modeling of the subgrid terms in the right hand side of (2.106).

2.4.5 Filtered Total Energy Equations

Applying the filtering operator to the total energy definition (2.4) leads to the following equation

$$\bar{\rho}\tilde{E} = \frac{\bar{p}}{\gamma - 1} + \frac{1}{2}\bar{\rho}\widetilde{u_i u_i} \tag{2.107}$$

which is not directly computable. This issue has been addressed in the literature using different techniques:

- Ragab and Sreedhar [226], Piomelli [216], Kosovic et al. [147], and Dubois et al. in a simplified form [61] write an equation of evolution for

$$\bar{\rho}\tilde{E} = \frac{\bar{p}}{\gamma - 1} + \frac{1}{2}\bar{\rho}\tilde{u}_i\tilde{u}_i + \frac{\tau_{ii}}{2} \tag{2.108}$$

  which implies that the pressure and the temperature are computed as

$$\bar{p} = (\gamma - 1)\left[\bar{\rho}\tilde{E} - \frac{1}{2}\bar{\rho}\tilde{u}_i\tilde{u}_i - \frac{\tau_{ii}}{2}\right] \tag{2.109}$$

  and

$$\tilde{T} = \frac{(\gamma - 1)}{R}\left[\tilde{E} - \frac{1}{2}\tilde{u}_i\tilde{u}_i - \frac{\tau_{ii}}{2\bar{\rho}}\right], \tag{2.110}$$

  respectively. The equation of state is not affected.
- Vreman in its "system II" [306] introduces a change of variable on the pressure

$$\breve{P} = \bar{p} + \frac{(\gamma - 1)}{2}\tau_{ii} \tag{2.111}$$

  by which the trace of the SGS tensor in the energy equation disappears

$$\bar{\rho}\tilde{E} = \frac{\breve{P}}{(\gamma - 1)} + \frac{1}{2}\bar{\rho}\tilde{u}_i\tilde{u}_i. \tag{2.112}$$

  The temperature is also modified as

$$\breve{T} = \tilde{T} + \frac{\tau_{ii}}{2C_v\bar{\rho}}, \tag{2.113}$$

  leaving the equation of state formally unchanged

$$\breve{P} = \bar{\rho}R\breve{T}. \tag{2.114}$$

- Comte and Lesieur [47, 174] have proposed a different change of variable on the pressure which results in the so-called macro-pressure

$$\bar{\mathcal{P}} = \bar{p} + \frac{1}{3}\tau_{ii}. \tag{2.115}$$

This change of variable is motivated by an analogy with incompressible flows where the isotropic part of the SGS tensor is added to the pressure. The temperature is modified according to (2.113) so that the SGS stress tensor does not appear in the definition of the filtered energy equation if computed with $\breve{T}$. The equation of state takes the form

$$\bar{\mathcal{P}} = \bar{\rho} R \breve{T} - \frac{3\gamma - 5}{6}\tau_{ii}. \tag{2.116}$$

For a monoatomic gas such as argon or helium (for which $\gamma = 5/3$) equation (2.116) recovers the classical form.

- Vreman in its "system I" establishes an equation for the computable energy $\breve{E}$

$$\breve{E} = \frac{\bar{p}}{\gamma - 1} + \frac{1}{2}\bar{\rho}\tilde{u}_i\tilde{u}_i. \tag{2.117}$$

This system does not require any modification of the thermodynamic variables.

A System for $\tilde{E}$, $\bar{p}$, $\tilde{T}$

The system can be written as

$$\frac{\partial \bar{\rho}\tilde{E}}{\partial t} + \frac{\partial (\bar{\rho}\tilde{E} + \bar{p})\tilde{u}_j}{\partial x_j} - \frac{\partial \breve{\sigma}_{ij}\tilde{u}_i}{\partial x_j} + \frac{\partial \breve{q}_j}{\partial x_j}$$
$$= -\frac{\partial}{\partial x_j}[(\overline{\rho u_j E} - \bar{\rho}\tilde{u}_j\tilde{E}) + (\overline{u_j p} - \tilde{u}_j\bar{p}) - (\overline{\sigma_{ij}u_j} - \breve{\sigma}_{ij}\tilde{u}_j) - (\overline{q_j} - \breve{q}_j)]. \tag{2.118}$$

It is possible to regroup the first two SGS terms of the right hand side of (2.118) in the following form

$$(\overline{\rho u_j E} - \bar{\rho}\tilde{u}_j\tilde{E}) + (\overline{u_j p} - \tilde{u}_j\bar{p}) = C_p Q_j + \mathcal{J}_j, \tag{2.119}$$

where

$$\mathcal{J}_j = \frac{1}{2}(\bar{\rho}\widetilde{u_j u_i u_i} - \bar{\rho}\tilde{u}_j\widetilde{u_i u_i}) = \frac{1}{2}(\bar{\rho}\widetilde{u_j u_i u_i} - \bar{\rho}\tilde{u}_j\tilde{u}_i\tilde{u}_i - \tau_{ii}) \tag{2.120}$$

is the SGS turbulent diffusion.

Introducing the SGS viscous diffusion

$$\mathcal{D}_j = \overline{\sigma_{ij}u_j} - \breve{\sigma}_{ij}\tilde{u}_j, \tag{2.121}$$

(2.118) can be rewritten as

$$\frac{\partial \bar{\rho}\tilde{E}}{\partial t} + \frac{\partial (\bar{\rho}\tilde{E} + \bar{p})\tilde{u}_j}{\partial x_j} - \frac{\partial \breve{\sigma}_{ij}\tilde{u}_i}{\partial x_j} + \frac{\partial \breve{q}_j}{\partial x_j} = -\frac{\partial}{\partial x_j}[C_p Q_j + \mathcal{J}_j - \mathcal{D}_j - (\overline{q_j} - \breve{q}_j)]. \quad (2.122)$$

A System for $\tilde{E}$, $\breve{p}$, $\breve{T}$

With the aforementioned change of variable on pressure and temperature, Vreman, in its system II, writes the energy equation as

$$\frac{\partial \bar{\rho}\tilde{E}}{\partial t} + \frac{\partial (\bar{\rho}\tilde{E} + \breve{p})\tilde{u}_j}{\partial x_j} - \frac{\partial \breve{\sigma}_{ij}\tilde{u}_i}{\partial x_j} + \frac{\partial \breve{q}_j}{\partial x_j} = -\frac{\partial}{\partial x_j}(C_p Q_j + \mathcal{J}_j - D_3 - D_4 + D_5), \quad (2.123)$$

with

$$\breve{q}_j = -\kappa(\breve{T})\frac{\partial \breve{T}}{\partial x_j}. \quad (2.124)$$

The term

$$D_3 = \frac{\partial}{\partial x_j}\left(\frac{\gamma - 1}{2}\tau_{ii}\tilde{u}_j\right) \quad (2.125)$$

results from the difference between $\bar{p}$ and $\breve{p}$.

The difference between

$$D_4 = \frac{\partial}{\partial x_j}(\overline{\sigma_{ij} u_i} - \breve{\sigma}_{ij}\tilde{u}_i) \quad (2.126)$$

and

$$D_5 = \frac{\partial}{\partial x_j}(\overline{q_j} - \breve{q}_j) \quad (2.127)$$

and their counterparts in (2.122) arise from the (inexact) replacement of $\tilde{T}$ by $\breve{T}$ in the computable heat flux ($\breve{q}_j$)) and strain of rate tensor $\breve{\sigma}_{ij}$ which introduces additional terms involving τ_{ii}.

A System for $\tilde{E}$, $\bar{\mathcal{P}}$, $\breve{T}$

With this set of variables Comte and Lesieur [47] derive the following energy equation

$$\frac{\partial \bar{\rho}\tilde{E}}{\partial t} + \frac{\partial (\bar{\rho}\tilde{E} + \bar{\mathcal{P}})\tilde{u}_j}{\partial x_j} - \frac{\partial \breve{\sigma}_{ij}\tilde{u}_i}{\partial x_j} + \frac{\partial \breve{q}_j}{\partial x_j} = -\frac{\partial}{\partial x_j}(\mathcal{Q}_j - D_4 + D_5) \quad (2.128)$$

with

$$\mathcal{Q}_j = \overline{(\rho E + p)u_j} - (\bar{\rho}\tilde{E} + \bar{\mathcal{P}})\tilde{u}_j, \quad (2.129)$$

which can be recast in a form similar to (2.119)

$$\mathcal{Q}_j = C_p Q_j + \mathcal{J}_j - \frac{1}{3}\tilde{u}_j \tau_{ii}. \quad (2.130)$$

A System for $\breve{E}$, $\bar{p}$, $\tilde{T}$

Vreman establishes the equation for the computable total energy adding the filtered internal energy equation to the filtered kinetic energy equation

$$\frac{\partial \breve{E}}{\partial t}+\frac{\partial(\breve{E}+\bar{p})\tilde{u}_j}{\partial x_j}-\frac{\partial \breve{\sigma}_{ij}\tilde{u}_i}{\partial x_j}+\frac{\partial \breve{q}_j}{\partial x_j}=-B_1-B_2-B_3+B_4+B_5+B_6-B_7. \quad (2.131)$$

The SGS terms B_i can be written as

$$B_1=\frac{1}{\gamma-1}\frac{\partial}{\partial x_j}(\overline{pu_j}-\bar{p}\tilde{u}_j)=\frac{\partial C_v Q_j}{\partial x_j}, \quad (2.132)$$

$$B_2=\overline{p\frac{\partial u_k}{\partial x_k}}-\bar{p}\frac{\partial \tilde{u}_k}{\partial x_k}=\Pi_{dil}, \quad (2.133)$$

$$B_3=\frac{\partial}{\partial x_j}(\tau_{kj}\tilde{u}_k), \quad (2.134)$$

$$B_4=\tau_{kj}\frac{\partial}{\partial x_j}\tilde{u}_k, \quad (2.135)$$

$$B_5=\overline{\sigma_{kj}\frac{\partial}{\partial x_j}u_k}-\overline{\sigma_{kj}}\frac{\partial}{\partial x_j}\tilde{u}_k=\epsilon_v, \quad (2.136)$$

$$B_6=\frac{\partial}{\partial x_j}(\overline{\sigma_{ij}}\tilde{u}_i-\breve{\sigma}_{ij}\tilde{u}_i)=\frac{\partial \mathcal{D}_j}{\partial x_j}, \quad (2.137)$$

$$B_7=\frac{\partial}{\partial x_j}(\overline{q_j}-\breve{q}_j). \quad (2.138)$$

The terms B_3 and B_4 are regrouped in the original work of Vreman [307]. The terms B_4 and B_5 can not be written in a conservative form. This might have some consequences for the treatment of flows with discontinuities.

2.4.6 Momentum Equations

In the vast majority of the published results, the selected system of equation is based on the filtered pressure $\bar{p}$. The filtered momentum equation is

$$\frac{\partial \bar{\rho}\tilde{u}_i}{\partial t}+\frac{\partial \bar{\rho}\tilde{u}_i\tilde{u}_j}{\partial x_j}+\frac{\partial \bar{p}}{\partial x_i}-\frac{\partial \breve{\sigma}_{ij}}{\partial x_j}=-\frac{\partial \tau_{ij}}{\partial x_j}+\frac{\partial}{\partial x_j}(\overline{\sigma_{ij}}-\breve{\sigma}_{ij}). \quad (2.139)$$

In the particular case of the change of variables introduced by Vreman [306] in its system II, replacing $\bar{p}$ by $\breve{P}$ in (2.139), an additional term $\frac{(\gamma-1)}{2}\tau_{ii}$ occurs on the right hand side of (2.139).
Using the change of variable proposed by Comte and Lesieur [47], $\mathcal{P}$ substitutes $\bar{p}$, and $\tau_{ii}/3$ is subtracted from the right hand side of (2.139). It is equivalent to a replacement of τ_{ij} by its deviatoric part τ_{ij}^d defined as

$$\tau_{ij}^d=\tau_{ij}-\delta_{ij}\tau_{kk}/3. \quad (2.140)$$

For the sake of completeness, one has to mention that in the case of the Vreman's system I $\breve{\sigma}_{ij}$ should be replaced by $\breve{\sigma}_{ij}$ in (2.139).

2.4.7 Simplifying Assumptions

SGS Force Terms

The different forms of the energy equation involve a large number of subgrid terms. Unfortunately, there are only two studies in which the forces of all subgrid terms have been computed using *a priori* tests.[6] The first one by Vreman et al. [306] and Vreman [307] is based on a 2D temporal shear layer. The convective Mach number effect has been investigated in the range 0.2–1.2. The 2D character of these simulations may limit the relevance of the conclusions drawn from this study. The other one is due to Martin et al. [192] who have carried out DNS of freely decaying homogeneous isotropic turbulence at a turbulent Mach number equal to 0.52.

In their study, Vreman et al. have compared the amplitude of the terms associated to resolved and SGS fields for their formulations I and II. Their conclusions are summarized in Table 2.3. The classification (large, medium, small, negligible) is based on the L_2 norm of the different terms of the filtered equations. One order of magnitude separates the norm of each class of terms. One can then expect that this classification may not hold locally.

Table 2.3. Classification of terms in the filtered energy equations

Influence of the term	System I	System II
Large	convective $\overline{NS}$	convective $\overline{NS}$
Medium	diffusive $\overline{NS}$, A_1, B_1, $B_2 = \Pi_{dil}$, B_3	diffusive $\overline{NS}$, C_1, D_1, D_2
Small	B_4, $B_5 = \epsilon_v$	D_3, D_4, D_5
Negligible	$\frac{\partial}{\partial x_j}(\overline{\sigma_{ij}} - \breve{\sigma}_{ij})$, B_6, B_7	$\frac{\partial}{\partial x_j}(\overline{\sigma_{ij}} - \breve{\sigma}_{ij})$

Martin et al. [192] have compared the main SGS terms appearing in the internal energy (2.100) and enthalpy (2.95) equations on the one hand and in the total energy equation (2.122) on the other hand. In the former cases they concluded that Π_{dil} is negligible, ϵ_v is one order of magnitude smaller than the divergence of the SGS heat flux ($C_v Q_j$). In the total energy equation (2.122), the SGS turbulent diffusion is comparable with the divergence of the SGS heat flux ($C_p Q_j$), and the SGS viscous diffusion ($\mathcal{D}_j$) is one order of magnitude smaller than the other terms.

From these two studies one can conclude that the non-linear terms occurring in the viscous terms and in the heat fluxes are small or negligible, depending on the chosen system. In practice, they are neglected by every authors. Specifically, this is equivalent to assume that $\overline{\sigma_{ij}} = \breve{\sigma}_{ij}$ and $\overline{\sigma_{ij} u_i} = \breve{\sigma}_{ij} \tilde{u}_i$. These two studies disagree on the importance of the $B_2 = \Pi_{dil}$ term. However, the respective conclusions are obtained for two different configurations

[6] Each term is computed on a DNS field and filtered on a LES grid. The forces of all terms are then compared.

and for two different systems of equations. In practice, this disagreement appears to have no significance. Martin et al. neglect this term and we will see in Chap. 4 that Vreman et al. model it together with B_1 so that eventually there is no model specific to this term published in the literature.

Small Scales Incompressibility

In order to present the different approaches developed in the literature, it is necessary to introduce the triple decomposition (adapted for compressible flows). A product of filtered terms can be decomposed

$$\overline{\rho\phi\psi} = \overline{\rho(\tilde{\phi} + \phi'')(\tilde{\psi} + \psi'')} \tag{2.141}$$

which can be recast as

$$\bar{\rho}\widetilde{\phi\psi} = \bar{\rho}(\widetilde{\tilde{\phi}\tilde{\psi}} + \widetilde{\tilde{\phi}\psi''} + \widetilde{\tilde{\psi}\phi''} + \widetilde{\phi''\psi''}). \tag{2.142}$$

A subgrid term can be expressed using the triple decomposition

$$\bar{\rho}(\widetilde{\phi\psi} - \tilde{\phi}\tilde{\psi}) = \mathbf{L} + \mathbf{C} + \mathbf{R}, \tag{2.143}$$

where one can distinguish:

- The Leonard term which relates only filtered quantities

$$\mathbf{L} = \bar{\rho}(\widetilde{\tilde{\phi}\tilde{\psi}} - \tilde{\phi}\tilde{\psi}). \tag{2.144}$$

- A cross term which represents the interactions between resolved scales and subgrid scales

$$\mathbf{C} = \bar{\rho}(\widetilde{\tilde{\phi}\psi''} + \widetilde{\tilde{\psi}\phi''}) \tag{2.145}$$

- A Reynolds term which accounts for interactions between subgrid scales

$$\mathbf{R} = \bar{\rho}(\widetilde{\phi''\psi''}). \tag{2.146}$$

The last two terms require modeling.

Restricting now the analysis to the subgrid scale tensor τ_{ij} ($\phi = u_i$ and $\psi = u_j$) and decomposing R_{ij} into an isotropic part R^i_{ij} and a deviatoric part R^d_{ij}, Erlebacher et al. [75] show that

$$R^i_{ij} = -\frac{1}{3}\gamma M^2_{sgs}\bar{p}\delta_{ij}, \tag{2.147}$$

where the subgrid Mach number M_{sgs} is defined as $M_{sgs} = \sqrt{q^2_{sgs}/\gamma R\tilde{T}}$ with $q^2_{sgs} = R_{ii}/\bar{\rho}$. Using DNS of isotropic turbulence these authors have found that the thermodynamic pressure is by far more important than R^i_{ij} for subgrid Mach numbers less than 0.4. The subgrid Mach number being lower than the turbulent Mach number, they also find that it is possible to neglect R^i_{ij} up to turbulent Mach numbers M_t as large as 0.6. This condition is valid for most of supersonic flows.

This subgrid-scales incompressibility hypothesis has been widely used. It has been extended from R^i_{ij} to τ^i_{ij} by many authors [47, 307, 226]. The main argument is that most of the compressibility effects are assumed to affect essentially the large scales. They are accounted for by resolved quantities. In this respect, it is much less restrictive to neglect the compressibility effect on sugbrid scales quantities than on quantities representing the whole turbulence spectrum (as in RANS) since subgrid scales fluctuations contain only a small part (typically a tenth of percent) of the fluctuating energy. One can anticipate that the limit usually taken equal to $M_t = 0.2$ [167] by which the compressibility effect must by taken into account in RANS simulations is not relevant in the framework of LES.

If the isotropic part of the SGS tensor is neglected, $\bar{\mathcal{P}}$ and $\check{P}$ degenerates towards $\bar{p}$, and $\check{T}$ degenerates towards $\tilde{T}$. Additionally, D_3 cancels out and $\mathcal{Q}_j$ can be identified as $C_p Q_j + \mathcal{J}_j$. Consequently, the systems based on $(\tilde{E}, \bar{p}, \tilde{T})$, $(\tilde{E}, \bar{\mathcal{P}}, \check{T})$, $(\tilde{E}, \check{P}, \check{T})$ become identical. The system based on $(\check{E}, \bar{p}, \tilde{T})$ preserves its character, and Vreman et al. use the Table 2.3 to argue that the latter system can be preferred since the contributions coming from the non-linearity in the viscous stresses and the heat fluxes are more important in the formulation II (D_4 and D_5) than in the formulation I (B_6 and B_7). Nevertheless, as already mentioned, these terms being both weak in intensity and difficult to model, they are commonly neglected in practical simulations. For the rest of this textbook, we will assume that this approximation holds. Furthermore, one has to notice that non conservative terms are present in system II. For B_1 and B_2 this is not an issue since we will see in Sect. 4.4 that this terms will be modeled with a conservative approximation. But once a model for τ_{ij} is chosen, B_4 can be computed explicitly and its non-conservative character remains. This consideration has motivated some authors to neglect also B_4. This in agreement with Vreman recommendation of modeling at least terms of "medium" importance (of the same order that the Navier-Stokes diffusive fluxes). According to Table 2.3, this remark concerns B_1, B_2 and B_3. This latter term is in conservative form and its modeling is not an issue since it results directly from the choice of τ_{ij}.

Comte and Lesieur justify their approach noticing that it is less restrictive to assume that the term $\frac{3\gamma-5}{6}\tau_{ii}$ is negligible in the state equation (2.116) than to assume that τ_{ii} is negligible. This statement has been motivated by the fact they used a global model for $\mathcal{Q}_j$ without making the decomposition (2.130), which depends explicitly on τ_{ii} [47].

2.5 Additional Relations for LES of Compressible Flows

This section is devoted to some additional relations which can result in further constraints on subgrid models.

2.5.1 Preservation of Original Symmetries

Governing equations of compressible flow dynamics have one-parameter symmetries which constitute a Lie group. Since LES is assumed to converge continuously towards DNS as the scale separation length vanishes, it is reasonable to require that LES governing equations should have the same symmetries as the unfiltered Navier–Stokes equations.[7] This will lead to a twofold constraint, since both the scale separation operator and subgrid models should be designed to preserve symmetries. Such a constraint has been devised in the incompressible case in a few articles (see Ref. [244] for a comprehensive review), the main conclusion being that symmetry-preserving scale separation operators are rare and that most existing subgrid models for incompressible flows violate at least one of the fundamental symmetries of the incompressible Navier–Stokes equations.

Such an analysis so far has not been performed for compressible LES. The scope of the present section is not to provide an extensive analysis, but to state the symmetries of compressible Navier–Stokes equations,[8] each symmetry being an additional constraint for the design of compressible LES models and theoretical filters. Let us also note that, numerical methods should also preserve symmetries of the continuous equations. This point, however will not be further discussed here, but let us mention the fact that many numerical schemes violate very fundamental symmetries such as Galilean invariance.

We restrict ourself to the case of a perfect gas. The symmetries are summarized in Table 2.4. The different cases correspond to possible choices of physical variables with respect to the symmetry. In the most general case, viscosity (and therefore diffusivity) is considered as an autonomous variable. As simplification, it can be considered as a function of temperature (e.g. through the Sutherland law), or a constant parameter. The ultimate simplification consists in considering inviscid fluids, i.e. the symmetries of the Euler equations.

[7] Note that the use of statistical averaging operator may result in a change of the fundamental symmetries of a system, an illustrative example being the loss of time reversal symmetry in statistical thermodynamics: while individual molecule behavior may be time-reversed, the mean behavior of an ensemble of molecules obeys the second law of thermodynamics.

[8] The full set of one-parameter symmetries of compressible Navier–Stokes equations is unpublished to the knowledge of the authors. The full Lie group of symmetries displayed in this section was determined by D. Razafindralandy and A. Hamdouni [227], whose contribution is gratefully acknowledged.

Table 2.4. One-parameter Lie group of symmetries of compressible Navier–Stokes equations. In the Basic Case, time, space, velocity, pressure, density and molecular viscosity are assumed to vary independently. Diffusivity is tied to viscosity assuming that the Prandtl number is constant. This case can be simplified assuming that the viscosity is a temperature-dependent variable. The problem is further simplified assuming that $\mu = \kappa = 0$. The parameter of the transformation is denoted a in the scalar case and $\mathbf{a}$ is the vector case. $\mathbf{R}$ is a 3D time-independent rotation matrix

Symmetry name	Definition	Basic case	$\mu = \mu(T)$	Constant μ	$\mu = \kappa = 0$
Time shift	$(t, \mathbf{x}, \mathbf{u}, p, \rho, \mu) \longrightarrow (t + a, \mathbf{x}, \mathbf{u}, p, \rho, \mu)$	yes	yes	yes	yes
Space shift	$(t, \mathbf{x}, \mathbf{u}, p, \rho, \mu) \longrightarrow (t, \mathbf{x} + \mathbf{a}, \mathbf{u}, p, \rho, \mu)$	yes	yes	yes	yes
Galilean transform	$(t, \mathbf{x}, \mathbf{u}, p, \rho, \mu) \longrightarrow (t, \mathbf{x} + \mathbf{a}t, \mathbf{u} + \mathbf{a}, p, \rho, \mu)$	yes	yes	yes	yes
3D rotation	$(t, \mathbf{x}, \mathbf{u}, p, \rho, \mu) \longrightarrow (t, \mathbf{Rx}, \mathbf{Ru}, p, \rho, \mu)$	yes	yes	yes	yes
Scaling 1	$(t, \mathbf{x}, \mathbf{u}, p, \rho, \mu) \longrightarrow (e^{2a}t, e^{a}\mathbf{x}, e^{-a}\mathbf{u}, e^{-2a}p, \rho, \mu)$	yes	no	yes	yes
Scaling 2	$(t, \mathbf{x}, \mathbf{u}, p, \rho, \mu) \longrightarrow (t, e^{a}\mathbf{x}, e^{a}\mathbf{u}, p, e^{-2a}\rho, \mu)$	yes	no	yes	yes
(Scaling 1) ∘ (Scaling 2)	$(t, \mathbf{x}, \mathbf{u}, p, \rho, \mu) \longrightarrow (e^{a}t, e^{a}\mathbf{x}, \mathbf{u}, p, e^{-a}\rho, \mu)$	yes	yes	yes	yes
Scaling 3	$(t, \mathbf{x}, \mathbf{u}, p, \rho, \mu) \longrightarrow (t, \mathbf{x}, \mathbf{u}, e^{a}p, e^{a}\rho, e^{a}\mu)$	yes	no	no	no
Scaling 4	$(t, \mathbf{x}, \mathbf{u}, p, \rho, \mu = 0) \longrightarrow (t, \mathbf{x}, \mathbf{u}, e^{a}p, e^{a}\rho, \mu = 0)$	no	no	no	yes

2.5.2 Discontinuity Jump Relations for LES

Shock Modeling and Jump Relations

The present discussion will be restricted to the inviscid case for the sake of simplicity. The rationale for that is that viscous effects are negligible compared to other physical mechanisms during the interaction (as can be proved a posteriori by comparing theoretical results with DNS and experimental results), and that relaxation times associated to vibrational, rotational and translational energy modes of the molecules are very small with respect to macroscopic turbulent time scales. Therefore, the shock is modeled as a surface discontinuity with zero thickness. An important consequence is that the shock has no intrinsic time or length scale, and its corrugation is entirely governed by incident fluctuations. Its effects are entirely represented by the Rankine–Hugoniot jump conditions for the mass, momentum and energy:

$$[[\rho u_n]] = 0, \tag{2.148}$$

$$[[\rho u_n^2 + p]] = 0, \tag{2.149}$$

$$[[\boldsymbol{u}_t]] = 0, \tag{2.150}$$

$$\left[\left[e + \frac{p}{\rho} + u^2\right]\right] = [[H]] = 0, \tag{2.151}$$

where H is the stagnation enthalpy and $\boldsymbol{u}$ is the velocity in the reference frame of the shock wave, i.e. $\boldsymbol{u} = \boldsymbol{v} - \boldsymbol{u}_s$ where $\boldsymbol{v}$ and $\boldsymbol{u}_s$ are the fluid velocity and the shock speed in the laboratory frame, respectively. Subscripts n and t are related to the normal and tangential components of vector fields with respect to the shock wave, respectively

$$u_n \equiv \boldsymbol{u} \cdot \boldsymbol{n}, \qquad \boldsymbol{u}_t \equiv \boldsymbol{n} \times (\boldsymbol{u} \times \boldsymbol{n}), \qquad \boldsymbol{u} = u_n \boldsymbol{n} + \boldsymbol{u}_t, \tag{2.152}$$

where $\boldsymbol{n}$ is the shock normal unit vector.

An exact general jump condition for the vorticity can be derived from the relations given above [110]. First noting that the vorticity vector $\boldsymbol{\Omega} = \nabla \times \boldsymbol{u}$ can be decomposed as $\boldsymbol{\Omega} = \Omega_n \boldsymbol{n} + \boldsymbol{\Omega}_t$ with

$$\Omega_n = (\nabla \times \boldsymbol{u}_t)_n \tag{2.153}$$

and

$$\boldsymbol{\Omega}_t = \boldsymbol{n} \times \left(\frac{\partial \boldsymbol{u}_t}{\partial n} + \boldsymbol{u}_t \cdot \nabla \boldsymbol{n} - \nabla_{||} u_n \right) \tag{2.154}$$

where $\nabla_{||}$ denotes the tangential (with respect to the shock surface) part of the gradient operator, one obtains the following vorticity jump conditions in unsteady flows in which the shock experiences deformations

$$[[\Omega_n]] = 0, \tag{2.155}$$

$$[[\boldsymbol{\Omega}_t]] = \boldsymbol{n} \times \left(\nabla_{||}(\rho u_n) \left[\left[\frac{1}{\rho}\right]\right] - \frac{1}{\rho u_n}[[\rho]](D_{||}\boldsymbol{u}_t + u_s D_{||}\boldsymbol{n}) \right), \qquad (2.156)$$

with

$$D_{||}\boldsymbol{u}_t = \left(\frac{d\boldsymbol{u}_t}{dt}\right)_t + \boldsymbol{u}_t \cdot \nabla_{||}\boldsymbol{u}_t = \left(\frac{\partial \boldsymbol{u}_t}{\partial t} + u_s \frac{\partial \boldsymbol{u}_t}{\partial n}\right)_t + \boldsymbol{u}_t \cdot \nabla_{||}\boldsymbol{u}_t \qquad (2.157)$$

and

$$D_{||}\boldsymbol{n} = \frac{d\boldsymbol{n}}{dt} + \boldsymbol{u}_t \cdot \nabla_{||}\boldsymbol{n} = -\nabla_{||}u_s + \boldsymbol{u}_t \cdot \nabla_{||}\boldsymbol{n}. \qquad (2.158)$$

It can be seen that the normal component of the vorticity is continuous across the shock, while the jump of the tangential component depends on the density jump, the tangential velocity and the shock wave deformation. In steady flows, the jump condition for the tangential vorticity simplifies as

$$[[\boldsymbol{\Omega}_t]] = \boldsymbol{n} \times \left(\nabla_{||}(\rho u_n) \left[\left[\frac{1}{\rho}\right]\right] - \frac{1}{\rho u_n}[[\rho]]\, \boldsymbol{u}_t \cdot \nabla_{||}\boldsymbol{u}_t \right). \qquad (2.159)$$

Filtered Jump Relations and Associated Constrains on Subgrid Terms

The question of deriving pseudo-jump relation for coarse resolution simulations, such as LES, is not a trivial task since several fundamental issues arise.

First, one has to decide if the LES solution can exhibit discontinuities. If it is assumed that LES governing equations originate from the application of a smoothing (i.e. regularizing) kernel to the exact equations, discontinuities are transformed into regions with large gradients but finite thickness. Therefore, jump conditions no longer hold, and must be replaced by classical global conservation laws. Such global relations are obtained by performing a volume integration of the LES governing equations over a control cell that encompasses the initial discontinuity.

The second issue comes from the relation between the grid size and the scale separation length. In almost all published works, authors have considered these lengthscales to be equal or very close. As a consequence, the large gradient region cannot be accurately computed on the grid, due to numerical errors. Therefore, jump relations are explicitly or implicitly used in practice to design shock-capturing techniques which yield entropic solutions. Here, the coupling between numerical discretization and the continuous LES formalism is obvious. It is worth noting that in the case of reacting flows with flames, the *thickened flame* approach has been proposed (see e.g. [220]) to allow for an accurate description of the dynamics inside the flame front, but the approach is not fully consistent in the sense that the thickened flame and the filtered turbulent field are not be obtained using a unique filtering operator. Extension of this approach to general discontinuities remains to be done.

A third issue is that pseudo-jump relations introduce additional constraints on subgrid models, which are not taken into account in most subgrid model derivations. To illustrate this point, let us consider the simple case in which LES pseudo-jump relations are obtained in a straightforward manner by applying the scale separation operator to the exact jump relations (2.148)–(2.151). One obtains (different forms can be obtained selecting other sets of filtered variables)

$$[[\overline{\rho u_n}]] = [[\overline{\rho \boldsymbol{u}} \cdot \bar{\boldsymbol{n}}]] + \underbrace{[[\overline{\rho u_n} - \overline{\rho \boldsymbol{u}} \cdot \bar{\boldsymbol{n}}]]}_{\text{subgrid}} = 0, \tag{2.160}$$

$$[[\overline{\rho u_n^2} + \bar{p}]] = \left[\left[\frac{(\overline{\rho \boldsymbol{u}} \cdot \bar{\boldsymbol{n}})^2}{\bar{\rho}} + \bar{p}\right]\right] + \underbrace{\left[\left[\overline{\rho u_n^2} - \frac{(\overline{\rho \boldsymbol{u}} \cdot \bar{\boldsymbol{n}})^2}{\bar{\rho}}\right]\right]}_{\text{subgrid}} = 0, \tag{2.161}$$

$$[[\overline{\boldsymbol{u}_t}]] = [[\overline{\rho \boldsymbol{u}} \cdot \bar{\boldsymbol{t}}]] + \underbrace{[[\overline{\rho u_t} - \overline{\rho \boldsymbol{u}} \cdot \bar{\boldsymbol{t}}]]}_{\text{subgrid}} = 0, \tag{2.162}$$

$$\left[\left[\bar{e} + \overline{\left(\frac{p}{\rho}\right)} + \overline{u^2}\right]\right] = \left[\left[\bar{e} + \left(\frac{\bar{p}}{\bar{\rho}}\right) + \bar{\boldsymbol{u}} \cdot \bar{\boldsymbol{u}}\right]\right]$$
$$+ \underbrace{\left[\left[\left(\overline{\left(\frac{p}{\rho}\right)} - \left(\frac{\bar{p}}{\bar{\rho}}\right)\right) + (\overline{u^2} - \bar{\boldsymbol{u}} \cdot \bar{\boldsymbol{u}})\right]\right]}_{\text{subgrid}} = 0. \tag{2.163}$$

It is observed that subgrid terms contribute to the jump relations which hold for the resolved scales. Therefore, the subgrid jump terms should ideally be taken into account to recover a fully satisfactory behavior of the computed solution in the vicinity of shock waves. It is worthy noting that subgrid jump terms differ from those found in the governing filtered equations.

2.5.3 Second Law of Thermodynamics

Compressible flow simulations raise the question of the compatibility of the computed solution with fundamental laws of thermodynamics. We now discuss the case of the second law of thermodynamics. It is worth noting that it is a non linear relation. Therefore, the filtered field does not a priori obey it, in the same way that it does not fulfill the original Navier–Stokes equations since nonlinearities give rise to subgrid residuals. As a consequence, the LES solution obeys new extended thermodynamic constraints, which are obtained by applying the scale-separation operator to the classical thermodynamic laws.

The Clausius-Duhem entropy inequality, using (2.15), can be recast as

$$\frac{1}{T}\left[\Phi + \frac{\partial}{\partial x_j}\left(\kappa \frac{\partial T}{\partial x_j}\right)\right] \geq 0. \tag{2.164}$$

Multiplying this relation by T and applying a *positive* scale-separation operator, one obtains

$$\bar{\Phi} + \overline{\frac{\partial}{\partial x_j}\left(\kappa \frac{\partial T}{\partial x_j}\right)} \geq 0 \tag{2.165}$$

which appears as an exact extension of the second law of thermodynamics for LES. It can be further refined by including the resolved and subgrid contributions

$$\breve{\Phi} + \epsilon_v + \frac{\partial \breve{q}_j}{\partial x_j} + B_7 \geq 0. \tag{2.166}$$

This last equation shows that the subgrid viscous dissipation ϵ_v (defined in (2.98)) and the subgrid viscous heat flux B_7 (see (2.138)) cannot be computed independently, since they are bounded by the second law of thermodynamics. Subgrid models which satisfy (2.166) can be referred to as *thermodynamically consistent* subgrid models, by analogy with previous works carried out within the RANS framework [9, 237, 238].

2.6 Model Construction

In the previous sections, we have shown that the reduction of the solution complexity (number of degree of freedom) in space and time by the filtering process results in coupling terms between resolved scales and subfilter scales that must be closed by an appropriate form of modeling. The modeling process consists in approximating the coupling terms on the basis of the information contained solely in the resolved scales. Among all the SGS terms, τ_{ij} possesses a particular status since it is the only term which appears in the equations for an incompressible isothermal fluid. One can anticipate that it will also play an important role for compressible flows.

2.6.1 Basic Hypothesis

Subgrid modeling usually is based on the following hypothesis: *If subgrid scales exist, then the flow is locally (in space and time) turbulent.* Consequently, the subgrid models will be built on the known properties of turbulent flows that will be summarized in chapter 3. Before discussing the various ways of modeling the subgrid terms, we have to set some constraints [244]. The subgrid modeling must be done in compliance with two types of constraints:

- Physical constraints. The model must be consistent from the viewpoint of the phenomenon being modeled, i.e.:
  - Conserve the basic properties of the underlying equations, such as Galilean invariance and asymptotic behavior;
  - Vanish wherever the exact solution exhibits no small scales corresponding to the subgrid scales;
  - Induce an effect of the same kind (dispersive or dissipative, for example) as the modeled terms;
  - Not destroy the dynamics of the resolved scales, and thus especially not inhibit the flow driving mechanisms.

- Numerical constraints. A subgrid model can only be thought of as part of a numerical simulation method, and must consequently:
  - Be of acceptable algorithmic cost, and especially be local in time and space;
  - Not destabilize the numerical simulation;
  - Be insensitive to discretization, i.e. the physical effects induced theoretically by the model must not be inhibited by the discretization.

2.6.2 Modeling Strategies

The problem of subgrid modeling consists in taking the interaction with the fluctuating field ϕ' into account in the evolution equation of the filtered field $\bar{\phi}$. Two modeling strategies exist [244]:

- *Structural modeling* of the subgrid term, which consists in making the best approximation of the modeled terms by constructing from an evaluation of $\bar{\phi}$ or a formal series expansion.
- *Functional modeling*, which consists in modeling the action of the subgrid terms on the quantity $\bar{\phi}$ and not the modeled term itself, i.e. introducing a dissipative or dispersive term, for example, that has a similar effect but not necessarily the same structure.

The structural approach requires no knowledge of the nature of the scale interaction, but does require sufficient knowledge of the structure of the small scales, and one of the two following conditions has to be met:

- The dynamics of the equation being computed leads to a universal form of the small scales (and therefore to their structural independence from the resolved motion, as all that remains to be determined is their energy level).
- The dynamics of the equation induces a sufficiently strong and simple scale correlation for the structure of the subgrid scales to be deduced from the information contained in the resolved field. This require both a knowledge of the nature of the scale interaction and an universal character of the small scales.

The distinction between these two types of modeling is fundamental and structures the presentation of subgrid models in Chaps. 4 and 5.

3

Compressible Turbulence Dynamics

3.1 Scope and Content of This Chapter

The scope of this chapter is to present the essential features of compressible turbulence dynamics which are relevant for subgrid modeling. It is not intended to provide the reader with a general discussion of compressibility effects on turbulence. The emphasis is rather placed on differences in scale interactions compared to the incompressible case. The interested reader is referred to relevant books and review articles [245, 266, 40] for a detailed survey of compressible turbulence dynamics.

A first conclusion drawn from works dealing with compressibility effects on turbulence is that three different flow categories must be distinguished for the above purpose:

- Shear flows without shocks. In most free shear flows (e.g. mixing layers, jets, wakes, ...) the governing mechanisms such as kinetic energy production and anisotropy production are tied to large-scale structures, which must be directly captured in Large-Eddy Simulation to recover a reliable description of the flow. The key dynamic phenomena being directly captured, their possible modification by compressibility effects will also be directly taken into account. For this class of flows, one may therefore expect that a simple, variable-density extension of subgrid models originally designed for incompressible flows may be sufficient.
- Flows with shocks. The shock phenomenon is associated with very small scales, which are much smaller than the scales of motion (a few tens of molecule mean free-path for strong shocks). The very concept of Direct Numerical Simulation is questionable for such flows, since the Navier–Stokes equations are not always relevant to describe interior dynamics of the shock wave. The Navier–Stokes solution gives an accurate shock profile for weak shock waves with $M_1 \leq 1.2$–1.3. In flows including shocks, at least two new "small" lengthscales must be considered: the shock thickness and a characteristic shock corrugation length. In practical LES, it appears

E. Garnier et al., *Large Eddy Simulation for Compressible Flows*,
Scientific Computation,

that these two lengthscales are much smaller than the turbulence integral lengthscale, and even than the Taylor microscale. The following empirical estimate for the shock thickness δ was given by Moin and Mahesh [202]

$$\frac{\eta}{\delta} \simeq 0.13 \frac{M_1 - 1}{M_t} \sqrt{Re_\lambda}, \tag{3.1}$$

where M_1 and M_t are the upstream Mach number and the upstream turbulent Mach number, respectively. η and Re_λ are the Kolmogorov lengthscale and the Taylor-micro-scale Reynolds number, respectively. A detailed analysis of the shock-turbulence interaction dynamics along with its consequences for subgrid modeling will be given in this chapter (see Sect. 3.4).

- Flows without shock and turbulence production. In such a flow, the main dynamic mechanism is the nonlinear transfer of kinetic energy between the scales of motion. Compressibility may alter the dynamics because new physical mechanisms arise which are not present in the incompressible case. Following Kovazsnay's decomposition [148, 42], these new mechanisms are represented by the acoustic mode and the entropy mode. The vorticity mode exists in both the compressible and the incompressible cases. The sensitivity of these flows to compressibility effects may be considerable. The possible modifications of scale interactions along with mode couplings must therefore be carefully analyzed to be able to draw relevant conclusions on the requirements for subgrid models for compressible flows. This point is illustrated in Sect. 3.5 considering compressible isotropic turbulence.

This chapter is organized as follows. First, the Kovasznay decomposition for turbulent compressible fluctuations is introduced in Sect. 3.2. This decomposition is the corner stone of the analysis of compressibility effects in most cases. Subsequently, shock-turbulence interaction is analyzed. It is worth noticing that the emphasis will be placed on flows in which turbulent structures cross the shock wave. Other cases can be considered as turbulence affected by a continuous pressure/density gradient in a region with finite thickness which can be computed unlike true shock-turbulence interaction. Shock-isotropic-turbulence interaction will be considered to analyze this class of flows. The last part of the chapter is dedicated to flows without shock and turbulence production. Here again, isotropic compressible turbulence will be considered. It will be shown that several physical regimes can be identified, in which scale and modal interactions may vary with possible impact on subgrid modeling.

3.2 Kovasznay Decomposition of Turbulent Fluctuations

3.2.1 Kovasznay's Linear Decomposition

The first question which arises when dealing with compressible turbulent flows is how to characterize the compressibility effects on turbulence. Before answering this question it is important to remark that incompressible flows are fully described by the velocity field as the pressure is directly coupled to the velocity by the divergence-free condition. For compressible turbulence this is no longer true since pressure is now given by an equation of state and at least one additional state variable is required to describe the solution.

A common way to answer this question is to split the observed fluctuations as the sum of a compressible part and an incompressible part, the latter being often understood as the part of the solution which satisfies the incompressible Navier–Stokes equations, the former being defined as the remainder. Unfortunately, no general decomposition based on this approach leading to tractable and useful analysis has been proposed up to now. The main reason is that such a decomposition does not explicitly distinguish between acoustic waves and other compressible phenomena.

To remedy this problem and to provide a useful decomposition of compressible fluctuations into physical modes Kovasznay [148] introduced a small-parameter expansion, which is based on the assumption that the turbulent fluctuations are small with respect to a uniform mean flow. As will be seen below, this decomposition leads to considerable insight, but its validity is restricted since it relies on a linearized theory. For flows in which nonlinear mechanisms are dominant another approach is to use the exact Helmholtz decomposition of the compressible velocity field. Since this decomposition does not rely on any assumption dealing with the amplitude of the turbulent fluctuations it is valid in all flows. But its weakness is that it does not allow a direct splitting of other flow variables such as density, pressure, or entropy. Therefore, it must be supplemented with to some extent arbitrary splitting procedures for these variables.

The first step in Kovasznay's approach consists in expanding the turbulent field as

$$\boldsymbol{u} = \boldsymbol{u}_0 + \epsilon \boldsymbol{u}_1 + \epsilon^2 \boldsymbol{u}_2 + \cdots , \tag{3.2}$$

$$\rho = \rho_0 + \epsilon \rho_1 + \epsilon^2 \rho_2 + \cdots , \tag{3.3}$$

$$p = p_0 + \epsilon p_1 + \epsilon^2 p_2 + \cdots , \tag{3.4}$$

$$s = s_0 + \epsilon s_1 + \epsilon^2 s_2 + \cdots , \tag{3.5}$$

where ϵ is a small parameter related to the amplitude of the perturbation field and $(\boldsymbol{u}_0, \rho_0, p_0, s_0)$ are related to a uniform mean field. Assuming that the source terms in the Navier–Stokes equations scale as ϵ and inserting the above expansions into these equations, one obtains the following linearized set of equations for the first-order fluctuating field $(\boldsymbol{u}_1, \rho_1, p_1, s_1)$ (the subscript 1 will be omitted hereafter for the sake of clarity)

$$\nabla \cdot \boldsymbol{u} + \frac{\partial p}{\partial t} - \frac{\partial s}{\partial t} = \frac{m}{\rho_0}, \tag{3.6}$$

$$\frac{\partial \boldsymbol{u}}{\partial t} + a_0^2 \nabla p - \nu_0 \nabla^2 \boldsymbol{u} - \frac{1}{3} \nu_0 \nabla (\nabla \cdot \boldsymbol{u}) = \boldsymbol{f}, \tag{3.7}$$

$$\frac{\partial s}{\partial t} - \kappa_0 \nabla^2 s - \kappa_0 (\gamma - 1) \nabla^2 p = \frac{Q}{\rho_0 c_p T_0}, \tag{3.8}$$

where $\nu_0 = \mu_0 / \rho_0$, and $\gamma = c_p / c_v$, c_p is the specific heat at constant pressure and c_v at constant volume. The volume source terms m, $\boldsymbol{f}$ and Q correspond to mass source/sink, external forces and heat source/sink, respectively. It is should be noted that the pressure and the entropy have been normalized by γp_0 and c_p, respectively (the notation has not been changed for the sake of simplicity). The speed of sound in the undisturbed medium, a_0, is computed as $a_0 = \sqrt{\gamma p_0 / \rho_0}$. The Prandtl number $\mu c_p / \kappa$ has been taken equal to 3/4 for the sole purpose of simplifying the algebra. This linear system can be rewritten by introducing the fluctuating vorticity $\boldsymbol{\Omega} = \nabla \times \boldsymbol{u}$. After some algebra, one obtains

$$\frac{\partial \boldsymbol{\Omega}}{\partial t} - \nu_0 \nabla^2 \boldsymbol{\Omega} = \nabla \times \boldsymbol{f}, \tag{3.9}$$

$$\frac{\partial s}{\partial t} - \kappa_0 \nabla^2 s = \kappa_0 (\gamma - 1) \nabla^2 p + \frac{Q}{\rho_0 c_p T_0}, \tag{3.10}$$

$$\begin{aligned} \frac{\partial^2 p}{\partial t^2} - a_0^2 \nabla^2 p - \kappa_0 \gamma \frac{\partial}{\partial t} (\nabla^2 p) = & \left[\left(\frac{\partial}{\partial t} - \kappa_0 \nabla^2 \right) \frac{m}{\rho_0} - \nabla \cdot \boldsymbol{f} \right. \\ & \left. + \frac{\partial}{\partial t} \left(\frac{Q}{\rho_0 c_p T_0} \right) \right]. \end{aligned} \tag{3.11}$$

This set of equations is supplemented by additional relations obtained by linearizing the perfect gas law and the entropy definition

$$\gamma p - \frac{\rho}{\rho_0} - \frac{T}{T_0} = 0, \tag{3.12}$$

$$p + \frac{1}{\gamma - 1} \left(s - \frac{T}{T_0} \right) = 0. \tag{3.13}$$

Using these equations, Kovasznay defines three *modes*, each mode corresponding to the solution of a subsystem extracted from (3.9)–(3.11):

- The *vorticity mode*, whose fluctuating field is denoted as $(\boldsymbol{\Omega}_\Omega, p_\Omega, s_\Omega, \boldsymbol{u}_\Omega)$, is defined as follows

$$\frac{\partial \boldsymbol{\Omega}_\Omega}{\partial t} - \nu_0 \nabla^2 \boldsymbol{\Omega}_\Omega = \nabla \times \boldsymbol{f}, \tag{3.14}$$

$$p_\Omega = 0, \qquad s_\Omega = 0, \qquad \nabla \times \boldsymbol{u}_\Omega = \boldsymbol{\Omega}_\Omega, \qquad \nabla \cdot \boldsymbol{u}_\Omega = 0. \tag{3.15}$$

  The vorticity mode is associated with a solenoidal rotational velocity field, and it can be interpreted as the "incompressible" part of the solution. But it is worth noting that there is no corresponding pressure disturbance because it is expected to be of order ϵ^2.

- The *entropy mode*, whose corresponding perturbation field is $(\boldsymbol{\Omega}_e, p_e, s_e, \boldsymbol{u}_e)$, is defined as

$$\frac{\partial s_e}{\partial t} - \kappa_0 \nabla^2 s_e = \kappa_0(\gamma - 1)\nabla^2 p_e + \frac{Q}{\rho_0 c_p T_0}, \tag{3.16}$$

$$\boldsymbol{\Omega}_e = 0, \qquad p_e = 0, \qquad \nabla \times \boldsymbol{u}_e = 0, \qquad \nabla \cdot \boldsymbol{u}_e = \frac{\partial s_e}{\partial t}. \tag{3.17}$$

  The velocity perturbation is purely dilatational and is induced by viscous effects, so that $\boldsymbol{u}_e = 0$ in the inviscid case.
- The *acoustic mode*, which is characterized by $(\boldsymbol{\Omega}_p, p_p, s_p, \boldsymbol{u}_p)$. The governing relations for this mode are

$$\frac{\partial^2 p_p}{\partial t^2} - a_0^2 \nabla^2 p_p - \kappa_0 \gamma \frac{\partial}{\partial t}(\nabla^2 p_p) = \left[\left(\frac{\partial}{\partial t} - \kappa_0 \nabla^2\right)\frac{m}{\rho_0} - \nabla \cdot \boldsymbol{f} + \frac{\partial}{\partial t}\left(\frac{Q}{\rho_0 c_p T_0}\right)\right], \tag{3.18}$$

$$\frac{\partial s_p}{\partial t} - \kappa_0 \nabla^2 s_p = \kappa_0(\gamma - 1)\nabla^2 p_p, \tag{3.19}$$

$$\nabla \times \boldsymbol{u}_p = 0, \qquad \nabla \cdot \boldsymbol{u}_p = \frac{\partial s_p}{\partial t} - \frac{\partial p_p}{\partial t} + \frac{m}{\rho_0}. \tag{3.20}$$

  It can be observed that the viscous effects lead to a dispersive solution. Acoustic entropy fluctuation originate from the viscous dissipation of the pressure waves.

Disturbances associated with the entropy mode and the vorticity mode are passively advected by the mean field (velocity $\boldsymbol{u}_0$ in a reference frame at rest), whereas acoustic disturbances travel at the speed of sound relative to the mean flow. Using this three-mode decomposition, all turbulent fluctuations can be decomposed as

$$p' = p_p, \tag{3.21}$$

$$s' = s_e + s_p, \tag{3.22}$$

$$\boldsymbol{u}' = \boldsymbol{u}_\Omega + \boldsymbol{u}_e + \boldsymbol{u}_p, \tag{3.23}$$

$$\boldsymbol{\Omega}' = \boldsymbol{\Omega}_\Omega. \tag{3.24}$$

3.2.2 Weakly Nonlinear Kovasznay Decomposition

The linear decomposition presented above makes it possible to define the three physical modes, but it does not provide any insight into the interactions between them since the modes evolve independently. Information dealing with the creation/destruction of fluctuations due to the modal interactions are recovered by considering terms of order ϵ^2 resulting from square interactions [42]. The full analysis brings in 18 terms and also involves another non-dimensional parameter $\epsilon' = \nu_0 k/a_0$ which measures the ratio of the characteristic length scale of the perturbation, $1/k$, and the intrinsic scale of the medium ν_0/a_0. Second-order terms scale as ϵ^2 or $\epsilon^2\epsilon'$. Since for turbulent flows at atmospheric pressure and density one has $\epsilon' \ll \epsilon$, one can neglect terms of order $\epsilon^2\epsilon'$. Remaining terms and associated production mechanisms are shown in Table 3.1. It is important to note that these second-order corrections lead to a full mode coupling since even self-interactions can affect other modes.

Table 3.1. Source terms associated with second-order square interactions according to the Kovasznay decomposition

Modal interaction	Acoustic production	Vorticity production	Entropy production
Acoustic-Acoustic	Steepening and self-scattering $\nabla \cdot \nabla \cdot (\boldsymbol{u}_p \boldsymbol{u}_p) + a_0^2 \nabla^2 p_p^2 + \frac{\gamma-1}{2} \frac{\partial^2 p_p^2}{\partial t^2}$	none	none
Vorticity-Vorticity	Generation $2\nabla \cdot \nabla \cdot (\boldsymbol{u}_\Omega \boldsymbol{u}_\Omega)$	Self-convection and stretching $-\boldsymbol{u}_\Omega \nabla \boldsymbol{\Omega}_\Omega + \boldsymbol{\Omega}_\Omega \nabla \boldsymbol{u}_\Omega$	none
Entropy-Entropy	none	none	none
Acoustic-Vorticity	Scattering $2\nabla \cdot \nabla \cdot (\boldsymbol{u}_\Omega \boldsymbol{u}_p)$	Vorticity convection and stretching $-\boldsymbol{u}_p \nabla \boldsymbol{\Omega}_\Omega + \boldsymbol{\Omega}_\Omega \nabla \boldsymbol{u}_p - \boldsymbol{\Omega}_\Omega \nabla \cdot \boldsymbol{u}_p$	none
Acoustic-Entropy	Scattering $\frac{\partial}{\partial t} \nabla \cdot (s_e \boldsymbol{u}_p)$	Baroclinic source $-a_0^2 (\nabla s_e) \times (\nabla p_p)$	Heat convection $-\boldsymbol{u}_p \cdot \nabla s_e$
Vorticity-Entropy	none	none	Heat convection $-\boldsymbol{u}_\Omega \cdot \nabla s_e$

3.3 Statistical Description of Compressible Turbulence

The statistical description of compressible turbulence is significantly more complex than that of incompressible turbulence as can be inferred from the existence of the different Kovasznay modes. In the previous chapter the statistical description of compressible turbulent fluctuations was performed using the density-weighted Favre-averaging procedure, leading to a definition of fluctuating quantities with respect to that average, denoted by a double prime. In this chapter we consider homogeneous flows only. For such flows it can be shown that the usual ensemble-averaging and the Favre averaging procedures give identical results (up to a Galilean transformation). Therefore, in the present chapter, no distinction will be made between these two averaging procedures.

It is common to split the mean kinetic energy dissipation $\bar{\varepsilon}$ into to two contributions:

$$\bar{\varepsilon} = \bar{\varepsilon}_s + \bar{\varepsilon}_d, \tag{3.25}$$

where $\bar{\varepsilon}_s$ and $\bar{\varepsilon}_d$ are referred to as the solenoidal and the dilatational dissipation rate, respectively. They are defined as

$$\bar{\varepsilon}_s = 2\frac{\bar{\mu}}{\bar{\rho}}\overline{W'_{ij}W'_{ij}} = \frac{\bar{\mu}}{\bar{\rho}}\overline{\Omega'_i\Omega'_i}, \quad W'_{ij} = \frac{1}{2}\left(\frac{\partial u'_i}{\partial x_j} - \frac{\partial u'_j}{\partial x_i}\right), \tag{3.26}$$

$$\bar{\varepsilon}_d = \frac{4}{3}\frac{\bar{\mu}}{\bar{\rho}}\overline{\left(\frac{\partial u'_i}{\partial x_i}\right)^2} = \frac{4}{3}\frac{\bar{\mu}}{\bar{\rho}}\overline{\left(\frac{\partial u'_{di}}{\partial x_i}\right)^2}, \tag{3.27}$$

where $\boldsymbol{\Omega} \equiv \nabla \times \boldsymbol{u}$. It can be observed that $\bar{\varepsilon}_s$ (resp. $\bar{\varepsilon}_d$) does not depend on the dilatational (resp. solenoidal) field. Restricting the analysis to the linear Kovasznay splitting the solenoidal dissipation $\bar{\varepsilon}_s$ is associated entirely with the vorticity mode, whereas the dilatational dissipation is mainly due to the acoustic mode in the absence of significant entropy source.

To get a deeper insight into compressible turbulence dynamics, one also introduces the following spectra:

- The solenoidal kinetic energy spectrum $E_{ss}(k)$, which is such that the solenoidal turbulent kinetic energy $\mathcal{K}_s$ is recovered as

$$\mathcal{K}_s = \int_0^{+\infty} E_{ss}(k)dk. \tag{3.28}$$

- The dilatational kinetic energy spectrum $E_{dd}(k)$, which is such that the solenoidal turbulent kinetic energy $\mathcal{K}_d$ is recovered as

$$\mathcal{K}_d = \int_0^{+\infty} E_{dd}(k)dk. \tag{3.29}$$

  The global turbulent kinetic energy is therefore defined as $\mathcal{K} = \mathcal{K}_s + \mathcal{K}_d$.
- The normalized pressure spectrum $E_{pp}(k)$, which is such that pressure variance is recovered as

$$\overline{p'p'}(t) = \rho_0^2 a_0^2 \int_0^{+\infty} E_{pp}(k,t)dk. \tag{3.30}$$

  The pressure spectrum can be further decomposed into the spectrum of the pressure fluctuations associated with the solenoidal velocity field, $E_{pp}^{inc}(k)$, and that associated with the dilatational motion, $E_{pp}^{acous}(k)$.

The two components of the turbulent kinetic energy dissipation are equal to

$$\bar{\varepsilon}_s = 2\nu \int_0^{+\infty} k^2 E_{ss}(k)dk, \qquad \bar{\varepsilon}_d = 2\kappa \int_0^{+\infty} k^2 E_{dd}(k)dk. \tag{3.31}$$

3.4 Shock-Turbulence Interaction

3.4.1 Introduction to the Linear Interaction Approximation Theory

We now briefly introduce the Linear Interaction Approximation (LIA), which is a very powerful tool pioneered in the 1950s [230, 203] to analyze shock-turbulence interactions. It relies on the following assumptions:

1. The shock wave has no intrinsic scale, and therefore it is enslaved by incident perturbations. It only enforces the jump conditions.
2. Both, mean and fluctuating parts of the upstream field (i.e. the field in the supersonic part of the flow) are given.
3. The downstream field is fully determined by the upstream field and the jump conditions. More precisely, it is assumed that the interaction process between turbulent fluctuations and the shock is essentially linear, so that
   (a) the mean flow obeys the usual Rankine–Hugoniot conditions, and
   (b) the fluctuating field obeys linearized jump conditions.

This scheme is illustrated in Fig. 3.1.

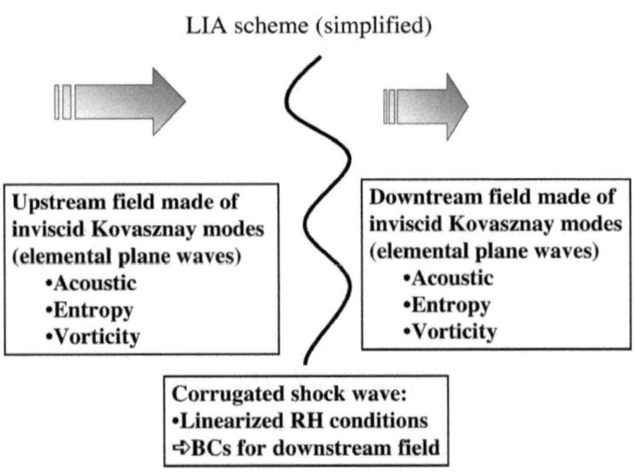

Fig. 3.1. Schematic view of the Linear Interaction Approximation for shock-turbulence interaction

Two conditions must be satisfied to ensure that the linear approximation is relevant:

1. The fluctuations must be weak in the sense that the distorted shock wave must remain well defined. Numerical experiments led Lee and coworkers [159] to propose the following empirical criterion for the linear regime

$$M_t^2 < \alpha(M_1^2 - 1), \tag{3.32}$$

where M_t and M_1 are the upstream turbulent and mean Mach numbers respectively, and $\alpha \approx 0.1$.

2. The time required for turbulent events to cross the shock must be small compared to the turbulence time scale $\mathcal{K}/\bar{\varepsilon}$, so that nonlinear mechanisms have no significant effects during the interaction.

The LIA analysis can be made more accurate decomposing the fluctuating field using the Kovasznay decomposition: both the upstream and downstream fluctuating fields are split as sums of individual modes, each mode being characterized by its nature (acoustic, vorticity or entropy mode) and wave number or frequency. Since linearized jump conditions are utilized, all cross interactions between modes are precluded, and the downstream fluctuating field is obtained by a simple superposition of the LIA results obtained for each upstream fluctuating mode. Let us emphasize here that the fact that interactions are precluded does not mean that an upstream perturbation wave is associated to an emitted downstream wave of the same nature (as a matter of fact, all physical modes exist in the downstream region in the general case), but that the interaction process is not sensitive to shock deformations induced by other upstream fluctuations.

The resulting LIA scheme is the following: one considers two semi-infinite domains separated by the shock wave. Both the mean and fluctuating fields in the upstream domain are arbitrarily prescribed. Since the flow is hyperbolic in this domain, it is not sensitive to the presence of the shock wave. The mean downstream field is computed using the mean upstream field and the usual Rankine–Hugoniot jump relations (2.148)–(2.151). The emitted fluctuating field is then computed using the linearized jump relations as boundary conditions. It is important to note that the wave vectors of the emitted waves are computed using the dispersion relation associated with each physical mode, the frequency and the tangential component of the wave vector being the same as the upstream perturbation. The linearized jump conditions are used only to compute the amplitudes of the emitted waves.

3.4.2 Vortical Turbulence-Shock Interaction

We first address the case in which the incident turbulence is isotropic and composed of vorticity modes only. This case was investigated by several researchers [159, 160] using LIA and direct numerical simulation (DNS). The trends found by DNS and LIA are corroborated by wind tunnel experiments, but a strict quantitative agreement cannot be expected since the nature of the incident turbulence in experiments cannot be controlled due to experimental limitations. The main observations are the following:

1. *Velocity fluctuations.* The streamwise distributions of the kinetic energy of the three velocity components given both by direct numerical simulation and by LIA are shown in Fig. 3.2. it is observed that all velocity components are amplified, leading to a global increase in the turbulent

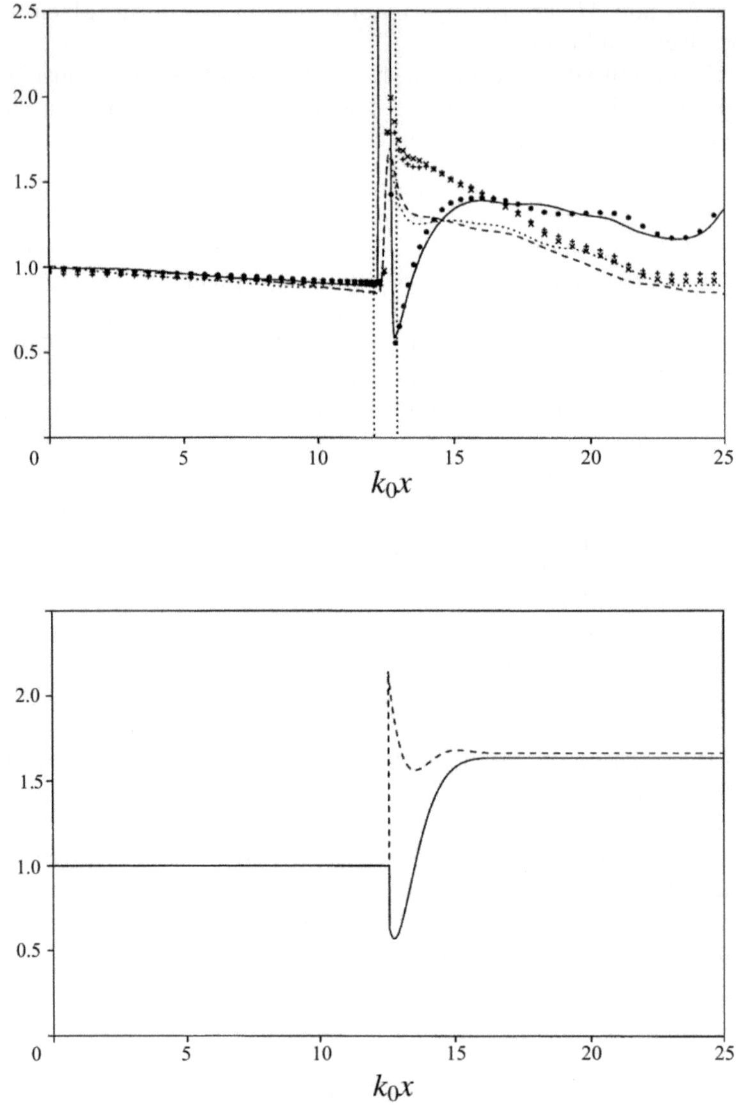

Fig. 3.2. Streamwise evolution of normalized Reynolds stresses. *Top*: DNS results (*lines* for ($M_1 = 2, M_t = 0.108, Re_\lambda = 19.0$) and *symbols* for ($M_1 = 3, M_t = 0.110, Re_\lambda = 19.7$)); streamwise component $R_{11} = \overline{u'u'}$: *solid line* and dots; spanwise component $R_{22} = \overline{v'v'}$: *dashed line* and '×'; spanwise component $R_{33} = \overline{w'w'}$: *dotted line* and '+'. *Bottom*: LIA results for ($M_1 = 2, M_t = 0.108$); streamwise component R_{11}: *solid line*; spanwise components R_{22} and R_{33}: *dashed line*. *Vertical dotted line* shows the limit of the shock displacement region. Reproduced from Ref. [160] with permission

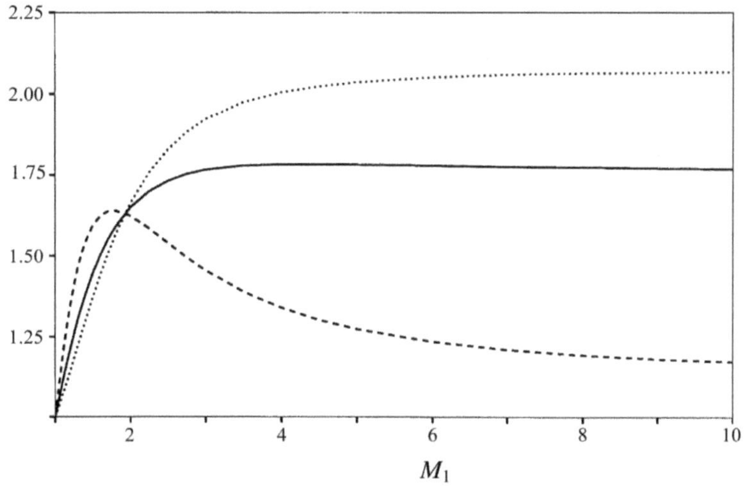

Fig. 3.3. LIA prediction of far-field Reynolds stress amplification versus the upstream Mach number. *Solid line*: turbulent kinetic energy; *dashed line*: streamwise Reynolds stress $R_{11} = \overline{u'u'}$; *dotted line*: spanwise Reynolds stresses $R_{22} = \overline{v'v'}$ and $R_{33} = \overline{w'w'}$. Reproduced from Ref. [160] with permission

kinetic energy. The amplification rate is well recovered by the LIA calculation, showing that the amplification is mainly due to linear mechanisms. In agreement with LIA, the velocity field behind the shock wave is axisymmetric. Both LIA and DNS predict that the amplification is Mach number dependent. The amplification level is plotted as a function of the upstream Mach number, M_1, in Fig. 3.3. It is interesting to note that the amplification of the transverse velocity components is an increasing monotonic function, while the shock-normal velocity component amplification exhibits a maximum near $M_1 = 2$. The transverse components are more amplified than the streamwise component for $M_1 > 2$, and the amplification of the total turbulent kinetic energy tends to saturate beyond $M_1 = 3$. The streamwise DNS profiles reveal that the velocity field experiences a rapid evolution downstream from the shock, leading to the existence of two different regions behind the shock wave. This observation is in full agreement with the LIA analysis, which predicts the existence of a near field region where the acoustic waves emitted during the interaction are not negligible. Comparing the LIA and DNS profiles (see Fig. 3.2) once again leads to the conclusion that the process is mainly governed by linear mechanisms. The rapid evolution in the near field region is due to the exponential decay of acoustic waves, which are responsible for the anti-correlation of the (acoustic) dilatational and (vortical) solenoidal field just downstream the shock. A close examination of DNS data shows that the near-field evolution is associated with an energy transfer from the

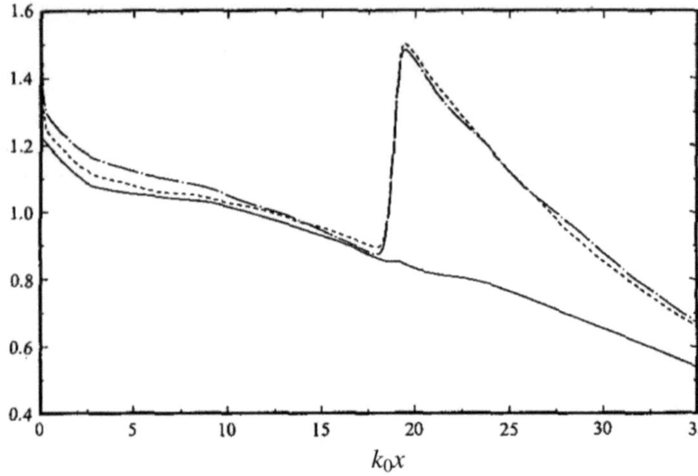

Fig. 3.4. Vorticity amplification across shock (DNS data, $M_1 = 1.2, Re_t = 84.8$). *Solid line*: streamwise component $\overline{\Omega_1'\Omega_1'}$; *Dashed line*: spanwise component $\overline{\Omega_2'\Omega_2'}$; *Dash-dot line*: spanwise component $\overline{\Omega_3'\Omega_3'}$. Reproduced from Ref. [159] with permission

acoustic potential energy in the form of density or pressure fluctuations to turbulent kinetic energy. Outside of the near-field region, the global behavior results from the competition between viscous decay and return to isotropy. In low-Reynolds number DNS the viscous effect is dominant: the turbulent kinetic energy balance simplifies as an equilibrium between the convection term and the viscous term, showing that the main effects are convection of turbulent velocity fluctuations by the mean field and their destruction by viscous effects.

2. *Vorticity field.* The vorticity is also strongly affected by the interaction with the shock wave. The streamwise evolution of the vorticity components computed in two different simulations are presented in Figs. 3.4 and 3.5. Several typical features are observed. First, the streamwise (i.e. shock normal) vorticity component is not affected, in agreement with the conclusion drawn from the jump condition (2.155). The two other components are amplified, in a symmetric way, leading to the definition of a statistically axisymmetric vorticity field behind the shock wave. This behavior is predicted by the LIA analysis. The amplification of the transverse component is Mach number dependent, and the LIA analysis presented in Fig. 3.6 shows that it increases monotonically and tends to saturate at very high Mach numbers. Since the vorticity has no contribution from the acoustic modes it does not exhibit a near field. But it is interesting to note that a Reynolds-number dependent behavior of the streamwise vorticity component is observed downstream the shock: it decreases monotonically at low Reynolds numbers, whereas it has a local maximum at higher

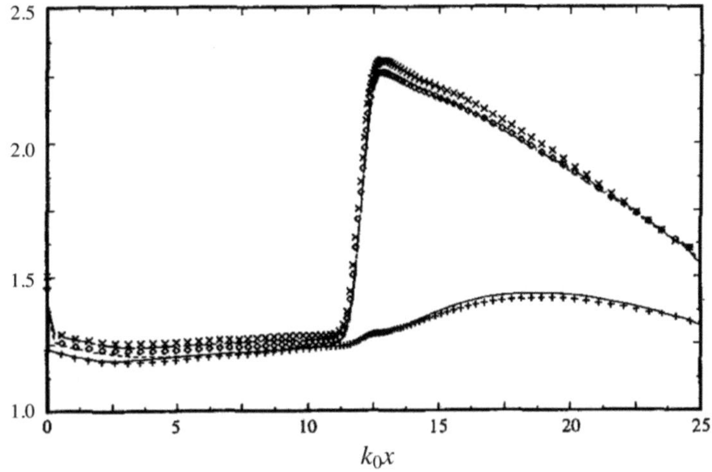

Fig. 3.5. Vorticity amplification across shock (DNS data on two computational grids, $M_1 = 1.2, Re_t = 238$). *Solid line* and '+': streamwise component $\overline{\Omega_1'\Omega_1'}$; *Dashed line* and '×': spanwise component $\overline{\Omega_2'\Omega_2'}$; *Dash-dot line* and *diamonds*: spanwise component $\overline{\Omega_3'\Omega_3'}$. Reproduced from Ref. [159] with permission

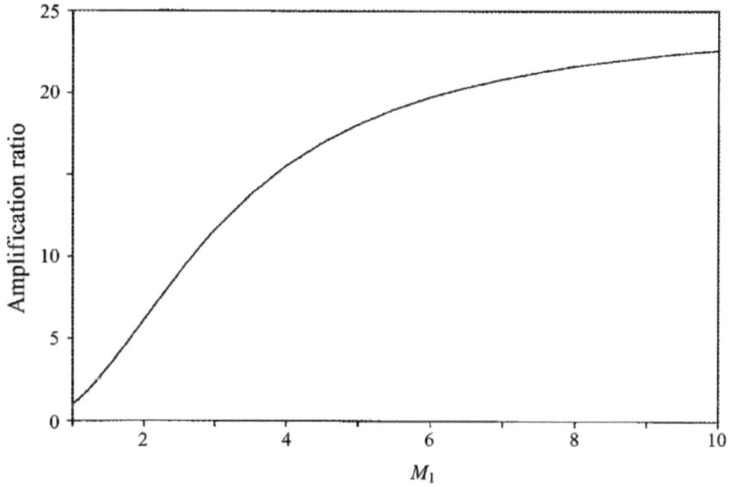

Fig. 3.6. Amplification of transverse vorticity components $\overline{\Omega_2'\Omega_2'}$ and $\overline{\Omega_3'\Omega_3'}$ across shock versus the upstream Mach number M_1: LIA results. Reproduced from Ref. [160] with permission

Reynolds numbers. The explanation for this difference is found by looking at the evolution equation of the vorticity component variances using DNS data. Inside of the shock wave, the transverse component evolution which leads to the existence of the jump in the LIA theory is dominated by the vorticity-compression mechanisms. Downstream of the shock region, the vortex stretching mechanism is balanced by the viscous effects. In all DNS cases, both the baroclinic production term and the turbulent transport are negligible. The dynamics of the streamwise component is different. Inside the shock wave, vortex stretching and vorticity compression balance each other, resulting in a negligible influence of the shock, which is in agreement with inviscid jump relations. Downstream of the shock wave the streamwise vorticity variance is governed by the balance between turbulent vortex stretching and viscous effects. In the low Reynolds number case the viscous damping overwhelms the vortex stretching effects, leading to a monotonic decay. At higher Reynolds number, the turbulent stretching is large enough to yield the existence of a local downstream maximum.

3. *Turbulence lengthscales.* Characteristic scales of turbulence are modified during the interaction in a scale-dependent manner. Let us first discuss the behavior of the one dimensional spectra $E_\alpha(k_\beta)$ which are defined such that (without summation over Greek indices)

$$\overline{u'_\alpha u'_\alpha} = \int_0^{+\infty} E_\alpha(k_\beta)\, dk_\beta, \tag{3.33}$$

where u'_α and k_β are the αth component of $\boldsymbol{u}'$ and the βth component of $\boldsymbol{k}$, respectively. Both LIA and DNS results show that:
(a) For the longitudinal spectra $E_\alpha(k_1)$ small scales (i.e. large wave numbers) are more amplified than large scales (i.e. small wave numbers).
(b) The amplification pattern is more complex for transverse spectra: higher amplification at small scales is found for $E_1(k_2)$ and $E_2(k_2)$, while the large scales are the most amplified for $E_3(k_2)$.

This complex behavior makes it necessary to carry out a specific analysis for each characteristic length scale, since they are spectrum-dependent. Defining the integral scale for the dummy variable ϕ as

$$\Lambda_\phi(r, x) = \int_{r=0}^{r=+\infty} C_{\phi\phi}(r, x)\, dr, \tag{3.34}$$

where the transverse two-point correlation $C_{\phi\phi}(r, x)$ is given by (ϕ is assumed to be a centered random variable)

$$C_{\phi\phi}(r, x) = \frac{\overline{\phi(x, y, z, t)\phi(x, y + r, z, t)}}{\overline{\phi(x, y, z, t)}\ \overline{\phi(x, y, z, t)}}, \tag{3.35}$$

in which the statistical averaging is carried out over time and homogeneous directions y and z, both DNS and LIA show that (see Fig. 3.7)
(a) Λ_{u_1}, Λ_{u_2} and Λ_ρ exhibit a significant Mach-number dependent decrease across the shock, and
(b) Λ_{u_3} is largely increased by the interaction.

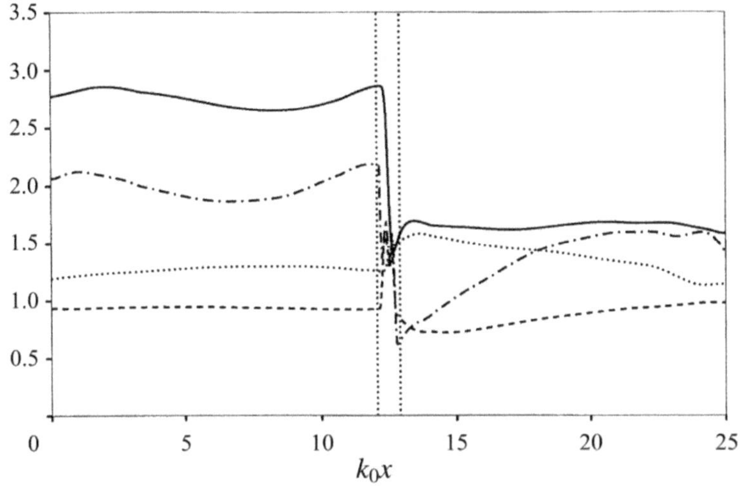

Fig. 3.7. Streamwise evolution of turbulence transverse integral scales (DNS, $M_1 = 2, M_t = 0.108, Re_\lambda = 19$). *Dashed line*: Λ_{u_1}; *solid line*: Λ_{u_2}; *dotted line*: Λ_{u_3}; *dashed-dotted line*: Λ_ρ. Reproduced from Ref. [160] with permission

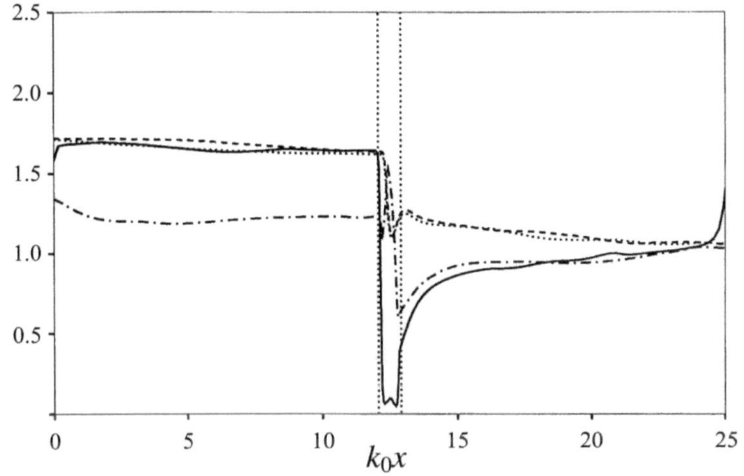

Fig. 3.8. Streamwise evolution of turbulence microscales (DNS, $M_1 = 2, M_t = 0.108, Re_\lambda = 19$). *Solid line*: λ_{u_1}; *dashed line*: λ_{u_2}; *dotted line*: λ_{u_3}; *dashed-dotted line*: λ_ρ. Reproduced from Ref. [160] with permission

Now looking at the Taylor microscales (see Fig. 3.8), it can be seen that they are all significantly reduced during the interaction, the reduction being more pronounced in the shock normal direction. We recall that the Taylor microscale λ_α associated with u'_α and the density microscale λ_ρ are computed as

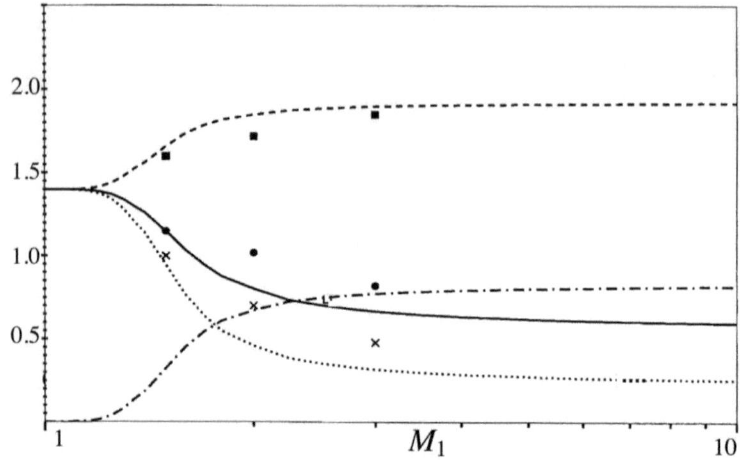

Fig. 3.9. Evolution of normalized correlation coefficients in the far field region downstream the shock versus the upstream Mach number M_1. $n_{\rho p}$: *solid line* (LIA) and *black circles* (DNS); $n_{\rho T}$: *dashed line* (LIA) and *black circles* (DNS); $C_{\rho T}$: *dotted line* (LIA) and '×' (DNS); i_s: *dashed-dotted line* (LIA). Reproduced from Ref. [160] with permission

$$\lambda_\alpha = \sqrt{\frac{\overline{u'_\alpha u'_\alpha}}{\overline{\frac{\partial u'_\alpha}{\partial x_\alpha}\frac{\partial u'_\alpha}{\partial x_\alpha}}}}, \qquad \lambda_\rho = \sqrt{\frac{\overline{\rho'\rho'}}{\overline{\frac{\partial \rho'}{\partial y}\frac{\partial \rho'}{\partial y}}}}. \tag{3.36}$$

4. *Thermodynamic quantities.* The thermodynamic properties of the flow downstream the shock are also modified by the interaction process. Both DNS and LIA results show that for an isentropic incident isotropic turbulence the downstream field remains isentropic for weak shocks with $M_1 < 1.2$. At higher upstream wave number the emitted entropy waves have significant energy since their magnitude becomes comparable to that of acoustic waves. This effect can be demonstrated by the normalized correlation coefficients (see Fig. 3.9):

$$n_{\rho p} \equiv \frac{\sqrt{\overline{p'p'}}}{\bar{p}^2}\frac{\bar{\rho}^2}{\sqrt{\overline{\rho'\rho'}}}, \qquad n_{\rho T} \equiv 1 + \frac{\sqrt{\overline{T'T'}}}{\bar{T}^2}\frac{\bar{\rho}^2}{\sqrt{\overline{\rho'\rho'}}}, \tag{3.37}$$

$$C_{\rho T} \equiv 1 + \frac{\bar{\rho}}{\tilde{T}}\frac{\sqrt{\overline{\rho' T'}}}{\sqrt{\overline{\rho'\rho'}}}, \tag{3.38}$$

along with the entropy fluctuation contribution

$$i_s \equiv \frac{\sqrt{\overline{s's'}}}{c_p^2}\frac{\bar{\rho}^2}{\sqrt{\overline{\rho'\rho'}}}. \tag{3.39}$$

One can observe that the entropy fluctuations are more significant than acoustic fluctuations for $M_1 > 1.65$. But it is worth noting that downstream of the shock neither the isentropic hypothesis (which states that the entropy fluctuations are negligible) nor the Strong Reynolds Analogy proposed by Morkovin [207] for shear flows (stating that the stagnation temperature is constant, which amounts to assuming that acoustic waves have negligible effect on the density fluctuations) are valid if $M_1 > 1.2$. We recall that the latter can be expressed as

$$\frac{\rho'}{\bar{\rho}} = -\frac{T'}{\bar{T}} = (\gamma - 1)M_1^2 \frac{u_1'}{\bar{u}_1}. \tag{3.40}$$

3.4.3 Mixed-Mode Turbulence-Shock Interaction

We now address the cases in which the incident turbulent field is composed of different types of Kovasznay modes: hybrid vortical/acoustic turbulence [185] and vortical/entropic turbulence [186, 187]. These cases are of great interest, since physical turbulence generated in wind tunnels or observed in natural flows is never strictly vortical or acoustic. It is worth recalling here the important conclusion that a Kovasnay mode will generate modes of different nature through non-linear self interactions. But since experimental data exhibit a significant dispersion it can be inferred that their sensitivity to incoming turbulence must be large (irrespective of the fact that such experiments are very difficult to perform). Another point is that, in real flows, the distribution of the total energy among the three Kovasznay modes is unknown and usually cannot be controlled. Therefore, we will hereafter put the emphasis on the theoretical results rather than giving an exhaustive presentation of experimental findings.

Let us begin by examining the two-dimensional linearized Euler equation for the vorticity fluctuation about a one-dimensional mean flow. We will use it as a simple phenomenological model to describe the amplification of the transverse vorticity components across the shock. The linearized evolution equation is

$$\frac{\partial \Omega'}{\partial t} + U\frac{\partial \Omega'}{\partial x} = \Omega' \frac{\partial U}{\partial x} - \frac{\partial \rho'}{\partial y}\frac{1}{\bar{\rho}^2}\frac{\partial \bar{p}}{\partial x} + \frac{\partial p'}{\partial y}\frac{1}{\bar{\rho}^2}\frac{\partial \bar{\rho}}{\partial x}. \tag{3.41}$$

The usual viscous model [151, 328] for the shock front shows that $(\partial \bar{u}/\partial x) < 0$, $(\partial \bar{p}/\partial x) > 0$ and $(\partial \bar{\rho}/\partial x) > 0$ in the shock region. The first term on the right-hand side of the above equation corresponds to the compression by the mean flow gradient. Since $\partial \bar{u}/\partial x$ is negative in the shock region the net effect of vorticity amplification by the bulk compression is recovered. The two last terms are related to the baroclinic mechanisms. The second term on the right hand side of (3.41) involves the fluctuating density and is therefore non-zero for both acoustic and entropy fluctuating modes, while the third one is non zero for acoustic perturbations only. This equation also shows that the baroclinic and the bulk compression contributions can have the same or opposite signs, depending on the respective signs of the vorticity, density and pressure fluctuations. If the contributions have the same sign the net amplification of vorticity increases by cooperative interaction, whereas the two mechanisms tend to cancel each other in the opposite case, yielding a decrease of the net vorticity fluctuation amplification. One can see that increased amplification is recovered if $\Omega' p' > 0$ or $\Omega' \rho' < 0$. The very important conclusion drawn from that very simple analysis is that the results of the shock/turbulence interaction will be quite sensitive to the correlation between the Kovasznay modes in the incident field.

Influence of the Upstream Entropy Fluctuations

Let us first consider the case of an incident field made of vorticity and entropy modes. We simplify the problem by considering a single plane entropy wave with amplitude A_e and a single vorticity wave with amplitude A_v with the same wave vector $\boldsymbol{k}$. The upstream field is composed of superimposed elementary plane waves

$$\frac{u'}{\bar{u}} \propto A_v e^{\imath(\boldsymbol{k}\cdot\boldsymbol{x}-\omega t)}, \qquad \frac{v'}{\bar{u}} \propto -A_v e^{\imath(\boldsymbol{k}\cdot\boldsymbol{x}-\omega t)}, \qquad \frac{s'}{c_p} \propto A_e e^{\imath(\boldsymbol{k}\cdot\boldsymbol{x}-\omega t)} \tag{3.42}$$

leading to

$$\frac{\Omega'}{\bar{u}} \propto -A_v e^{\imath(\boldsymbol{k}\cdot\boldsymbol{x}-\omega t)}, \qquad \frac{\rho'}{\bar{\rho}} = -\frac{T'}{\bar{T}} \propto A_e e^{\imath(\boldsymbol{k}\cdot\boldsymbol{x}-\omega t)}, \qquad \frac{p'}{\bar{p}} = 0. \tag{3.43}$$

It can be seen that Ω' and u' on the one hand and ρ' and T' on the other hand have opposite phases. Therefore, the condition for cooperative interaction $\Omega' \rho' < 0$ is equivalent to $u'T' < 0$. Introducing the complex ratio

$$\frac{A_e}{A_v} = \varrho e^{\imath\varphi} \tag{3.44}$$

cooperative interaction is observed for $\varphi \in]-\pi/2, \pi/2[$, whereas partial cancellation occurs for $\varphi \in]\pi/2, -\pi/2[$.

Direct numerical simulation and LIA results are presented in Fig. 3.10. They show that, in the case where the streamwise velocity component and temperature fluctuations are strongly anti-correlated ($\overline{u'T'}/u_{rms}T_{rms} \approx -1$), the amplification of all velocity components is significantly enhanced, the effect being more important on the streamwise component. The velocity field still exhibits a near field whose properties are similar to that of the near field generated by a pure vortical incident field. On the opposite, the amplification is reduced when they are correlated, i.e. $\overline{u'T'} > 0$. The evolution of the amplification of the far field velocity variances with respect to the upstream Mach number is displayed in Fig. 3.11. LIA predicts that the amplification saturates for $M_1 > 2$ with a remarkable exception: if the upstream fluctuations satisfy the Morkovin's hypothesis (3.40) the amplification factor does not saturate

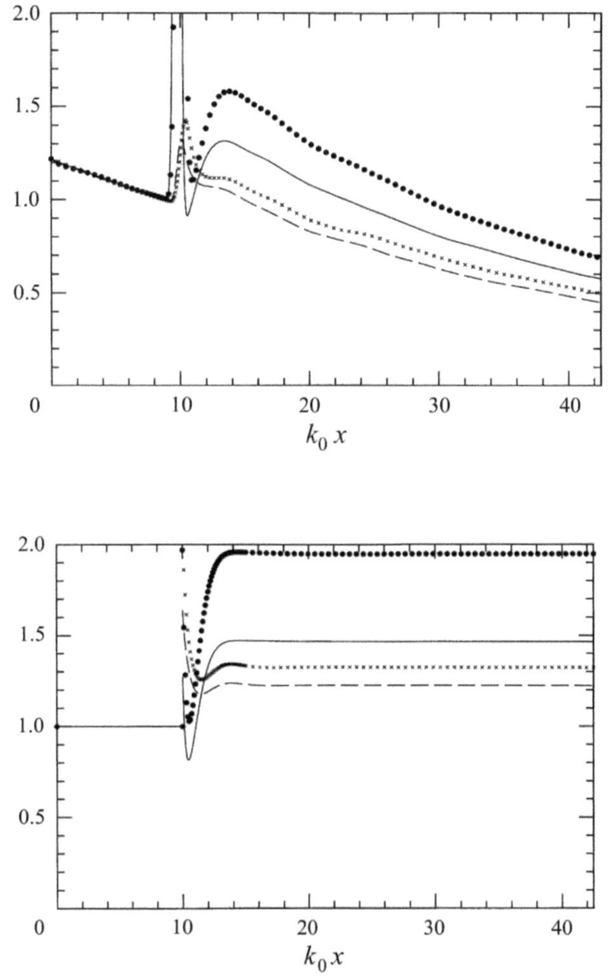

Fig. 3.10. Influence of upstream entropy fluctuations on the streamwise evolution of Reynolds stresses at $M_1 = 1.29$. *Top*: DNS data. Streamwise Reynolds stress $R_{11} = \overline{u'u'}$ for $\overline{u_1'T'}/u_{rms}T_{rms} = -0.06$ (*solid line*) and -0.84 (*full circles*); transverse Reynolds stress $R_{22} = \overline{v'v'}$ for $\overline{u_1'T'}/u_{rms}T_{rms} = -0.06$ (*dashed line*) and -0.84 ('×'). *Bottom*: LIA analysis, same cases and symbols as for DNS data. Reproduced from Ref. [187] with permission

and keeps growing with M_1. The main reason is that, if the Morkovin's hypothesis is assumed to hold in the upstream region, the relative importance of the entropy modes with respect to the vorticity mode scales as M_1^2.

Similar conclusions hold for the vorticity field: both DNS and LIA confirm the predictions drawn from the simplified model. The amplification factor of the transverse vorticity components is plotted versus the upstream Mach

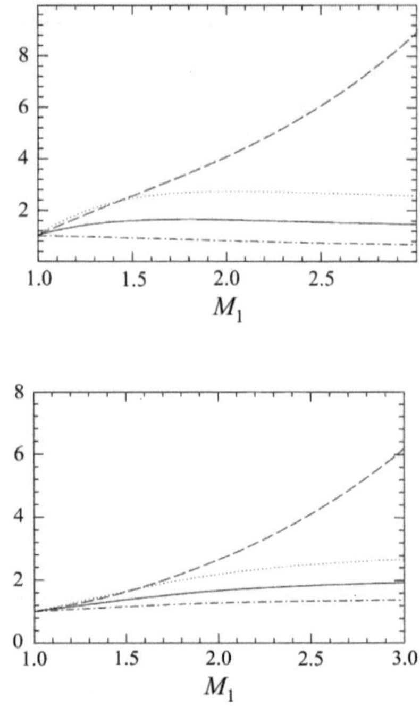

Fig. 3.11. LIA analysis of influence of upstream entropy fluctuations on the downstream evolution of Reynolds stresses versus the upstream Mach number M_1. *Top*: Streamwise Reynolds stress $R_{11} = \overline{u'u'}$. *Solid line*: pure vortical incident turbulence; *dotted line*: $\overline{u_1'T'} < 0$; *dashed-dotted line*: $\overline{u_1'T'} > 0$; *dashed line*: Morkovin's hypothesis satisfied upstream. *Bottom*: transverse Reynolds stress $R_{22} = \overline{v'v'}$, same symbols as above. Reproduced from Ref. [187] with permission

number in Fig. 3.12. Once again, the amplification is enhanced if $\overline{u_1'T'} < 0$ and exhibits very strong values if the incident turbulent field satisfies Morkovin's hypothesis.

An interesting point is that the interaction with the shock results in a breakdown of the Morkovin hypothesis downstream the shock wave, even if it holds upstream of the shock. We recall that the fundamental assumption in Morkovin's hypothesis is that the stagnation temperature T^0 is constant in the flow. Decomposing the stagnation temperature as

$$T^0 = \bar{T} + T' + \frac{1}{2}\frac{(U+u')^2 + v'^2 + w'^2}{c_p} \tag{3.45}$$

the Rankine-Hugoniot jump relation for the energy (2.151) yields the continuity of T^0 across the shock wave

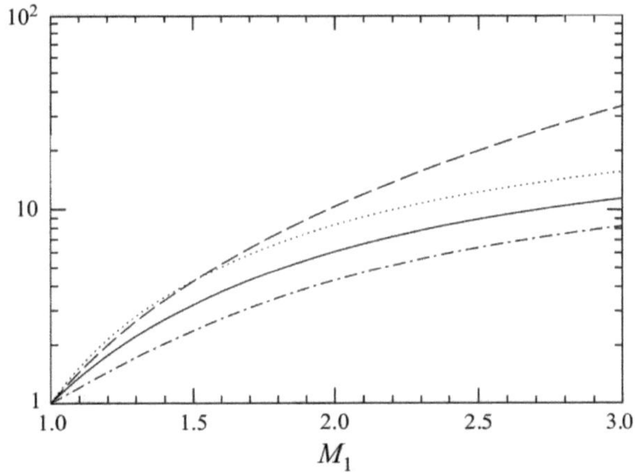

Fig. 3.12. LIA analysis of influence of upstream entropy fluctuations on the far field amplification of transverse vorticity components $\overline{\Omega_2'\Omega_2'} = \overline{\Omega_3'\Omega_3'}$ versus the upstream Mach number M_1. *Solid line*: pure vortical incident turbulence; *dotted line*: $\overline{u_1'T'} < 0$; *dashed-dotted line*: $\overline{u_1'T'} > 0$; *dashed line*: Morkovin's hypothesis satisfied upstream. Reproduced from Ref. [187] with permission

$$\bar{T}_1 + T_1' + \frac{1}{2}\frac{(U_1 + u_1' - u_s)^2 + {v_1'}^2 + {w_1'}^2}{c_p}$$
$$= \bar{T}_2 + T_2' + \frac{1}{2}\frac{(U_2 + u_2' - u_s)^2 + {v_2'}^2 + {w_2'}^2}{c_p}, \tag{3.46}$$

where u_s is the shock speed associated with the corrugation of the shock front by incident perturbations. Assuming that fluctuations are small enough one can linearize (3.46)

$$T_1' + \frac{U_1(u_1' - u_s)}{c_p} = T_2' + \frac{U_2(u_2' - u_s)}{c_p}. \tag{3.47}$$

Now assuming that the upstream flow satisfies the Morkovin hypothesis, one obtains the following expression for the linearized fluctuation of the stagnation temperature behind the shock wave

$$T_2' + \frac{U_2 u_2'}{c_p} = \frac{u_s(U_2 - U_1)}{c_p}. \tag{3.48}$$

Using this expression, one can write

$$\frac{T_2'}{\bar{T}_2} + (\gamma - 1)M_2^2\frac{u_2'}{U_2} = -(\gamma - 1)M_2(\mathfrak{C} - 1)\frac{u_s}{\bar{a}_2}, \tag{3.49}$$

where $\mathfrak{C}$ is the compression factor defined by

$$\mathfrak{C} = \frac{\bar{\rho}_2}{\bar{\rho}_1} = \frac{U_1}{U_2}, \tag{3.50}$$

Therefore, the Morkovin hypothesis holds downstream the shock wave if and only if the right hand side of (3.49) is zero, which is observed to be invalid for both, LIA and DNS results.

Influence of the Upstream Acoustic Fluctuations

The analysis of the influence of upstream acoustic waves is simpler than that for entropy waves, since these waves cannot be correlated with the vortical fluctuations, as their propagation speed is different. Therefore, the emitted far field is obtained by a simple superposition of the far fields corresponding to the vortical fluctuations and acoustic fluctuations considered separately.

3.4.4 Consequences for Subgrid Modeling

The above results show that the shock-isotropic-turbulence interaction, which is the simplest case of turbulence-shock interaction has the following features:

- The flow downstream of the shock is anisotropic.
- The interaction with the shock distributes energy among the three Kovasznay modes, independently of the character of the upstream flow.
- Important features of the downstream flow, such as the energy amplification ratio, strongly depend on both the relative importance of the three Kovasznay modes upstream of the shock, but also on the correlations between the entropy and the vorticity modes.
- The corrugation of the shock wave by incoming perturbations is a key element.
- The existence of decaying acoustic waves downstream the shock yields the existence of both a near field and a far field downstream the shock. The thickness of the near field region scales with the integral scale of incoming turbulence.

The last item in the list above is such that the issue of subgrid modeling for the shock/turbulence problem must be considered in three different ways, each way corresponding to a *level of resolution*. In analogy to the theory of interfaces between open flows and flows in porous media, one can distinguished between three resolution levels (see Fig. 3.13):

1. *The microscopic level*, in which (almost) all shock corrugation scales are directly captured on the grid. Since the shock responds to all perturbations, the microscopic level of description corresponds to a quasi-DNS of the shock-turbulence interaction. In this case, the main features of the interaction are directly captured, and special subgrid modeling is not needed. The shock is still represented as a 2D surface. The main practical problem is that of numerically capturing this discontinuity, by either

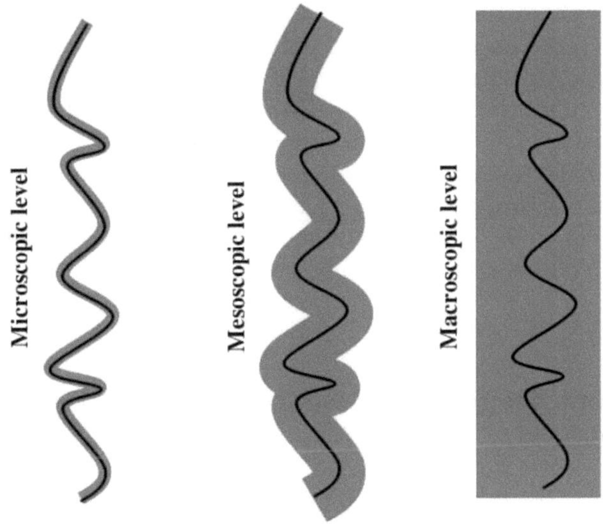

Fig. 3.13. Schematic view of the three LES resolution levels in flows with discontinuities. The *solid black line* denotes the original unfiltered shock front. The *gray area* is related to the thickened shock region obtained applying the smoothing operator

a shock tracking or a shock capturing technique. Available results show that for shock capturing techniques the subgrid contribution is masked by the effects of the artificial diffusion in cells located in the vicinity of the shock wave. An empirical criterion to design the mesh in the shock region is $\Delta x/\Delta y \sim u_{rms}/U$, where Δx and $\Delta y \sim \eta$ are the mesh size in the shock-normal and shock-tangential directions, respectively. Here, U and u_{rms} denote the upstream shock-normal mean and rms velocities, respectively.

2. *The mesoscopic level*, in which a significant part of the shock corrugation is now unresolved, but the grid is fine enough to represent the near-field downstream the shock. In this case, some of the subgrid terms appearing in the filtered jump conditions cannot be neglected and should therefore be taken into account by a special subgrid model. Practical experience shows that qualitatively correct results can be obtained even when such subgrid contributions are neglected.
3. *The macroscopic level*, in which the grid is so coarse that neither the shock corrugation nor the downstream near field are captured. Here, the full interaction must be taken into account by special subgrid models. The shock and the near field are fully embedded within one or two cells, and filtered jump relations are the basic model to be used.

Almost all published reliable LES results belong to the microscopic level. The possibility of deriving efficient subgrid models for the macroscopic level may be questioned. As a matter of fact, previous results dealing with the physics of the shock-turbulence interaction show that the decomposition of the upstream turbulence in terms of both scales and Kovasznay modes must be known to get a reliable prediction of the interaction. On the one hand, the scale decomposition is far from being trivial, but can be thought as a possible extension of multiscale/multiresolution subgrid modeling approaches. On the other hand, no efficient technique to split the instantaneous LES field given by conservative variables into Kovasznay modes has been proposed up to now. Therefore, the definition of general purpose, robust LES models at the macroscopic level seems to be out of reach for existing subgrid modeling strategies.

3.5 Different Regimes of Isotropic Compressible Turbulence

We now discuss the dynamics of compressible isotropic turbulence, the emphasis being placed on compressibility effects induced by a sufficiently high value of the Mach number based on the characteristic velocity of the turbulent motion. Dilatational effects observed in some low-speed flows in the presence of large temperature gradients (e.g. in flows dominated by natural convection) will not be considered. Numerical experiments and theoretical analyses show that several dynamical regimes exist in isotropic compressible turbulence, even in the freely decaying case where no external forcing is present. A major difficulty is that these regimes are very sensitive to a large number of parameters, such as the turbulent Mach number $M_t = \sqrt{\mathcal{K}}/a_0$ and the initial condition.

Three physical regimes will be distinguished, according to the influence of compressibility effects on the turbulence dynamics:

- The *low-Mach number quasi-isentropic regime*, in which the turbulent Mach number is low and the interactions between the solenoidal and dilatational components are weak. Moreover, the dilatational component is assumed to obey a quasi-linear acoustic dynamics. The vast majority of available studies are devoted to the case where the dilatational mode is restricted to the acoustic mode.
- The *nonlinear subsonic regime*, in which the turbulent Mach number is still less than one, but the fluctuations of the dilatational mode are strong enough to give rise to non-linear phenomena. In this case, some turbulence-induced very small shocks (referred to as *shocklets* or *eddy-shocklets*) are detected.
- The *supersonic regime*, in which the turbulent Mach number is larger than one. In this case, the dilatational mode is of great importance and shocklets have a considerable impact.

3.5.1 Quasi-Isentropic-Turbulence Regime

The quasi-isentropic regime is a regime in which the vorticity and the acoustic modes interact, the entropy mode having negligible influence. It occurs at relatively low turbulent Mach number ($M_t \leq 0.2$–0.4 can be considered as a reasonable upper bound). The corresponding dynamics has been investigated by many research groups, using both direct numerical simulation and theoretical analysis. It is worth noting that the high Reynolds number dynamics has been investigated only using extended versions of the EDQNM spectral model.

The main conclusions drawn from the existing literature are the following:

- There is no significant feedback of the acoustic modes on the vorticity modes. As a consequence, the solenoidal part of the velocity field evolves as in the strictly incompressible case. The interscale energy transfers are identical to those observed for incompressible isotropic turbulence.
- There is a one-way coupling between the vortical and the acoustic part of the solution. This coupling is associated with the physical mechanisms mentioned in Table 3.1, namely the production of acoustic waves by vortical mode self-interactions and the scattering of acoustic modes by vortical motion. The generation of acoustic waves, which represents a transfer of energy from the vortical modes towards the acoustic modes, must remain of small intensity for the vortical dynamics to be identical to the one observed for incompressible turbulence. The normalized acoustic-energy-production spectrum $\mathcal{P}^*_{acous}(k)$ deduced from advanced EDQNM analysis by Fauchet and Bertoglio [79, 80] is very similar to Lilley's model

$$\mathcal{P}^*_{acous}(\omega^*) = \frac{8}{3\pi S_t} \frac{\omega^*/(2S_t)^4}{(1 + (\omega^*/2S_t)^2)^3}, \quad S_t = 1.24, \tag{3.51}$$

where $\omega^* = a_0 k u'^2/\bar{\varepsilon}_s$ is the normalized frequency. The maximum production $\max(\mathcal{P}^*_{acous}(\omega^*)) \simeq 0.1$ is obtained for $\omega^* \simeq 3.5$. The total radiated acoustic power P_{tot} is then equal to

$$P_{tot} = \int_0^{k_\eta} \mathcal{P}_{acous}(k) dk \propto \bar{\varepsilon}_s M_t^5, \tag{3.52}$$

which is in very good agreement with usual estimates found by the Lighthill acoustic analogy.
- Some nonlinear acoustic phenomena, such as self-steepening and self-scattering, can occur.

The analysis can be further refined when a statistical equilibrium state is reached by isotropic turbulence. The spectral features of the solution at high Reynolds number has been investigated in detail by Fauchet and Bertoglio [79, 80] using an extended EDQNM approach. Main features of this equilibrium solution are illustrated in Fig. 3.14. As mentioned above $E_{ss}(k)$ is

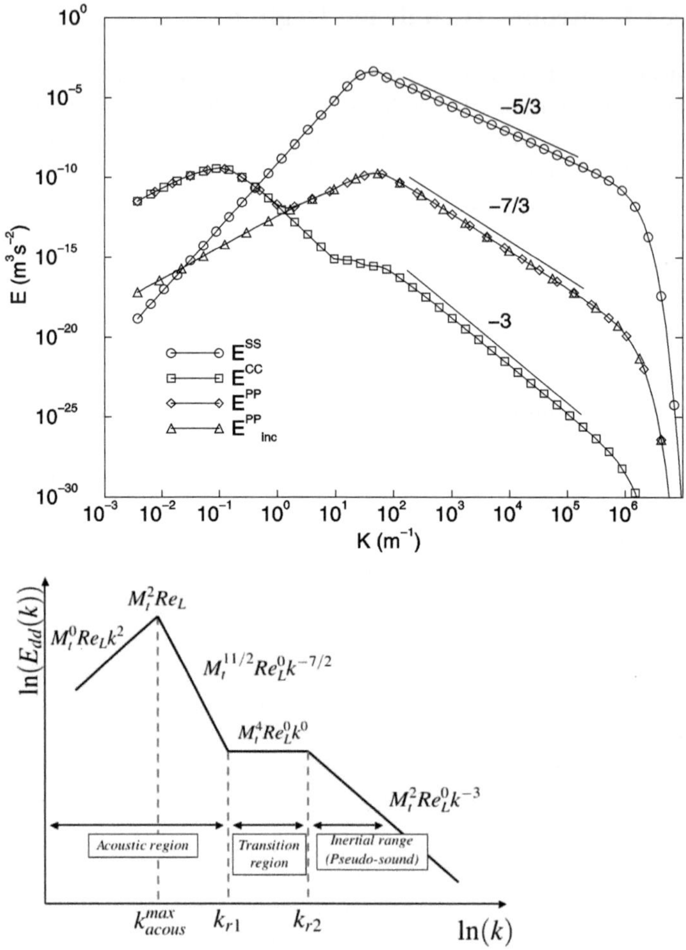

Fig. 3.14. *Top*: Spectra in the nonlinear equilibrium state computed using an extended EDQNM-type closure for compressible flows (the dilatational energy spectrum is denoted E^{cc} instead of E_{dd}). Courtesy of G. Fauchet and J.P. Bertoglio. *Bottom*: schematic view of the dilatational kinetic energy spectrum for $M_t \leq 0.1$–0.2

the same as for the incompressible case and exhibits an inertial range with a $-5/3$ exponent. The dilatational energy spectrum is much more complex. The analysis shows that three spectral bands must be distinguished, each band corresponding to a different type of dynamics:

- The *acoustic region*, which corresponds to modes such that $k < k_{r1}$, with

$$k_{r1} \simeq 1.8 k_L, \tag{3.53}$$

where k_L corresponds to the peak of $E_{ss}(k)$. In this region, the dilatational motion essentially obeys linear acoustic dynamics, and the so-called *strong acoustic equilibrium* is valid: the kinetic energy of the dilatational mode is equal to the potential energy of the pressure mode at all scales, i.e.

$$E_{dd}(k) = \frac{1}{\rho_0^2 a_0^2} E_{pp}(k). \tag{3.54}$$

One can show that in this region

$$E_{dd}(k) \propto \begin{cases} M_t^0 Re_L^0 k^2 & k < k_{r1}, k \ll k_{acous}^{\max}, \\ M_t^{11/2} Re_L^0 k^{-7/2} & k < k_{r1}, k \gg k_{acous}^{\max}, \end{cases} \tag{3.55}$$

where the peak of $E_{dd}(k)$ is found at $k_{acous}^{max} \propto M_t k_L$ and is proportional to $\mathcal{K}_s^{5/2} M_t^2 \bar{\varepsilon}_s^{-1}$. Integrating $E_{dd}(k)$ in this spectral band one obtains that the *turbulent acoustic kinetic energy* scales like $\mathcal{K}_s M_t^3$.

- The *transition region*, for modes $k_{r1} < k < k_{r2}$, where

$$k_{r2} \propto M_t^{7/11} Re_L^{2/11} k_L. \tag{3.56}$$

In this region, nonlocal nonlinear effects are not negligible, the strong linear acoustic equilibrium does not hold and $E_{dd}(k)$ exhibits a plateau whose amplitude scales like M_t^4.

- The *pseudo-sound region*, for $k_{r2} < k$, in which strong local nonlinear interactions lead to the definition of another inertial range in which $E_{dd}(k) \propto M_t^4 k^{-3}$. The compressible turbulent kinetic energy contained in this inertial range and the corresponding dilatational dissipation are found to be equal to

$$\mathcal{K}_d = \int_{k \geq k_{r2}} E_{dd}(k) dk \propto \mathcal{K}_s M_t^4, \tag{3.57}$$

and

$$\bar{\varepsilon}_d = \frac{4}{3} \nu \int_{k \geq k_{r2}} k^2 E_{dd}(k) dk \propto \bar{\varepsilon}_s M_t^4 \frac{\ln(Re_L)}{Re_L}, \tag{3.58}$$

respectively.

The energy balance associated with the statistical equilibrium state (at low Reynolds number) has been analyzed in both decaying and forced isotropic turbulence [139, 141, 200]. In these studies, the transfers between the dilatational turbulent kinetic energy $\mathcal{K}_d$, the solenoidal turbulent kinetic energy $\mathcal{K}_s$ and the fluctuating internal energy $\tilde{e}$ have been investigated. The main conclusions with respect to the global energy transfers at the equilibrium state are the following:

1. In the acoustic equilibrium state both $\mathcal{K}_d(t)$ and $\tilde{e}(t)$ fluctuate sinusoidally about a constant mean value (see Fig. 3.15). The two signals have opposite phases and similar amplitude, leading to $\mathcal{K}_d(t) + \tilde{e}(t) \simeq$ const.
2. The solenoidal kinetic energy $\mathcal{K}_s$ varies slowly with irregular fluctuations of small amplitude and exhibits phase locking with either $\mathcal{K}_d(t)$ or $\tilde{e}(t)$.

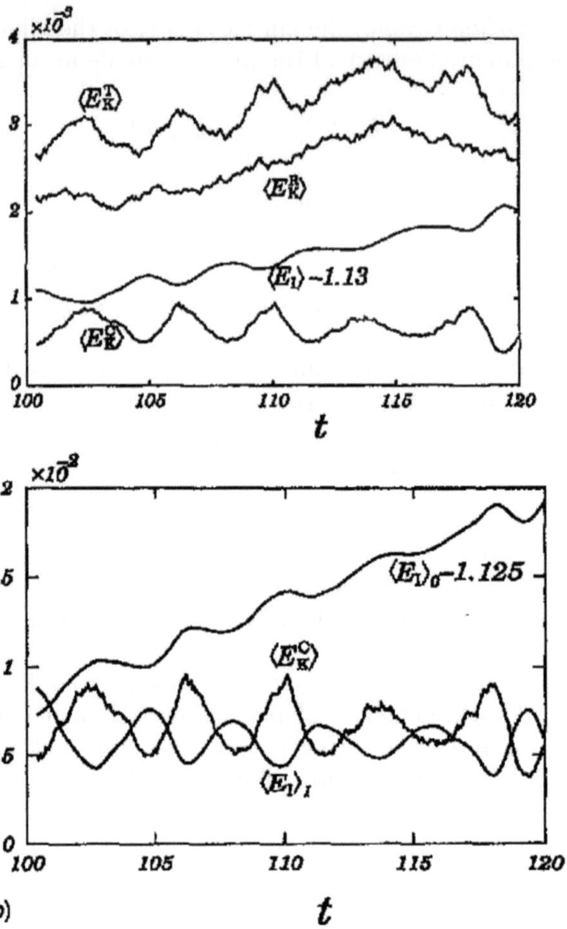

(b)

Fig. 3.15. Computed time history of volume-averaged energies in forced compressible isotropic turbulence. $\langle E_K^T \rangle, \langle E_K^R \rangle, \langle E_K^C \rangle$ and $\langle E_I \rangle$ are the full turbulent kinetic energy, solenoidal kinetic energy, dilatational kinetic energy and internal energy, respectively. Since a source term is present the mean internal energy undergoes a constant growth, and is split as the sum of a uniform part $\langle E_I \rangle_0$ and a turbulent part $\langle E_I \rangle_1$. Reproduced from Ref. [200] with permission

3. The interactions between the solenoidal and dilatational components of the turbulent kinetic energy are weaker than self-interactions of the respective components.
4. The pressure-dilatation term governs the coupling between $\mathcal{K}_d(t)$ and $\tilde{e}(t)$ (see Fig. 3.16). It is also observed to overwhelm other terms which appears in the evolution equations for $\mathcal{K}_d(t)$, the total mean turbulent kinetic energy $\mathcal{K} = \mathcal{K}_d + \mathcal{K}_s$ and the internal energy $\tilde{e}$. It exhibits a periodic

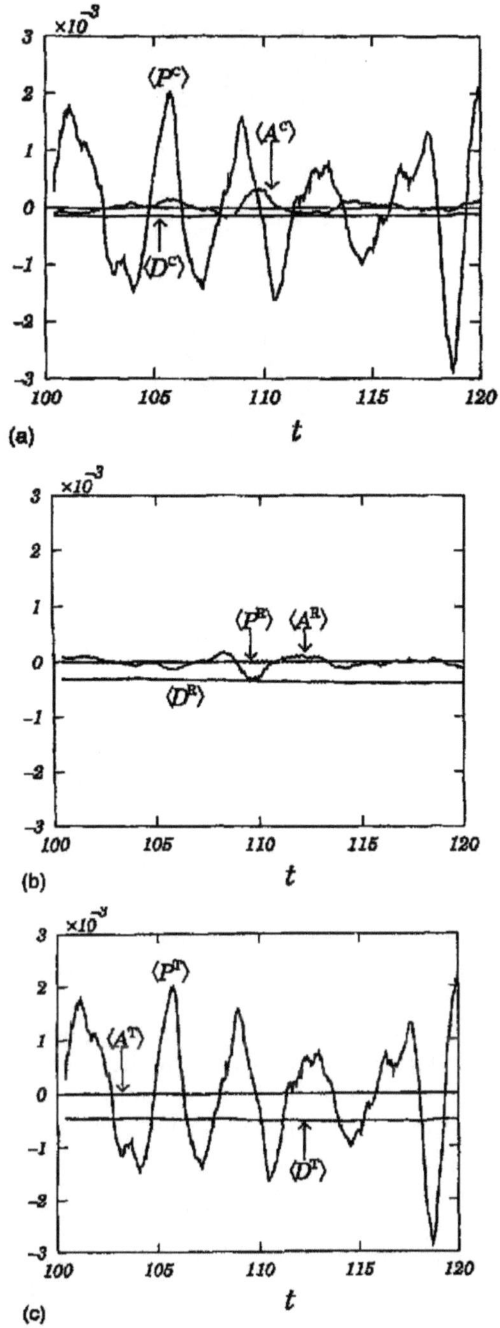

Fig. 3.16. Time histories of the volume-averaged budget terms for the dilatational turbulent kinetic energy (*top*), the solenoidal turbulent kinetic energy (*middle*) and the internal energy (*bottom*). Terms A, D and P denote the advection, viscous diffusion and pressure terms, respectively. Reproduced from Ref. [200] with permission

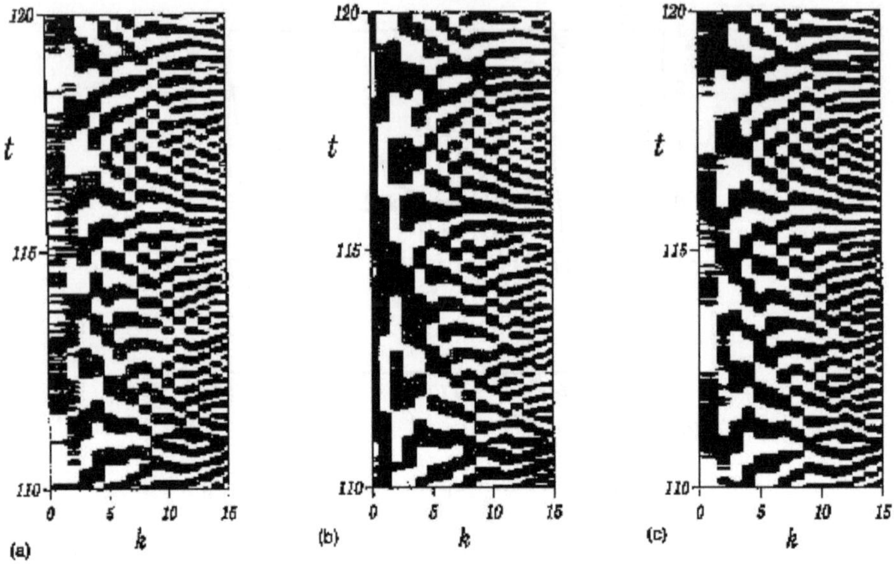

Fig. 3.17. Time histories of the wave number spectra of dilatational turbulent kinetic energy (*left*), internal energy (*middle*) and dilatational pressure (*right*). *White* (resp. *dark*) *regions* are regions where the instantaneous spectrum coefficients are decreasing (resp. increasing) in time. Reproduced from Ref. [200] with permission

behavior with the same period as $\mathcal{K}_d(t)$ and $\tilde{e}(t)$, and is caused by acoustic pressure fluctuations.

This dynamical picture can be further refined looking at energy exchanges at individual wave numbers. The main findings of Miura and Kida [200] are:

1. Periodic behavior of the compressible kinetic energy and the internal energy is observed at each wave number in the spectra associated to these quantities, $E_{dd}(k,t)$ and $E_e(k,t)$, respectively (see Fig. 3.17). The same observation holds for the compressible pressure spectrum $E_{pp}(k,t)$.
2. The period of oscillation $\tau(k)$ depends on the wave number and is the same for the three spectra at each wave number. The measured period corresponds almost exactly to that of acoustic waves

$$\tau(k) = \frac{\pi}{a_0 k}. \tag{3.59}$$

At every wave number it is found that the phase of oscillation of $E_{pp}(k,t)$ precedes that of $E_{dd}(k,t)$ and lags behind that of $E_e(k,t)$ by a quarter of period.

3.5.2 Nonlinear Subsonic Regime

The regime discussed above is expected to occur in the limit of nearly incompressible turbulence, i.e. at very low turbulent Mach number. When M_t cannot longer be considered as a small parameter the asymptotic analyses presented above are theoretically no longer valid, since one expects that the non-linearities arising from the convective terms can play a major role. As a matter of fact, numerical simulations show that small shocks, referred to as *shocklets* or *eddy-shocklets* can develop.

Conditions for Occurrence of Shocklets

Before discussing the properties of these shocklets and analyzing their influence on the dynamics and the statistical properties of compressible isotropic turbulence, it is worth noting that they can occur at very low Mach numbers depending on the initial condition. No exact threshold value for the turbulent Mach number M_t associated to the occurrence of shocklets is known since this phenomenon also depends on other parameters. It seems that all simulations carried out for $M_t \geq 0.4$ exhibit shocklets.

Energy Budget and Shocklet Influence

Shocklets are weak normal shocks which satisfy the Rankine–Hugoniot jump conditions. They induce sharp pressure and density gradients and, because they are associated with a compression, a negative velocity divergence. As a consequence, one can reasonably expect that they should have a non-negligible influence on the energy balance. We summarize below observations retrieved from direct numerical simulations. Since all these simulations were carried out at small Reynolds number viscous effects are important and they damp the effect of the shocklets. Therefore, exact values of the quantities given below must be interpreted as a qualitative description of high-Reynolds number flows rather than quantitative.

Numerical experiments show that the probability density function of the dilatation (i.e. the divergence of the velocity field) is strongly skewed: about 2/3 of the volume is associated with an expansion ($\nabla \cdot \boldsymbol{u} > 0$), while only 1/3 corresponds to compression. On the average, the expansion regions are responsible for 80–90% of the solenoidal dissipation $\bar{\varepsilon}_s$ and 50–60% of the total dissipation ($\bar{\varepsilon}_d + \bar{\varepsilon}_s$). The global dilatational dissipation $\bar{\varepsilon}_d$ is found to be small with respect to the global solenoidal dissipation $\bar{\varepsilon}_s$: Lee and coworkers [157] report that $\bar{\varepsilon}_d$ is less or equal than 10% of the total dissipation for M_t up to 0.6.

The shocklets fill only a few percent of the total volume: Pirozzoli and Grasso [218] found that they represent only 1.4% of the volume at $M_t = 0.8$ while Samtaney et al. [247] report a fraction smaller than 2% in their set of numerical experiments. Nevertheless, the shocklets strongly modify the local relative importance of the physical mechanisms: near shocklets, the dilatational dissipation is up to 10 times larger than the solenoidal dissipation. Although they fill only a very small part of the fluid domain shocklets are responsible for about 20% of the global dilatational dissipation.

The shocklets perturb the dilatation field. This perturbation can be roughly estimated from the jump condition for the dilatation for a normal shock moving into a two-dimensional inviscid steady flow provided by Kida and Orszag [139, 141]

$$[[\nabla \cdot \boldsymbol{u}]] \simeq \frac{2}{R(\gamma+1)} \left(\frac{(\gamma-1)M_s^2+2}{(\gamma+1)M_s^2} - \frac{3M_s^2+1}{M_s^2-1} \tan^2\theta \right) u_n, \tag{3.60}$$

where R, u_n, M_s and θ are the radius of curvature of the shock, relative velocity normal to the shock, the shocklet Mach number defined by the ratio of u_n about the upstream speed of sound, the angle between the fluid velocity and the shock normal, respectively. It can be seen that the sign of the induced dilatation depends on both R and θ, and that its amplitude is a function of the square of the normal Mach number. The use of a two-dimensional simplified model was proved to be qualitatively relevant by Kida and Orszag, since the three-dimensional curved shock can be locally projected on a two-dimensional space. Lee and coworkers observed that the correlation between pressure fluctuations and dilatation fluctuations is large near shocklets, leading to a local enhancement of the transfer between the internal energy and the turbulence kinetic energy. These authors also report that the overall effect of the pressure-dilatation term $\overline{p'd}$ on the evolution of kinetic energy in freely decaying isotropic turbulence is comparable to the overall dilatational dissipation $\bar{\varepsilon}_d$. This effect is typical of the presence of the shocklets, since this term is theoretically and experimentally found to be negligible in the pseudo-acoustic regime.

Enstrophy Budget and Shocklet Influence

The shocklets also have a large impact on the dynamics of vorticity and enstrophy. It is important to emphasize that the main trends and the relative importance of the different physical mechanisms are not the same in the shocklet case as in the strong-shock case discussed above. The main reason for this is a scale effect: shocklets are small shock waves which form when turbulent eddies allow for the local steepening of pressure waves, and their size is comparable to that of the turbulent eddies, whereas the extent of strong shocks is much larger than that of turbulent vortical structures.

We first recall an estimate related to vorticity creation by a normal shock moving in a steady, inviscid two-dimensional flow [140]

$$[[\boldsymbol{\Omega}]] \simeq \frac{4(M_s^2-1)\sin\theta}{R(\gamma+1)M_s^2\left((\gamma-1)M_s^2+2\right)} |\boldsymbol{u}|, \tag{3.61}$$

where the nomenclature is the same as in the previous paragraph. The sign of the created vorticity is seen to depends on the local shock curvature and the angle of incidence. The created vorticity is zero for normal shocks ($\theta = 0$) and Mach waves ($\theta = \pm \cos^{-1}(1/M_s)$). The effect of the baroclinic term $-(\nabla p \times \nabla \rho)/\rho^2$ is evaluated as

$$[[u_n \boldsymbol{\Omega}]] \simeq \frac{4(M_s^2 - 1)^2}{R(\gamma + 1)^2 M_s^4} u_n^2 \tan\theta \tag{3.62}$$

and is observed to depend on M_s^4 instead of M_s^2 for the global vorticity creation.

Numerical experiments show that the volume-averaged enstrophy budget is governed by the vortex stretching term $\boldsymbol{\Omega} \cdot \mathbf{S} \cdot \boldsymbol{\Omega}$ and the viscous dissipation. The former is positive and creates vorticity, while the latter is strictly negative. The baroclinic term is negligible, while the compression term $\Theta(\nabla \cdot \boldsymbol{u})$ exhibits an oscillatory behavior with a period very similar to that of the compressible kinetic energy and the internal energy. Therefore, this phenomenon is interpreted as a coupling between acoustic waves and the vorticity.

A more detailed analysis can be achieved by distinguishing between regions of negative dilatation and regions of positive dilatation. The compression term is observed to be dominant in shocklet areas, while the stretching term is the most important in expansion region. A careful look at direct numerical simulation data reveals that vorticity is created on shocklets through the baroclinic interaction and is enhanced in expansion regions by the vortex stretching phenomenon. The baroclinic production is relatively small because there is a clear trend for the pressure gradient and the density gradient to be aligned with each other: the global probability density function of the angle between these vectors exhibits a peak near 4 degrees, and is almost null for angles higher than 10 degrees, even for values of M_t as high as 0.74. It is also found that increasing the turbulent Mach number yields a stronger alignment of these vectors. Due to the weakness of the baroclinic production Kida and Orszag observed that the barotropic relation

$$\left(\frac{p}{\bar{p}}\right) = \left(\frac{\rho}{\bar{\rho}}\right)^\gamma \tag{3.63}$$

is overall valid.

A last observation is that the vorticity has a statistical preference to align with the density gradient $\nabla\rho$ near the shocklets and to be orthogonal to it outside shocklet areas. Since the shocklets fill a very small fraction of the fluid domain, the overall p.d.f. of the angle between $\boldsymbol{\Omega}$ and $\nabla\rho$ has a peak at 90 degrees. As in the incompressible case, the vorticity is observed to be aligned on the overall with intermediate eigenvector of the velocity gradient tensor.

3.5.3 Supersonic Regime

The supersonic regime, in which the turbulent Mach number is larger than unity, is much less known than the other regimes. The main reason why so little attention as been paid to this configuration is that it is encountered in astrophysics only, and that it escapes in-depth theoretical analysis since it does not allow small parameter expansion.

Only very few numerical experiments are available [221–223], which all reveal the existence of two distinct quasi-equilibrium phases separated by a short transition phase:

- The *quasi-supersonic phase*, whose typical duration is of the order of a few acoustic time scale. During this initial period, nonlinear phenomena yield the formation of a myriad of small but intense shock waves. No vortical structures are observed during this period. Then, the shocks interact, leading to the existence of vortex sheets which roll-up due to Kelvin–Helmholtz-type instabilities, generating vortex tubes. These vortex tubes experience vortex stretching, leading to the appearance of the usual kinetic energy cascade phenomenon. During this phase, which is dominated by shock formation and shock interaction, the evolution of the vorticity is governed by the baroclinic production and the linear terms (vortex stretching and dilatation terms), which are of equal amplitude. At the end of the quasi-supersonic phase, both dilatational velocity spectrum and solenoidal velocity spectrum exhibit an inertial range with a -2 slope

$$E_{dd}(k) \propto k^{-2}, \qquad E_{ss}(k) \propto k^{-2}. \tag{3.64}$$

  It is worth noting that most of turbulent kinetic energy is contained in the solenoidal mode once the vortical structures have been created.
- The *immediate post-supersonic phase* which is governed by vortex interaction and vortical decay. The main processes involved in subsonic vortex dynamics are present, but shocks are still present and very active. As a consequence, the following inertial range scalings are observed:

$$E_{dd}(k) \propto k^{-2}, \qquad E_{ss}(k) \propto k^{-1}. \tag{3.65}$$

  The vorticity dynamics is dominated by the vortex stretching and the dilation term during this phase, the baroclinic production being now much weaker due to the decrease of the turbulent Mach number.

At much longer times, an equilibrium state similar to the subsonic regime is recovered in which the shocks are much weaker and the solenoidal velocity dynamics is decoupled (at the leading order approximation) from the acoustic field. The measured inertial range behaviors are $E_{dd}(k) \propto k^{-2}$ and $E_{ss}(k) \propto k^{-5/3}$.

3.5.4 Consequences for Subgrid Modeling

The subgrid modeling issue must be discussed separately in the three regimes described in the preceding sections.

In the quasi-isentropic low-Mach-number regime, almost the entire turbulent kinetic energy is contained in the solenoidal modes since $\mathcal{K}_d/\mathcal{K}_s \propto M_t^4$, and the kinetic energy dissipation is mostly due to the solenoidal dissipation, since $\bar{\varepsilon}_d/\bar{\varepsilon}_s \propto M_t^4 \ln(Re_L)/Re_L$. Therefore, it is expected that it will not be necessary to account for compressibility effects to derive an efficient subgrid model, and that using variable-density extensions of subgrid models designed for incompressible flows will be sufficient. One can also note that the difference in the location of the peaks of E_{ss} and E_{dd} scales as M_t. Therefore, the size of the computational domain must be very large compared to that used for incompressible isotropic turbulence to capture all acoustic modes. But since there is no significant feedback of acoustic modes on the solenoidal motion the use of a mesoscale model for unresolved very large scales is not mandatory.

Let us now consider the subsonic nonlinear regime. The main difference with the quasi-isentropic regime is the occurrence of shocklets. The dynamics of the solenoidal motion is still very similar to that observed in strictly incompressible turbulence, and most of the total turbulent kinetic energy is carried by vorticity modes. Shocklets are rare (they usually fill less than 2% of the total volume) but represent up to 20% of the global dilatational dissipation. The most probable value of their thickness is about 5η, where η is the Kolmogorov length. One can therefore expect that most of the shocklets will be part of the subgrid motion. Their main effect is an increase of the dilatational dissipation $\bar{\varepsilon}_d$, and mostly of its subgrid contribution. But since $\bar{\varepsilon}_d$ is much smaller than the solenoidal dissipation $\bar{\varepsilon}_s$, even at $M_t = 0.6$, it does not appear necessary to develop new subgrid models for them. A slight modification of the dissipation induced by usual incompressible subgrid models might be enough. And since the LES results are not very sensitive to a change of the subgrid dissipation of the order of one or two percent it is likely that a simple variable-density extension of incompressible subgrid models may lead to satisfactory results.

The case of the supersonic regime is very different. In this regime, usual inertial ranges are not present, since the dynamics are not driven by the usual physical mechanisms. Interscale energy transfers are mostly due nonlinear acoustic phenomena and shock wave interactions. It is closer to the dynamics generated by the Burgers equation than to the classical Navier–Stokes turbulence. Therefore, usual subgrid models are not relevant, and their successful use should be considered as fortuitous. But it must be borne in mind that the simulation of such a turbulent field necessarily requires the use of shock capturing schemes, and that the Implicit LES approach seems to be the most adequate way to address this problem.

Let us conclude this section by a last comment. The subgrid terms appear as source terms in filtered governing equations for resolved variables, which are formally identical to the original Navier–Stokes equations. Therefore, the

Kovasznay system (3.6)–(3.8) shows that subgrid terms will act on the three physical modes at the same time. A straightforward conclusion is that subgrid models should ideally be designed to recover the correct effect on the three modes at the same time. Let us illustrate this point on the example of the subgrid term in the momentum equations. This term can be considered as an extension of its incompressible counterpart. Whereas in the incompressible case most subgrid models are designed to recover the correct balance of resolved turbulent kinetic energy,[1] in the compressible case it should also recover the correct acoustic subgrid contribution. In a similar way, the subgrid terms that appear in the energy equation should be modeled in such a way that subgrid production/destruction of both resolved energy and resolved acoustic field will be accurately taken into account. This new multiphysics constraint has been ignored in almost all works dealing with compressible subgrid scale development. The main reasons are that many compressible LES applications put the emphasis on the aerodynamic field, and that in existing aeroacoustics-oriented LES applications the subgrid contribution to noise generation is very small compared to that of resolved scales. Nevertheless, an ideal, physically fully consistent closure should account for the existence of different physical modes at all scales.

[1] This is the case for almost all subgrid-viscosity-based models.

4

Functional Modeling

According to the conclusions of Chap. 3, in the quasi-isentropic low-Mach regime, almost the entire turbulent kinetic energy is contained in the solenoidal modes, and the kinetic energy dissipation is mostly due to the solenoidal dissipation. Therefore, for this regime it is reasonable to assume that compressibility effects can be neglected when deriving an effective subgrid-scale model. This fundamental assumption is the basis of functional modeling for compressible flows being rooted in a incompressible framework. After introducing the classical concept of eddy-viscosity, we present some functional subgrid-scale (SGS) models that have been used for compressible flows. The modeling of the isotropic part of the SGS tensor and of the turbulent heat flux is discussed later. Before concluding with a brief introduction to implicit modeling from the functional perspective, some techniques improving the performance of eddy-viscosity models are detailed.

4.1 Basis of Functional Modeling

4.1.1 Phenomenology of Scale Interactions

Most of available theoretical tools have been developed in the framework of homogeneous turbulence, and it is in this framework that most of the SGS models have been set-up. These theories usually are developed in the spectral space [173, 245] and aim at quantifying the interaction between the different scales of turbulence.

The scenario proposed to describe the inter-scale transfer is the following [302]. The largest eddies (associated to the smallest wave numbers) extract kinetic energy from the mean flow. Such structures which can be found in high-mean-strain areas, are initially relatively thick ("pancake-like") and flatten into vortex sheets. These sheets are unstable and roll-up into vortex tubes which stretch producing filamentary structures. From an energetic point of view, this results in a direct energy transfer from the largest to the smallest

E. Garnier et al., *Large Eddy Simulation for Compressible Flows*,
Scientific Computation,

scales, until the molecular viscosity dissipates the kinetic energy into internal energy.

A spectrum typical for homogeneous turbulence is sketched in Fig. 4.1. It encompasses a production zone (at the smallest wave numbers), a zone in which energy is transferred from the large scales to the small ones (outscatter), and eventually a dissipation zone at the largest wavenumbers. The transfer zone is also called inertial zone in which the spectral energy density decreases following the Kolmogorov law (with a $k^{-5/3}$ slope). Moreover, there exists an energy transfer from the small scales to the large ones (inverse-cascade phenomenon, backscatter) but its intensity is much weaker than outscatter. For the small wavenumbers, the slope of the spectrum ranges between k^1 and k^4 (see [245], Sect. 3.2).

Considering a sharp cut off filter,[1] SGS models are constructed by assuming that the cut-off wave number ($k_c = \pi/\Delta$) is located within the inertial range (see Fig. 4.1) and that the smallest scales are isotropic. This means that they are independent from the large energetic scales (which are anisotropic in the general case). The small-scales isotropy implies some universality of SGS models which are based on this hypothesis. Additionally, the fact that the cut-off is located in the inertial zone allows to assume that the energy transfer through the cut-off range is equal to the dissipation rate of the smallest scales. Finally, these smallest scales are assumed to be in energetic equilibrium with the large ones and to adjust instantaneously to their level of energy. This is reasonable considering the eddy turn-over time of such small structures.

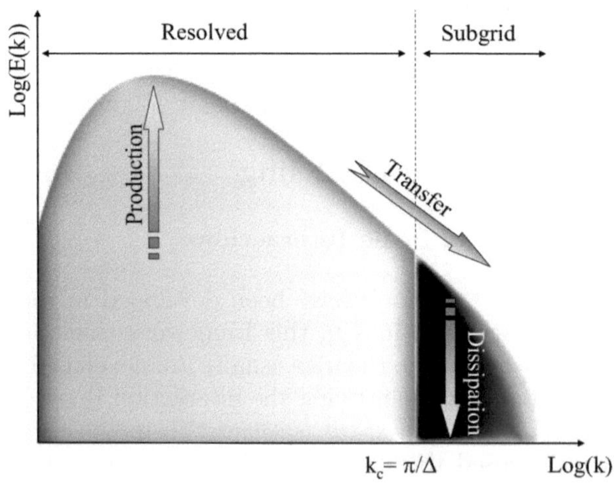

Fig. 4.1. Spectrum of homogeneous turbulence including the definition of the filter (case of a sharp cut-off filter)

[1] It is not a necessary condition and a formalism has been derived for Gaussian filters.

4.1.2 Basic Functional Modeling Hypothesis

All functional subgrid models make a more or less implicitly use of the following hypothesis:

Hypothesis 4.1. *The action of the subgrid scales on the resolved scales is essentially an energetic action, so that the sole balance of the energy transfer between the two scale ranges is sufficient to describe the action of the subgrid scales.*

Using this hypothesis as a basis for modeling we neglect a part of the information contained in the small scales, such as the structural information related to the anisotropy. As was seen above, the energy transfer between subgrid scales and resolved scales mainly exhibits two mechanisms: a forward energy transfer toward the subgrid scales and a backward transfer to the resolved scales which is much weaker in intensity. All the approaches existing today for numerical simulation at high Reynolds numbers consider the energy lost by the resolved scales, while for compressible flows, no attempt has been made to consider the backward energy cascade. Following the above hypothesis, modeling consists in modifying the different evolution equations of the system in such a way as to integrate the desired dissipation or energy production effects into them. Two different approaches can be found in current work:

- *Explicit modeling* of the desired effects, i.e. including them by adding additional terms to the equations: the actual subgrid models.
- *Implicit inclusion* by the numerical scheme which is adjusted so that the truncation error induces the desired effects.

Let us note that while the explicit approach is what would have to be called the classical modeling approach, the implicit one appears generally only as an *a posteriori* interpretation of dissipative properties for certain numerical methods used. This issue is discussed in detail in Chap. 6 where the interaction of the numerical method with SGS modeling is treated.

4.2 SGS Viscosity

This section is dedicated to the modeling of the SGS tensor which is the only term present in the incompressible equations. Therefore, this term deserves particular attention. Most of the material presented here can also be found in Ref. [244].

4.2.1 The Boussinesq Hypothesis

The forward energy cascade is modeled explicitly using the following hypothesis:

Hypothesis 4.2. *The energy transfer mechanism from the resolved to the subgrid scales is analogous to molecular mechanisms represented by the diffusion term, in which the molecular viscosity appears.*

This hypothesis is equivalent to assuming that the behavior of the subgrid scales is analogous to Brownian motion superimposed onto the motion of the resolved scales. In the kinetic theory of gases, molecular motion draws energy from the flow by way of molecular viscosity. Accordingly, the energy cascade mechanism is modeled by a term having a mathematical structure similar to that of molecular diffusion, but in which the molecular viscosity is replaced by a subgrid viscosity denoted ν_{sgs}. Following a Boussinesq type hypothesis [31], this choice of mathematical form of the subgrid model is written[2] as

$$\tau_{ij}^{d} = \tau_{ij} - \frac{1}{3}\delta_{ij}\tau_{kk} = -2\bar{\rho}\nu_{sgs}\left(\tilde{S}_{ij} - \frac{1}{3}\delta_{ij}\tilde{S}_{kk}\right). \tag{4.1}$$

A scalar subgrid viscosity requires the adoption of the following hypothesis: *A characteristic length l_0 and a characteristic time t_0 are sufficient for describing the subgrid scales.* By dimensional reasoning we arrive at

$$\nu_{sgs} = \frac{l_0^2}{t_0}. \tag{4.2}$$

Models of the form (4.1) are local in space and time, which is necessary to be practically useful. This local character, similar to that of the molecular diffusion terms, implies [244]:

Hypothesis 4.3 (Scale Separation Hypothesis). *There exists a total separation between the subgrid and resolved scales.*

Filtering associated to large-eddy simulation does not introduce such a separation between resolved and subgrid scales because the turbulent energy spectrum is continuous. The characteristic scales of the smallest resolved scales are consequently very close to those of the largest subgrid scales. This continuity gives rise to the existence of a spectrum region near the cutoff where the effective viscosity varies rapidly as a function of wave number. The result of this difference in nature with the molecular viscosity is that the subgrid viscosity is not a characteristic quantity of the fluid but of the flow. It is worth noting that subgrid-viscosity based models for the forward energy cascade induce an unphysical alignment of the eigenvectors for resolved strain rate tensor and subgrid-scale tensor.

Eventually, the local equilibrium hypothesis is employed to calibrate the model constant:

Hypothesis 4.4 (Local Equilibrium Hypothesis). *The flow is in constant spectral equilibrium, so there is no accumulation of energy at any frequency and the shape of the energy spectrum remains invariant with time.*

---

[2] In the particular case of high-pass filtered eddy-viscosity models, the strain rate tensor is computed using a filtered velocity field, see (4.29).

This implies an instantaneous adjustment of all scales of the solution to the turbulent-kinetic-energy production mechanism, and therefore a balance between the production, dissipation, and energy flux through the cutoff.

In the following, we restrict our discussion to models expressed in physical space and to models that have been used for computations in the compressible regime. Under these two restrictions, it follows that, to our knowledge, authors who have contributed to this field have limited themselves to a small subset of existing explicit functional models, namely the Smagorinsky model, the structure function model, and the mixed scales models. These models are no more than variable-density extensions of their incompressible counterparts.

4.2.2 Smagorinsky Model

The simplest expression of (4.2) is the Smagorinsky model [264]. The time scale is obtained from the resolved strain rate tensor which gives

$$\nu_{sgs} = C_s^2 \Delta^2 |\tilde{S}|, \tag{4.3}$$

with

$$|\tilde{S}| = (2\tilde{S}_{ij}\tilde{S}_{ij})^{1/2}. \tag{4.4}$$

Using the local equilibrium hypothesis and assuming a Kolmogorov spectrum, the constant can be evaluated as

$$C_s = \frac{1}{\pi}\left(\frac{3K_0}{2}\right)^{-3/4} \sim 0.18, \tag{4.5}$$

with the Kolmogorov constant K_0 equal to 1.4. Using the same kind of hypothesis and an evaluation of the averaged SGS viscosity with EDQNM theory, the value of 0.148 can also be found. Finite Reynolds number and spectrum shape effects can be taken into account leading to a complex definition of C_s [197]. These evaluations of the Smagorinsky constant are based on an average rate of dissipation, whereas the SGS viscosity expression here is localized in space and time for practical reasons. This constant can be interpreted as the ratio of the mixing length associated to the non-resolved scale to the filter cut-off length. The latter one is generally evaluated as $\Delta = (\Delta x \Delta y \Delta z)^{1/3}$.[3] Practically, the Smagorinsky constant takes value ranging from 0.1 to 0.2 depending on the flow.

In addition to this flow dependency of the constant, this model is generally over-dissipative in regions of large mean strain. This is particularly true in the transition region between laminar and turbulent flows. Moreover, the limiting behavior near the wall is not correct, and the model predictions correlate poorly with the exact SGS tensor (a correlation coefficient of 0.3 is obtained at best). Despite all these identified drawbacks, its simplicity makes this model still very popular. Most of the time it is used with a damping function to correct its erroneous near-wall behavior.

[3] Other expressions more adapted to anisotropic filtering have been proposed, see for example [254].

4.2.3 Structure Function Model

This model is a transposition of Métais and Lesieur constant effective viscosity model into the physical space, and can consequently be interpreted as a model based on the energy at cutoff, expressed in physical space. The authors [196] propose evaluating the energy at cutoff $E(k_c)$ by means of the second-order velocity structure function. This is defined as

$$F_2(\mathbf{x}, r, t) = \int_{|x'|=r} [\bar{\mathbf{u}}(x, t) - \bar{\mathbf{u}}(x + x', t)]^2 d^3\mathbf{x}'. \tag{4.6}$$

In the case where $r = \Delta$, the model takes the simplified form

$$\nu_{sgs}(x, \Delta, t) = 0.105 K_0^{-3/2} \Delta \sqrt{F_2(x, \Delta, t)}. \tag{4.7}$$

The function F_2 is homogeneous with respect to a norm of the resolved velocity gradient. If this function is evaluated in the simulation in a way similar as the resolved strain rate tensor is computed for the Smagorinsky model, we can in theory expect the Structure Function model to suffer some of the same weaknesses: the information contained in the model is local in space, therefore non-local in frequency, which induces a poor estimation of the kinetic energy at cutoff and a loss of precision of the model in the treatment of large-scale intermittency and spectral nonequilibrium. The discrete form of the function $F_2(\mathbf{x}, r, t)$ can be found in [47].

4.2.4 Mixed Scale Model

Models based solely on large scales (such as the Smagorinsky model) cannot be expected to vanish in areas where all dynamic scales are resolved. Moreover, noticing that the indication of an under-resolved simulation is the accumulation of kinetic energy near the cut-off (in spectral space), Sagaut and Loc [239] propose to base their model on the turbulent kinetic energy at cut-off which acts as a sensor to detect areas where SGS modeling is necessary. A one parameter model is proposed. The SGS viscosity is written as a non-linear combination of the second invariant $|\widetilde{S}|$ of the stress tensor, of the characteristic length Δ and of the kinetic energy of the smallest resolved scales q_c^2

$$\nu_{sgs} = C_m |\tilde{S}|^{\alpha} (q_c^2)^{\frac{(1-\alpha)}{2}} \Delta^{(1+\alpha)}. \tag{4.8}$$

In the compressible framework, the kinetic energy of the smallest resolved scales is defined as

$$q_c^2(x, t) = \frac{1}{2} (\tilde{u}_i(x, t))' (\tilde{u}_i(x, t))'. \tag{4.9}$$

The test field $(\tilde{u})'$ represents the high-frequency part of the resolved velocity field, defined using a test filter, indicated by the hat symbol and associated with the cutoff length $\hat{\Delta} > \Delta$

$$(\tilde{u})' = \tilde{u} - \hat{\tilde{u}}. \tag{4.10}$$

The following expression is employed for q_c^2

$$q_c{}^2 = \frac{1}{2}(\tilde{u}_i - \hat{\tilde{u}}_i)^2. \tag{4.11}$$

In practical implementations the test filter is derived from a three-point explicit filter, such as the trapezoidal rule (2.49), and the α parameter is taken equal to 0.5. This implies that the constant $C_m = C_m(\alpha) = 0.06$ (computed on the basis of an equilibrium production/dissipation assumption for homogeneous isotropic turbulence). Assuming that the cut-off lengths of the two filters are both in the inertial range it is possible to show [244] that the SGS kinetic energy is equal to the kinetic energy of smallest scales if the test filter cut-off length is $\sqrt{8}$ times larger than the primary filter.

4.3 Isotropic Tensor Modeling

Yoshizawa [320] has proposed a modeling for the isotropic part of the SGS tensor as

$$\tau_{kk} = 2C_I\bar{\rho}\Delta^2|\tilde{S}|^2. \tag{4.12}$$

This model has been derived from the Direct Interaction Approximation framework for shear flows. It relies on an asymptotic expansion about an incompressible state, and for the model derivation an equilibrium hypothesis between production and dissipation is used.

Erlebacher et al. [73] have performed direct simulations of compressible homogeneous turbulence at different Mach numbers ranging from 0.1 to 0.6 with this model. The value maximizing the correlation of the modeled stress with the exact stress is $C_I = 0.0066$. This value was determined with a linear least-squares regression technique on the vector level. It was found to depend weakly on the Mach number.

Additionally, Zang et al. [323] have studied the influence of this constant in LES of compressible homogeneous decaying turbulence at a turbulent Mach number of 0.1. C_I was varied from its standard value (0.0066) to a value 50 times larger. The results obtained with C_I ranging from 0.0066 to 0.066 were indistinguishable. However, a 50-fold increase in C_I leads to a significant overestimation of the compressible energy decay rate while leaving the incompressible energy unaffected so that the C_I effect on the simulation remains very limited.

Isotropic-tensor modeling has been assessed by Speziale et al. [273]. They have shown that it correlates poorly with the exact isotropic part of the SGS tensor (the correlation coefficient is about 15%) and they suggested to add a model for the Leonard and cross terms which improves drastically the correlation coefficient. Later, the same authors [75] have neglected τ_{kk} explaining that this term is generally weak with respect to the thermodynamic pressure.

It is worthwhile to note that with this type of modeling, τ_{kk} is expected to assume large values in areas of high compression/dilatation where the underlying numerical scheme introduces a large amount of numerical dissipation, the latter probably overwhelming the model effect.

4.4 SGS Heat Flux

Whatever the chosen formulation of the energy equation, a model is needed for the SGS heat flux. To this end, commonly authors follow the hypothesis of Eidson [69] who assumes that the energy transfer from the resolved scales to the subgrid scales is proportional to the gradient of resolved temperature. The proportionality coefficient is the SGS conductivity κ_{sgs} which is linked to the SGS viscosity through the relationship

$$\kappa_{sgs} = \frac{\mu_{sgs} C_p}{Pr_{sgs}}, \tag{4.13}$$

where Pr_{sgs} is the SGS Prandtl number. It is chosen in the interval [0.3; 0.9] and from a strict point of view it is a function of the considered wave number. The EDQNM theory [173] gives a value of 0.6 for this quantity. The term Q_j is modeled as

$$Q_j = -\frac{\bar{\rho}\nu_{sgs}}{Pr_{sgs}} \frac{\partial \tilde{T}}{\partial x_j}. \tag{4.14}$$

Following his classification (Table 2.3), Vreman [307] emphasizes the need to model $B_2 = \Pi_{dil}$. He proposes to use the same hypothesis of SGS conductivity for the sum of B_1 and B_2

$$B_1 + B_2 = \frac{\partial C_v Q_j}{\partial x_j} + \Pi_{dil} = -\frac{\partial}{\partial x_j}\left(\frac{C_p \bar{\rho}\nu_{sgs}}{Pr_{sgs}} \frac{\partial \tilde{T}}{\partial x_j} \right), \tag{4.15}$$

whereas $\tilde{u}_j \frac{\partial \bar{p}}{\partial x_j} - \overline{u_j \frac{\partial p}{\partial x_j}}$ is neglected. The relative magnitude of this term with respect to Π_{dil} is unknown and one can consider that it is simpler to neglect directly this latter term, keeping only B_1. This choice is significant within the framework of dynamic and dynamic mixed models where Vreman's formulation of the model is more complicated. Moreover, other authors, following Erlebacher et al. [75] and Moin et al. [201], have neglected Π_{dil}, invoking the incompressibility of the smallest scales. This approximation is supported by the more recent work of Martin et al. [192] (see Sect. 2.4.7). Furthermore, Lenormand et al. [168] have shown in the framework of *a priori* tests realized in a plane channel configuration that the correlation between terms $B_1 + B_2$ computed from filtered DNS data of Coleman et al. [43] and their model based on a turbulent conductivity hypothesis almost vanishes.

4.5 Modeling of the Subgrid Turbulent Dissipation Rate

Ghosal et al. [98] propose to model this term as

$$B_5 = C_\epsilon \bar{\rho} k^{3/2} / \Delta, \tag{4.16}$$

where $k = \frac{1}{2}\tau_{ii}$ is the subgrid turbulent kinetic energy and C_ϵ a dynamic constant. Ghosal et al. argue that this constant cannot be modeled with the standard dynamic procedure (see Sect. 4.6.5) since turbulent dissipation is essentially a small-scale phenomenon and no dissipation is left at large (resolved) scales for high Reynolds number flows, making the Germano identity useless in this context. The authors have proposed a procedure based on a local balance of the terms of the k equation. This procedure was simplified by Vreman [307] who computes the time dependent value of C_ϵ as

$$C_\epsilon = \frac{\int_\Omega (B_2 + B_3 + B_4 - \partial_t(\bar{\rho} k)) d\mathbf{x}}{\int_\Omega (\bar{\rho} k^{3/2} / \Delta) d\mathbf{x}}, \tag{4.17}$$

involving integration over the computational domain Ω. In practice, this procedure has been used infrequently, and only for applications with two directions of homogeneity such as shear layers [309] and channel flows [168, 312].

4.6 Improvement of SGS models

4.6.1 Structural Sensors and Selective Models

In order to improve the prediction of intermittent phenomena occurring for example in transitional flows, a sensor based on structural information can be introduced. This is done by incorporating a selection function into the model, based on the local angular fluctuations of the vorticity, developed by David [48, 172]. The idea here is to modulate the subgrid model in such a way as to apply it only when the assumptions underlying the model are satisfied, i.e. when not all the scales of the exact solution are resolved and the flow corresponds to fully-developed turbulence type. The problem therefore consists in determining if these two hypotheses are valid at each point and each time step. David's structural sensor tests the second hypothesis. For this purpose it is assumed that, if the flow is turbulent and developed, the highest resolved frequencies have certain characteristics specific to isotropic homogeneous turbulence, and particularly structural properties. David observed in direct numerical simulations that the probability density function of the local angular fluctuation of the vorticity vector exhibit a peak around the value of 20 degrees. As a consequence, he proposes to identify the flow as being locally under-resolved and turbulent at those points for which the local angular fluctuations of the vorticity vector corresponding to the highest resolved frequencies are larger than or equal to a threshold value θ_0. The selection criterion is therefore based on an estimation of the angle θ between the instantaneous vorticity vector ω and the local average vorticity vector $\hat{\omega}$, which is computed by applying a test filter to the vorticity vector. The angle θ is given by the following relation

$$\theta = \arcsin\left(\frac{\|\hat{\omega} \times \omega\|}{\|\hat{\omega}\|.\|\omega\|}\right). \tag{4.18}$$

A selection function which damps the subgrid model when the angle θ is less than a threshold angle θ_0 is defined. In the original version developed by David this function f_{θ_0} is a Boolean operator

$$f_{\theta_0}(\theta) = \begin{cases} 1 & \text{if } \theta \geq \theta_0, \\ 0 & \text{otherwise.} \end{cases} \tag{4.19}$$

This function is discontinuous by definition, which may pose problems in the numerical solution. One variant that exhibits no discontinuity for the threshold value is defined as [239]

$$f_{\theta_0}(\theta) = \begin{cases} 1 & \text{if } \theta \geq \theta_0, \\ r(\theta)^n & \text{otherwise,} \end{cases} \tag{4.20}$$

where θ_0 is the chosen threshold value, and r is the function

$$r(\theta) = \frac{\tan^2(\frac{\theta}{2})}{\tan^2(\frac{\theta_0}{2})} \tag{4.21}$$

with positive exponent n. In practice, it is taken to be equal to 2. Considering the fact that we can express the angle θ as a function of the norms of the vorticity vector ω, the average vorticity vector $\hat{\omega}$, and the norm ω' of the fluctuating vorticity vector defined as $\omega' = \omega - \hat{\omega}$, by the relation

$$\omega'^2 = \hat{\omega}^2 + \omega^2 - 2\hat{\omega}\omega\cos(\theta), \tag{4.22}$$

and the trigonometric relation:

$$\tan^2(\theta/2) = \frac{1 - \cos\theta}{1 + \cos\theta}, \tag{4.23}$$

the quantity $\tan^2(\theta/2)$ is estimated as:

$$\tan^2(\theta/2) = \frac{2\tilde{\omega}\omega - \tilde{\omega}^2 - \omega^2 + \omega'^2}{2\tilde{\omega}\omega + \tilde{\omega}^2 + \omega^2 - \omega'^2}. \tag{4.24}$$

The selection function is used as a multiplicative factor of the subgrid viscosity, leading to the following general definition of selective models

$$\nu_{sgs}^{(s)} = \nu_{sgs} f_{\theta_0}(\theta), \tag{4.25}$$

where ν_{sgs} is calculated by an arbitrary subgrid viscosity model. It should be noted that, in order to keep the same average subgrid viscosity value over the entire fluid domain, the constant that appears in the subgrid model has to be multiplied by a correction factor which has been estimated to be 1.65 with the structure function model on the basis of isotropic homogeneous turbulence simulations [48]. The selection function has been used successfully with functional model like the structure function [48], the Smagorinsky model and the mixed scales model [168].[4] In particular, it allows the treatment of transitional flows.

4.6.2 Accentuation Technique and Filtered Models

The Gabor-Heisenberg generalized principle of uncertainty stipulates that the precision of the information cannot be improved in space and in frequency at the same time because the contribution of the low frequencies excludes any precise determination of the energy at the cutoff. This quantity is nonetheless necessary to determine if the exact solution is entirely resolved.

In order to be able to detect the existence of the subgrid modes better, Ducros et al. [62] propose an accentuation technique which consists in applying the subgrid models to a modified velocity field obtained from a high-pass filter to the resolved velocity field. This filter, denoted as HP^n, is defined recursively as

$$HP^1(\bar{u}) \simeq \Delta^2 \nabla^2 (\bar{u}) \tag{4.26a}$$

$$HP^n(\bar{u}) = HP(HP^{n-1}(\bar{u})). \tag{4.26b}$$

This type of filter modifies the spectrum of the initial solution by emphasizing the contribution of the highest frequencies. The resulting field therefore represents mainly the high frequencies of the initial field and serves to compute the subgrid model. To remain consistent the subgrid model has to be modified. Such models are called filtered models. The case of the Structure Function model which has been used for computation of compressible boundary layer by Ducros et al. [62] is given as an example. Nevertheless, the technique is very general and filtered versions of the Smagorinsky and Mixed Scale models have been proposed by Sagaut et al. [241]. We define the second-order structure function of the filtered field as

$$F_2^{HPn}(\mathbf{x}, r, t) = \int_{|x'|=r} [HP^n(\bar{\mathbf{u}})(x,t) - HP^n(\bar{\mathbf{u}})(x+x',t)]^2 d^3\mathbf{x}'. \tag{4.27}$$

Ducros et al. [62] recommend to use a triply-iterated Laplacian which leads to the following expression of the model

$$\nu_{sgs}(x, \Delta, t) = 0.0014 K_0^{-3/2} \Delta \sqrt{F_2^{HP3}(x, \Delta, t)}. \tag{4.28}$$

[4] In these cases, the 1.65 factor has been kept.

4.6.3 High-Pass Filtered Eddy Viscosity

The high-pass filtered eddy viscosity model [282] shares with the former accentuation technique the idea of filtering the velocity field to highlight the high frequencies. This model is still based on the Boussinesq hypothesis but the SGS tensor is reconstructed with a filtered velocity

$$\tau_{ij}^d = \tau_{ij} - \frac{1}{3}\delta_{ij}\tau_{kk} = -2\bar{\rho}\nu_{sgs}(M \star \tilde{\mathbf{u}})(S_{ij}(M \star \tilde{\mathbf{u}}) - \frac{1}{3}\delta_{ij}S_{kk}(M \star \tilde{\mathbf{u}})). \quad (4.29)$$

The filtered velocity field is obtained by subtracting the low-pass filtered quantities from the unfiltered ones

$$M \star \tilde{\mathbf{u}} = (I - G) \star \tilde{\mathbf{u}}, \quad (4.30)$$

where G is typically computed in discrete form as in (2.40).[5] This technique is inspired by approximated deconvolution methods (ADM) presented in Chap. 5. Expressions for $\nu_{sgs}(M\star\tilde{\mathbf{u}})$ have been proposed for both Smagorinsky and Structure Function models in an earlier paper dedicated to incompressible flows [281]. It simply consist in substituting $\tilde{\mathbf{u}}$ by $M \star \tilde{\mathbf{u}}$ in (4.3) and (4.7) respectively. In the same way as for accentuation technique of Sect. 4.6.2, these models are suitable for application to transitional flows.

4.6.4 Wall-Adapting Local Eddy-Viscosity Model

Most of subgrid viscosity models do not exhibit the correct behavior in the vicinity of solid walls in equilibrium boundary layers on fine grids, resulting in a too high damping of fluctuations in that region and to a wrong prediction of the skin friction.[6] The common way to alleviate this problem is to add a damping function, which requires the distance to the wall and the skin friction as input parameters, leading to complex implementation issues. An elegant solution to solve the near-wall region problem on fine grids is proposed by Nicoud and Ducros [208], who found a combination of resolved velocity spatial derivatives that exhibits the expected asymptotic behavior $\nu_{sgs} \propto z^{+3}$, where z^+ is the distance to the wall expressed in wall units. The subgrid viscosity is defined as

$$\nu_{sgs} = (C_w\Delta)^2 \frac{(S_{ij}^d S_{ij}^d)^{3/2}}{(\tilde{S}_{ij}\tilde{S}_{ij})^{5/2} + (S_{ij}^d S_{ij}^d)^{5/4}}, \quad (4.31)$$

with $C_w \approx \sqrt{10.6}C_s$ and

$$S_{ij}^d = \frac{1}{2}\left(\frac{\partial \tilde{u}_i}{\partial x_l}\frac{\partial \tilde{u}_l}{\partial x_j} + \frac{\partial \tilde{u}_j}{\partial x_l}\frac{\partial \tilde{u}_l}{\partial x_i}\right) + \frac{1}{3}\frac{\partial \tilde{u}_m}{\partial x_l}\frac{\partial \tilde{u}_l}{\partial x_m}\delta_{ij}. \quad (4.32)$$

This model also possesses the interesting property that the subgrid viscosity vanishes when the flow is two-dimensional, in agreement with the physical analysis. This WALE model has been used in particular by Wollblad et al. [314] for shock/boundary layer interaction (see Chap. 10).

[5] The weights are chosen to impose a cut-off wave number of about $2\pi/3$.

[6] The asymptotic behavior of classical SGS models is discussed in Ref. [244], p. 159.

4.6.5 Dynamic Procedure

In order to adapt the models better to the local structure of the flow, Germano et al. [95] proposed an algorithm for adapting the Smagorinsky model by automatically adjusting the constant at each point in space and at each time step. This procedure, described below, is applicable to any model that makes explicit use of an arbitrary constant C_d, such that the constant now becomes time- and space-dependent: C_d becomes $C_d(x,t)$. The procedure is very general and can be applied to the determination of Pr_{sgs} and C_I following the original proposal of Moin et al. [201].

Computation of the Deviatoric SGS Tensor

The dynamic procedure is based on the Germano relationship which employs a second filter level F with cut-off wave number $k'_c = \pi/\hat{\Delta} < k_c$ (see Fig. 4.2). Using the property of the commutator (2.27), it is possible to write the SGS tensor at the filtering level FG (noted T_{ij}) in the following way

$$\begin{aligned} T_{ij} &= [(F\star) \circ (G\star), H](\rho, \rho u_i, \rho u_j) = [G\star, H] \circ (F\star) + (G\star) \circ [F\star, H] \\ &= \mathcal{L}_{ij} + \widehat{\tau_{ij}}, \end{aligned} \tag{4.33}$$

with

$$T_{ij} = \widehat{\overline{\rho u_i u_j}} - \frac{1}{\hat{\bar{\rho}}} \widehat{\overline{\rho u_i}}\, \widehat{\overline{\rho u_j}} \tag{4.34}$$

and

$$\mathcal{L}_{ij} = [G\star, H] \circ (F\star) = (\bar{\rho}\tilde{u}_i\tilde{u}_j)^{\widehat{}} - \frac{1}{\hat{\bar{\rho}}} \widehat{\bar{\rho}\tilde{u}_i} \widehat{\bar{\rho}\tilde{u}_j}, \tag{4.35}$$

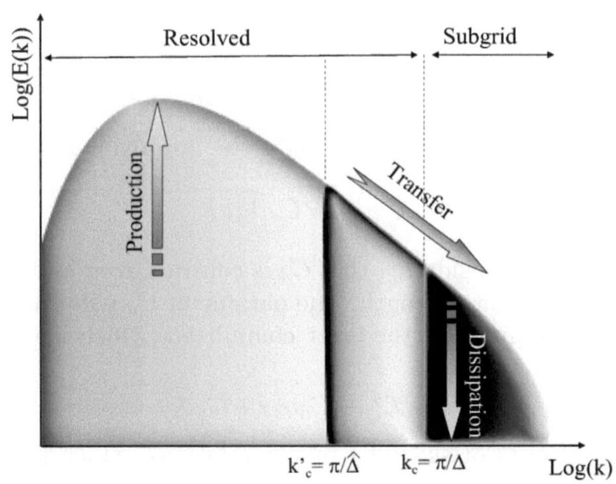

Fig. 4.2. Spectrum of homogeneous turbulence including the definition of the filter (case of a sharp cut-off filter)

where $\hat{\phi}$ is the filtered variable at the F level, and H is defined as in (2.81). The notation $(\phi)\widehat{}$ is equivalent to $\hat{\phi}$. $\hat{\Delta} = (\hat{\Delta}_1\hat{\Delta}_2\hat{\Delta}_3)^{1/3}$ ($\hat{\Delta}_i$ is the length of the test filter in the ith direction). Spyropoulos and Blaisdell [274] have studied the influence of the ratio $\hat{\Delta}/\Delta$ within the range [1.6; 8/3] on compressible homogeneous turbulence LES. The optimal value is found as 2.

It is assumed that the two subgrid tensors τ and T can be modeled by the same constant C_d for both filtering levels. Formally, this is expressed by

$$\tau_{ij} - \frac{1}{3}\tau_{kk}\delta_{ij} = C_d\beta_{ij}, \tag{4.36a}$$

$$T_{ij} - \frac{1}{3}T_{kk}\delta_{ij} = C_d\alpha_{ij}, \tag{4.36b}$$

where the tensors α and β designate the deviators of the subgrid tensors obtained using the subgrid model deprived of its constant. Some examples of subgrid model kernels for α_{ij} and β_{ij} are given in Table 4.1.

Table 4.1. Examples of subgrid model kernels for the dynamic procedure

Model	β_{ij}	α_{ij}
Eq. (4.3)	$-2\Delta^2\bar{\rho}\lvert\tilde{S}\rvert(\tilde{S}_{ij} - \frac{1}{3}\delta_{ij}\tilde{S}_{kk})$	$-2\hat{\Delta}^2\hat{\bar{\rho}}\lvert\hat{\tilde{S}}\rvert(\hat{\tilde{S}}_{ij} - \frac{1}{3}\delta_{ij}\hat{\tilde{S}}_{kk})$
Eq. (4.7)	$-2\Delta\bar{\rho}\sqrt{F_2(\Delta)}(\tilde{S}_{ij} - \frac{1}{3}\delta_{ij}\tilde{S}_{kk})$	$-2\hat{\Delta}\hat{\bar{\rho}}\sqrt{F_2(\hat{\Delta})}(\hat{\tilde{S}}_{ij} - \frac{1}{3}\delta_{ij}\hat{\tilde{S}}_{kk})$
Eq. (4.8)	$-2\bar{\rho}\lvert\tilde{S}\rvert^{\alpha}(q_c^2)^{\frac{(1-\alpha)}{2}}\Delta^{(1+\alpha)}$ $\times(\tilde{S}_{ij} - \frac{1}{3}\delta_{ij}\tilde{S}_{kk})$	$-2\hat{\bar{\rho}}\lvert\hat{\tilde{S}}\rvert^{\alpha}(\hat{q}_c^2)^{\frac{(1-\alpha)}{2}}\hat{\Delta}^{(1+\alpha)}$ $\times(\hat{\tilde{S}}_{ij} - \frac{1}{3}\delta_{ij}\hat{\tilde{S}}_{kk})$

Introducing the formulae (4.36a) and (4.36b) into the relation (4.33) one obtains

$$\mathcal{L}_{ij}^d = \mathcal{L}_{ij} - \frac{1}{3}\mathcal{L}_{kk}\delta_{ij} = C_d\alpha_{ij} - \widehat{C_d\beta_{ij}}. \tag{4.37}$$

For the further procedure we need to make the approximation

$$\widehat{C_d\beta_{ij}} = C_d\widehat{\beta_{ij}}, \tag{4.38}$$

which is equivalent to considering that C_d is constant over an interval at least equal to the test filter cutoff length. The parameter C_d will thus be computed in such a way as to minimize the error committed, which is evaluated using the residual E_{ij}

$$E_{ij} = \mathcal{L}_{ij}^d - C_d\alpha_{ij} + C_d\widehat{\beta_{ij}}. \tag{4.39}$$

This definition consists of six independent relations, which in theory makes it possible to determine six values of the constant and possibly to a tensorial SGS model. In order to maintain a single relation and thereby determine a single value of the constant, Germano et al. propose to contract the relation

(4.39) with the resolved strain rate tensor. The value sought for the constant is a solution of

$$\frac{\partial E_{ij}\tilde{S}_{ij}}{\partial C_d} = 0. \tag{4.40}$$

This method raises the problem of indetermination when the tensor S_{ij} cancels out. To remedy this problem, Lilly [178] proposes to calculate the constant C_d by a least-squares method, by which the constant C_d now becomes a solution of

$$\frac{\partial E_{ij}E_{ij}}{\partial C_d} = 0, \tag{4.41}$$

or more explicitly

$$C_d = \frac{M_{ij}\mathcal{L}^d_{ij}}{M_{kl}M_{kl}}, \tag{4.42}$$

where

$$M_{ij} = \alpha_{ij} - \widehat{\beta_{ij}}. \tag{4.43}$$

It is worth to note that the computed C_d has the two following properties:

- It can assume negative values, so the model can have an anti-dissipative effect locally. This is a characteristic that is often interpreted as a modeling of the backward energy cascade mechanism.
- In its implemented form, C_d is not bounded, since it appears in the form of a fraction whose denominator can cancel out without a strict cancellation of the numerator.

These two properties have important practical consequences on the numerical solution because they can both influence the stability of the simulation. The constant therefore needs an *ad hoc* process to ensure the models good numerical properties. There are a number of different ways of performing this process on the constant:

- statistical average in the directions of statistical homogeneity, in time or local in space

$$C_d = \left\langle \frac{M_{ij}\mathcal{L}^d_{ij}}{M_{kl}M_{kl}} \right\rangle, \tag{4.44}$$

- averaging the denominator and numerator separately, which is denoted symbolically as

$$C_d = \frac{\langle M_{ij}\mathcal{L}^d_{ij}\rangle}{\langle M_{kl}M_{kl}\rangle}, \tag{4.45}$$

- clipping the constant value imposing the following relations

$$\nu_{sgs} + \nu \geq 0, \tag{4.46a}$$

$$C_d \leq C_{max}. \tag{4.46b}$$

The latter technique can complement the two first. Additionally, it is worth noting that the averaging process reduces the locality of the model and subsequently a significant part of its self-adaptation capabilities. Many other variants of the dynamic procedure have been published in the literature (see [244] for a review), but, to our knowledge, none of them has been used for compressible flows.

Computation of the Isotropic Part of the SGS Tensor

Using the Germano identity for the trace of the tensor T and τ, and the model (4.12), the dynamic computation of C_I can been written according to Moin et al. [201] as

$$C_I^{dyn} = \frac{\langle \mathcal{L}_{kk} \rangle}{\langle 2\hat{\bar{\rho}}\hat{\Delta}^2|\hat{\tilde{S}}|^2 - 2\Delta^2(\bar{\rho}|\tilde{S}|^2)\hat{} \rangle}. \tag{4.47}$$

Computation of the Dynamic Prandtl Number

The principle is the same for the dynamic Prandtl number. The Germano identity for the SGS temperature flux writes

$$\begin{aligned} \mathcal{K}_j &= [(F\star)\circ(G\star), H](\rho, \rho u_j, \rho T) = [G\star, H]\circ(F\star) + (G\star)\circ[F\star, H] \\ &= \mathcal{L}_j^\theta + \widehat{Q}_j, \end{aligned} \tag{4.48}$$

with

$$\mathcal{K}_j = \widehat{\overline{\rho u_j T}} - \frac{1}{\hat{\bar{\rho}}}\widehat{\overline{\rho u_j}}\widehat{\overline{\rho T}}, \tag{4.49}$$

which is modeled as

$$\mathcal{K}_j = -\frac{\hat{\bar{\rho}}\hat{\nu}_{sgs}}{Pr_{sgs}}\frac{\partial \hat{\tilde{T}}}{\partial x_j} \tag{4.50}$$

and

$$\mathcal{L}_j^\theta = [G\star, H]\circ(F\star) = (\bar{\rho}\tilde{u}_j\tilde{T})\hat{} - \frac{1}{\hat{\bar{\rho}}}\widehat{\bar{\rho}\tilde{u}_j}\widehat{\bar{\rho}\tilde{T}}. \tag{4.51}$$

Assuming that the SGS Prandtl is constant over the wavenumber range between the cut-off wavenumbers of F and G, one can write the following equation

$$\frac{M_j^\theta}{Pr_{sgs}} = \mathcal{L}_j^\theta. \tag{4.52}$$

with

$$M_j^\theta = -\hat{\bar{\rho}}\hat{\nu}_{sgs}\frac{\partial \hat{\tilde{T}}}{\partial x_j} + \left(\bar{\rho}\nu_{sgs}\frac{\partial \tilde{T}}{\partial x_j}\right)\hat{}. \tag{4.53}$$

Following Vreman [307], the turbulent Prandtl number is computed from the least-squares minimization procedure proposed by Lilly. The Prandtl number is the solution of

$$\frac{\partial E_j^\theta E_j^\theta}{\partial Pr_{sgs}} = 0, \tag{4.54}$$

with

$$E_j^\theta = \mathcal{L}_j^\theta - \frac{M_j^\theta}{Pr_{sgs}}. \tag{4.55}$$

More explicitly, the turbulent Prandtl number is equal to

$$Pr_{sgs} = \frac{\langle M_j^\theta M_j^\theta \rangle}{\langle \mathcal{L}_j^\theta M_j^\theta \rangle}. \tag{4.56}$$

In the Vreman implementation of the model, the numerator and denominator are averaged separately.

4.6.6 Implicit Diffusion and the Implicit LES Concept

As mentioned in Chap. 1, the existence of strong gradients in compressible flows requires the use of special numerical methods which must be at least locally dissipative. This numerical dissipation is generally nonlinear and differentiation schemes can be represented as filters using the equivalent wave number formalism (see Sect. 2.3.4). Turbulent kinetic energy is then extracted from the resolved field by the numerical scheme which mimics, at least in the energetic sense, the action of a subgrid scale model. Employing numerical dissipation as model in this straightforward sense relies on the hypothesis that *the action of subgrid scales on the resolved scales is strictly dissipative*, and the SGS model can be considered to be implicitly contained in the numerical scheme. This notion essentially has been introduced by Boris *et al.* [30]. It is generally referred to as MILES (Monotone Integrated Large Eddy Simulation). This terminology comes from the fact that the authors have used a FCT (Flux Corrected Transport) scheme which ensures monotonicity of the solution. Nevertheless, the numerical scheme can be dissipative without being monotonicity preserving.[7] This approach is more generally called ILES (Implicit Large Eddy Simulation) [106] and also applies to incompressible flows following the work of Kawamura and Kuwahara [137] published in 1984. The connection between numerical method and modeling which is of crucial importance for compressible LES is reviewed in more detail in Chap. 6.

[7] High order filters used to stabilize centered compact schemes also enter in this category without being monotonicity preserving [303].

5

Explicit Structural Modeling

5.1 Motivation of Structural Modeling

Similarly as with structural models for incompressible flows [244, Chap. 6] for compressible flows structural models can be differentiated into different groups. The essence of structural modeling is that there is a attempt to reconstruct some of the subgrid-scale information directly by generating a new field from the filtered field (velocity, density etc.) which approximates the unfiltered field more closely than the filtered field [55]. Functional models represent the effect of the subgrid-scales in the filtered transport equations without first reconstructing subgrid-scales. Structural models commonly require some additional measures for small-scale energy removal so that the discretized transport equations for the filtered field can be advanced stably in time. This is mostly handled by including an additional functional model. One may ask what the benefit of structural modeling is if there is need to include an additional functional model. *A priori* analyses have unanimously shown that structural models give a much better correlation of the predicted and the real subgrid-scale stress tensor. Also, for well-resolved LES, i.e. LES whose resolution differs by no more than about one order of magnitude from the required DNS resolution in each spatial dimension, a large part of the energy transfer occurs on the range of scales which are represented on the numerical grid. The prediction of this energy transfer is greatly improved by structural models, alleviating the modeling load on the functional model. One could say that there is a symbiosis between structural and functional model, where the prediction accuracy of anisotropic energy redistribution gains from structural modeling, and a rather simple isotropic energy transfer provided by the additional functional model is sufficient.

In the following we will essentially distinguish between the different structural models as those based on the scale-similarity hypothesis, those based on approximate deconvolution, and those which follow a multi-resolution concept in reconstructing the subgrid-scale field. As not in all of these classes for compressible flows the same level of development has been reached as for in-

E. Garnier et al., *Large Eddy Simulation for Compressible Flows*,
Scientific Computation,

compressible flows [244] not the full scope of possible model variations will be given, but the focus will be placed on such models which have been employed for compressible LES.

Before addressing the different structural models it is useful to recall some formal framework. For simplicity we consider an equidistant mesh with mesh size h. The filter kernel is characterized by a cutoff-wavenumber k_C which for graded filters is a matter of definition. The largest wavenumber which can be represented by this mesh is the Nyquist wavenumber $k_N = \pi/h$, We distinguish between scales that can be *represented* on the underlying grid with wavenumbers $|k| \leq k_N$, scales that are considered as *resolved* by the filter kernel $|k| \leq k_C$, and the converse scale ranges, i.e. non-resolved scales $|k| > k_C$, non-represented scales $|k| > k_N$, and non-resolved represented scales $k_C < |k| \leq k_N$. The objective of structural modeling is to recover as much as possible of the unfiltered solution from the available scales so that an approximation of the non-truncated and unfiltered field is obtained. This task is illustrated in Fig. 5.1.

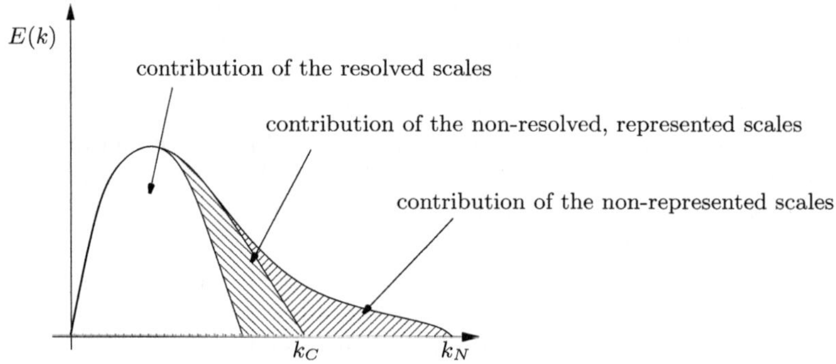

Fig. 5.1. Schematic energy spectrum and different contributions to an energy spectrum evolving in time by Navier–Stokes dynamics

Whereas the reconstruction of represented scales from the resolved scales for positive filters, i.e. filter kernels with positive transfer function on $|k| \leq k_N$, is possible by regularized inversion of the filter operation, the reconstruction of unrepresented scales requires further *a priori* knowledge and cannot be directly employed to compute nonlinear interactions on the underlying mesh. The former can be considered as an approximation problem, a *soft deconvolution* problem, whereas the latter is a modeling problem, a *hard deconvolution* problem. In the following we will distinguish between these two issues and consider defiltering procedures which apply to the soft deconvolution problem and regularization procedures which apply to the hard deconvolution problem.

5.2 Models Based on Deconvolution

In this section we consider approaches which operate only on the represented scales. The fact that scales with wavenumbers near k_N cannot be represented accurately by the underlying numerical scheme can be exploited. Numerical errors induced by scales near k_N can be reduced by mollifying the solution by explicit filtering. On the other hand, a realistic energy transfer near k_N requires that small scales have sufficient energy and follow Navier–Stokes dynamics, whereas their numerical accuracy is less significant. Therefore it is conceivable that the range of non-resolved but represented scales can be used for manipulation in order to achieve a stable computation of the resolved scales, and to model the proper energy transfer from the resolved to the non-resolved scales.

The concept of defiltering, deconvolution, or reconstruction approaches can be cast in the following way. If we consider for simplicity a general one-dimensional and one-component conservation law

$$\frac{\partial u}{\partial t} + \frac{\partial F(u)}{\partial x} = 0, \tag{5.1}$$

we project the solution onto a discrete, equidistantly spaced mesh with size h and obtain formally an equation for the grid function u_N

$$\frac{\partial u_N}{\partial t} + \frac{\partial F_N(u_N)}{\partial x} = 0, \tag{5.2}$$

where, aside of grid truncation no other approximations have been made yet. For simplicity we consider the computational domain as infinite and the grid points ordered as $x_j = jh$, for $j = \ldots, -1, 0, 1, \ldots$. The nonlinear flux function $F(u)$ now also is evaluated as grid function $F_N(u_N)$, implying that the flux is computed from the grid function u_N and projected onto the grid. The derivative operation $\partial/\partial x$ at this point does not yet involve a numerical approximation. Note that an equivalent continuous function $u_N(x)$ can be obtained from u_N, and accordingly for F_N, by interpolation using the Whittaker cardinal function [301], so that an exact derivative of the grid function can be defined.

At this point the *hard deconvolution problem* can be stated. The non-represented contribution to the solution $u'(x) = u(x) - u_N(x)$ with scales $|k| > k_N$ is lost and cannot be recovered by approximation methods. The effect of these scales on the represented part of the solution $u_N(x)$ requires *a priori* knowledge for modeling. From Fourier approximation theory it is known that the relative error of scales near k_N is of the same order as the magnitude of these scales, implying that scales near the Nyquist wavenumber cannot be computed accurately. The situation becomes certainly worse for other than spectral numerical discretization schemes. If the truncated conservation law is convolved with a filter the numerical approximation of the filtered solution can be made more accurate by mollifying the contribution of the marginally resolved scales at the expense of further constraining the range of physically relevant scales to $|k| \leq k_C$. For positive filters the definition of k_C is somewhat arbitrary, one often takes that wavenumber where the filter transfer function has the magnitude $\hat{G}(k_C) = 0.5$. Note also that at this point it does not make sense to consider a spectral cut-off filter since that would be the same as a grid projection with smaller wavenumber k_N.

By application of an explicit, translation-invariant filter we obtain

$$\frac{\partial \bar{u}_N}{\partial t} + G * \frac{\partial F_N(u_N)}{\partial x} = \frac{\partial \bar{u}_N}{\partial t} + \frac{\partial \overline{F_N(u_N)}}{\partial x} = 0. \tag{5.3}$$

Now the soft deconvolution problem can be stated in two alternative forms:

(a) the deconvolution form

$$\frac{\partial \bar{u}_N}{\partial t} + \frac{\partial \overline{F_N(u_N^*)}}{\partial x} = 0, \tag{5.4a}$$

(b) the residual-stress form

$$\frac{\partial \bar{u}_N}{\partial t} + \frac{\partial F_N(\bar{u}_N)}{\partial x} = \frac{\partial F_N(\bar{u}_N)}{\partial x} - \frac{\partial \overline{F_N(u_N^*)}}{\partial x} = \tau_{SGS} = 0, \tag{5.4b}$$

where we have introduced the approximate inverse Q of the filter G by $u^* = Q * \bar{u}$. The approximate inverse will be discussed in more detail in the following. Note that versions (a) and (b) differ in terms of satisfying Galilean invariance. As shown by Berselli et al. [19, Sect. 8.6.1], the generalized scale-similarity form leads to a Galilean invariant SGS tensor for the momentum equations. Formulation (a), however, violates Galilean invariance by an error which is on the order of the deconvolution error. In practical computations no significant differences between formulations (a) and (b) have been found. The soft-deconvolution problem amounts to reconstructing u_N from $\bar{u}_N$.

Due to the grid truncation, u_N is equivalent to a C^∞ function $u_N(x)$ on its support. $\hat{u}_N(k)$ can be obtained from $\hat{\bar{u}}_N(k)$ for a filter kernel with positive Fourier transform $\hat{G}(k)$ by

$$\hat{u}_N(k) = \hat{G}^{-1}(k)\hat{\bar{u}}_N(k), \tag{5.5}$$

and u_N from a subsequent inverse Fourier transform. This direct inversion, however, is neither numerically stable for large k nor desirable since it amplifies errors near k_N which have a relative error of order unity [38]. With deconvolution, energy in the range of represented non-resolved scales is amplified by the bounded operator Q, which is a regularized inverse of G. Several regularizations are possible. For frequent applications, however, the number of required computational operations should be low, which limits relevant regularizations to the most simple ones known from the theory of solutions of the Fredholm integral equation of the first kind, such as the van Cittert iterative scheme, see Ref. [20]. The van Cittert iterative scheme is the basis of the approximate deconvolution model (ADM) of Stolz and Adams [276], where Q is defined as

$$Q = \sum_{m=0}^{N} (I - G)^m \tag{5.6}$$

and I is the identity operator. This sum does not converge in general for $N \to \infty$, but approximates G^{-1}, where the error can be minimized by an optimum finite value of N which serves as regularization parameter. In applications $N = 5$ has been found as suitable choice [276–278]. Note that the concept of ADM requires that for wavenumbers up to k_C the underlying discretization scheme provides spectral-like resolution. The most efficient discretizations of this kind are compact finite-difference schemes for spatial discretization. Lower-order schemes result in a smaller k_C/k_N and can imply larger overall computation cost [279]. ADM has the properties of energy stability of the approximation [64] and consistency with theoretical scaling laws of turbulence for kinetic energy and helicity [156].

We have introduced the ADM framework with (5.4) since it provides a formal approach to other deconvolution models. The scale-similarity model $\tau_{ij} = \overline{\bar{u}_i\bar{u}_j} - \bar{u}_i\bar{u}_j$ corresponds to the residual-stress form of ADM, i.e. $\overline{G_N * \bar{u}_i G_N * \bar{u}_j} - \bar{u}_i\bar{u}_j$, as mentioned above, where $N = 0$. The tensor-diffusivity model, which commonly is derived from a formal series expansion of the inverse filter in terms of filter width [163], can be shown to correspond to ADM for a Gaussian filter and with a regularization parameter $N = 2$ up to fourth order in filter width. Regularization for this non-convergent expansion is achieved by truncating the expansion at the $\mathcal{O}(\Delta^4)$ term. The fact that regularization for the defiltering with the tensor-diffusivity model and ADM is achieved differently (series truncation vs. truncated summation of the approximate inverse) expresses itself in the difference at the higher-order terms.

For simplicity, we consider two velocity components $u(x)$ and $v(x)$ as function of only one spatial variable and suppress time dependency. Assuming sufficiently smooth $u(x)$ and $v(x)$ we can expand these around a grid node x_i as a Taylor series

$$u(x' - x_i) = \sum_{l=0}^{\infty} \frac{1}{l!}(x' - x_i)^l \left.\frac{\partial^l u}{\partial x^l}\right|_{x_i} .$$

On applying a filter kernel $G(x' - x_i) = G(\xi\Delta_i - x_i)$ with support Δ_i to this expansion and on evaluating the filtered expression one obtains

$$\bar{u}(x_i) = \sum_{l=0}^{\infty} \frac{1}{l!}\Delta_i^l M^l(x_i) \left.\frac{\partial^l u}{\partial x^l}\right|_{x_i} ,$$

where $M^l(x_i) = \int_{-1/2}^{1/2} \xi^l G(\xi, x_i) d\xi$ is the normalized l-th moment of the filter kernel at x_i. On Fourier transforming the grid function $\bar{u}(x_i)$ in terms of the wavenumber k, using the above expansion in terms of filter moments, one obtains

$$\hat{\bar{u}}(k) = \hat{G}(k)\hat{u}(k) = \sum_{l=0}^{\infty} \frac{1}{l!} \Delta_i^l M^l(x_i) \mathrm{i}^l k^l \hat{u}(k),$$

where $\mathrm{i} = \sqrt{-1}$. By comparison of like expressions, the filter transfer function can be identified as

$$\hat{G}(k) = \sum_{l=0}^{\infty} \frac{1}{l!} \Delta_i^l M^l \mathrm{i}^l k^l.$$

Applying ADM with $N = 2$ to this expression for a Gaussian filter, i.e. defining the filter moments accordingly,

$$\hat{G}(k) = e^{-\frac{-k^2\Delta^2}{24}} = 1 - \frac{1}{24}k^2\Delta^2 + \frac{1}{1152}k^4\Delta^4 + \mathcal{O}(\Delta^6)$$

gives

$$\hat{G}^*(k) = 1 + \frac{1}{24}k^2\Delta^2 + \frac{1}{1152}k^4\Delta^4 + \mathcal{O}(\Delta^6).$$

For u^* one obtains by inverse Fourier transform

$$u^* = \bar{u} - \frac{\Delta^2}{24}\frac{\partial^2 \bar{u}}{\partial x^2} + \frac{\Delta^4}{1152}\frac{\partial^4 \bar{u}}{\partial x^4} + \mathcal{O}(\Delta^6).$$

The same procedure applied to $v(x)$ and filtering the product u^*v^* results in

$$u^*v^* = \bar{u}\bar{v} + \frac{\Delta^2}{12}\frac{\partial^2 \bar{u}}{\partial x^2} + \frac{\Delta^4}{228}\frac{\partial^4 \bar{u}}{\partial x^4} + \mathcal{O}(\Delta^6).$$

For the subgrid-scale stress the following model is obtained

$$\overline{uv} - \bar{u}\bar{v} \approx \overline{u^*v^*} - \bar{u}\bar{v} = \frac{\Delta^2}{12}\frac{\partial^2 \bar{u}}{\partial x^2} + \frac{\Delta^4}{228}\frac{\partial^4 \bar{u}}{\partial x^4} + \mathcal{O}(\Delta^6)$$

which agrees to order $\mathcal{O}(\Delta^4)$ with the tensor-diffusivity model for a Gaussian filter.

5.2.1 Scale-Similarity Model

With respect to the scale-similarity model for incompressible flows the reader may want to consult Ref. [244, Sect. 6.3]. Adaptations of the scale-similarity model to compressible flows have been proposed by several authors [75, 273, 306–308] in terms of a mixed formulation. The scale-similarity SGS-stress tensor for compressible flow using Favre filtering is

$$\tau_{ij} = \bar{\rho}\big(\widetilde{\tilde{u}_i\tilde{u}_j} - \tilde{\tilde{u}}_i\tilde{\tilde{u}}_j\big). \tag{5.7}$$

According to the conservative form of the evolution equations, refer to Chap. 2.2, the model formulation is based on the filtered density $\bar{\rho}$ and the filtered momentum densities $\overline{\rho u_i}$. In terms of filter width the scale similarity model approximates the exact SGS stresses up to order $\mathcal{O}(\Delta^2)$. A parameterized scale-similarity model can be formulated by introducing a parameter of order unity as prefactor to the right-hand side of (5.7) [244], the parameter can also be determined dynamically. Such an extension of the scale-similarity model has been introduced by Liu et al. [179] for LES of incompressible flows. The nonlinearity of the viscous stresses due to the temperature-dependency of the viscosity is commonly neglected. A scale-similarity approach for such terms would amount to approximate the SGS residual by replacing the non-computable quantities by their computable counterparts using the filtered temperature and velocity.

The scale-similarity concept extends in a straight-forward fashion to the subgrid-scale terms of the energy equation, whether in conservative, enthalpy, temperature, or pressure formulation. The SGS-heat flux can be approximated as

$$Q_j = \bar{\rho}\big(\widetilde{\tilde{u}_j \tilde{T}} - \tilde{\tilde{u}}_j \tilde{\tilde{T}}\big). \tag{5.8}$$

Depending on the particular choice of energy-equation formulation, see Sect. 2.4, different SGS occur:

1. For the enthalpy formulation, (2.95), there is the divergence of the heat flux Q_j, the pressure dilatation Π_{dil}, the SGS heat flux $\partial(\bar{q}_j - \check{q}_j)/\partial x_j$, and the SGS viscous dissipation $\epsilon_v = \bar{\Phi} - \check{\Phi}$. For the pressure-dilatation the scale-similarity model is [308]

$$\Pi_{dil} = C_\Pi \left(\overline{\bar{p}\frac{\partial \tilde{u}_j}{\partial x_j}} - \bar{\bar{p}}\frac{\partial \tilde{\tilde{u}}_j}{\partial x_j} \right), \tag{5.9}$$

and for the viscous dissipation one obtains

$$\epsilon_v = C_{\epsilon_v} \left(\overline{\check{\sigma}_{ij}\frac{\partial \tilde{u}_i}{\partial x_j}} - \bar{\check{\sigma}}_{ij}\frac{\partial \tilde{\tilde{u}}_i}{\partial x_j} \right). \tag{5.10}$$

The SGS residual for the heat flux is commonly neglected, refer to Sect. 2.4.7. However, a scale-similarity model could be written in a similar fashion as shown above for the viscous-stress residual. As values for the modeling parameters Vreman et al. [308] propose $C_\Pi = 2.2$ and $C_{\epsilon_v} = 8$. Alternatively, the coefficients can also be computed dynamically from the generalized Germano identity [192, 308]. $\check{\mathcal{E}}_v$ is a scale-similarity estimate of ϵ_v on the test-filter level (the analog of the Leonard stress)

$$\check{\mathcal{E}}_v = \mathcal{E}_v - \hat{\epsilon}_v,$$

and ϵ_v is the viscous dissipation on the grid filter level. For $\mathcal{E}_v$ and for ϵ_v the above model with the same constant C_v is assumed so that the generalized Germano identity can be solved for C_v at each grid point. A similar procedure can be applied to Π_{dil}.

2. For the temperature formulation (2.100) the same terms occur as for the enthalpy formulation, and the same models can be used.
3. The only difference between the pressure formulation (2.105) and the enthalpy or temperature formulation is that the three unclosed terms for ideal gases have different constant factors but otherwise are the same. Again, the same closures can be used.
4. The entropy formulation has not been used with scale-similarity modeling to the authors' knowledge. The scale-similarity framework, however, gives a straight-forward way to proceed with this formulation by replacing the unfiltered variables by the filtered variables in all unclosed terms.
5. For the energy formulation (2.122) one has to distinguish between the different ways the energy, pressure and temperature are defined.
   (a) When the filtered total energy $\tilde{E}$, the filtered pressure $\bar{p}$, and the filtered temperature $\tilde{T}$ are used we obtain as SGS terms the heat flux Q_j, the SGS turbulent diffusion $\mathcal{J}_j$, and the SGS viscous dissipation $\mathcal{D}_j$. For SGS turbulent diffusion the scale-similarity model becomes

$$\mathcal{J}_j = \frac{1}{2}\left(\bar{\rho}\widetilde{\tilde{u}_j\tilde{u}_i\tilde{u}_i} - \bar{\rho}\tilde{\tilde{u}}_j\widetilde{\tilde{u}_i\tilde{u}_i}\right) \tag{5.11}$$

and the SGS viscous dissipation

$$\mathcal{D}_j = \overline{\check{\sigma}_{ij}\tilde{u}_i} - \bar{\check{\sigma}}_{ij}\tilde{\tilde{u}}_i, \tag{5.12}$$

using the resolved-stress tensor $\check{\sigma}_{ij}$. An alternative choice for the turbulent diffusion is motivated by analogy to RANS modeling [143]

$$\mathcal{J}_j = -\tau_{ij}\tilde{u}_i.$$

A dynamic modeling for turbulent diffusion was proposed by Martin et al. [192] by introducing a prefactor $C_{\mathcal{J}}$ to the left-hand side of (5.11) which can be determined dynamically from the generalized Germano identity, as mentioned above for ϵ_v.
   (b) When the filtered energy $\tilde{E}$, the pressure $\check{p}$, and the temperature $\check{T}$ are used the additional SGS residuals D_3, D_4, D_5 occur. The term D_3 is closed by the model for τ_{ij}, whereas $D_4 = \partial\mathcal{D}_j/\partial x_j$ and can be modeled by using (5.12). D_5 is the divergence of the SGS residual for the heat flux and can be treated as mentioned above.
   (c) When the SGS energy is absorbed into the pressure, one operates with $\tilde{E}$, $\bar{\mathcal{P}}$, and the temperature $\check{T}$. In this case new SGS terms occur for which no closure has been devised yet.
   (d) The most convenient choice is to use the resolved energy $\check{E}$, the filtered pressure $\bar{p}$, and the filtered temperature $\tilde{T}$ since this results in a straight-forward application of the definitions for the unfiltered energy and the unfiltered equation of state. Now the SGS terms B_1 through B_7 occur. The term B_1 essentially is the divergence of the SGS heat

flux and can be treated as given by (5.8). B_2 is the SGS pressure dilatation and can be modeled as

$$B_2 = \overline{\bar{p}\frac{\partial \tilde{u}_j}{\partial x_j}} - \bar{\bar{p}}\frac{\partial \tilde{\tilde{u}}_j}{\partial x_j}, \tag{5.13}$$

B_3 and B_4 are closed by the closure of τ_{ij}. For B_5 one can use

$$B_5 = \overline{\breve{\sigma}_{kj}\frac{\partial \tilde{u}_k}{\partial x_j}} - \bar{\breve{\sigma}}_{kj}\frac{\partial \tilde{\tilde{u}}_k}{\partial x_j}. \tag{5.14}$$

Term B_6 arises due to the nonlinearity of the viscosity and has found to be negligible in most cases. A scale-similarity model would amount to inserting $\breve{\sigma}_{ij}$ for σ_{ij}, i.e. $B_6 = 0$. The term B_7 corresponds to D_5.

Note that for any of such terms the scale-similarity approach gives a straight-forward estimate by simply replacing the unknown unfiltered variables in the respective expressions by the known filtered ones. Some terms can be argued to be negligible in many cases, refer to Sect. 2.4.7. For the mixed model, explained in Sect. 5.3, some of the effects of the above terms are modeled by an additional functional model. Also note that the scale-similarity model without further regularization in most cases will not result in stable time integration. One of the regularization techniques mentioned below has to be employed.

5.2.2 Approximate Deconvolution Model

Approximate deconvolution modeling (ADM) applies to any graded primary filter with positive transfer function over the range of represented wave numbers. The underlying modeling principle has been explained for a generic conservation law in the introduction to this section. It is important to note that ADM involves explicit filtering, so that the shape of the filter kernel is directly imposed on the solution. This has implications, e.g. for the near-wall turbulent-boundary-layer flow, which in the average solution should follow the law of the wall. Such properties transfer to the filtered averaged solution only if the filter kernel obeys sufficient moment-preservation conditions, which implies that for a given (inhomogeneous) computational grid the filter kernel has to be determined locally from the moment conditions. These issues are discussed in detail by Stolz et al. [277, 278] and Vasilyev et al. [299]. ADM has been formulated by Stolz et al. [278] for the conservative form of the compressible Navier-Stokes equations, using a transport equation for the total energy. In addition to the sets of derived thermodynamic properties given in Sect. 2.4, for ADM also the following definitions are introduced:

1. The deconvolved density

$$\rho^* = Q * \bar{\rho}. \tag{5.15}$$

2. The deconvolved momentum

$$(\rho u)_i^* = Q * \overline{\rho u}_i . \tag{5.16}$$

3. The deconvolved total energy

$$E^* = Q * \bar{E}. \tag{5.17}$$

With the deconvolved quantities, approximations of unfiltered derived thermodynamic properties and dynamic quantities can be obtained directly, in the same sense as for the resolved energy and resolved temperature mentioned above for the scale-similarity model. Also, the intrinsic nonlinearities due to the equation of state, the temperature dependence of the viscosity or the heat conductivity can be recovered to some extent. The deconvolved pressure is computed from

$$\breve{p}^* = (\gamma - 1)\left(E^* - (\rho u)_i^* (\rho u)_i^* / \rho^*\right). \tag{5.18}$$

The deconvolved temperature for an ideal gas results as

$$\breve{T}^* = \breve{p}^* / (\rho^* R). \tag{5.19}$$

One obtains the approximations for the viscous stress tensor $\breve{\tau}_{ij}^*$ and for the heat flux $\breve{q}^*$ by computing the viscosity from $\breve{T}^*$ and the strain rate S_{ij}^* from $(\rho u)_i^* / \rho^*$.

Given these deconvolved quantities they can be inserted directly into the convective and diffusive flux terms of the conservation equations instead of the unknown filtered quantities, and one obtains the continuity equation

$$\frac{\partial \bar{\rho}}{\partial t} + \overline{\frac{\partial (\rho u)_j^*}{\partial x_j}} = 0, \tag{5.20a}$$

the momentum equations

$$\frac{\partial \overline{(\rho u)}_i}{\partial t} + \overline{\frac{\partial}{\partial x_j}\left(\frac{(\rho u)_i^* (\rho u)_j^*}{\rho^*} + \breve{p}^* \delta_{ij} - \breve{\tau}_{ij}^*\right)} = 0, \tag{5.20b}$$

and the energy equation

$$\frac{\partial \bar{E}}{\partial t} + \overline{\frac{\partial}{\partial x_j}\left(\frac{(\rho u)_j^*}{\rho^*}(E^* + \breve{p}^*) - \breve{\tau}_{ji}^* \frac{(\rho u)_i^*}{\rho^*} + \breve{q}_j^*\right)} = 0. \tag{5.20c}$$

Note that these equations can also be written in the residual-stress form of (5.4b) where a SGS-tensor appears explicitly and can be closed by replacing all unfiltered quantities by the approximately deconvolved ones.

Without regularization this ADM formulation does not ensure the proper SGS energy transfer and in most cases leads to numerical instability, a property shared with the scale-similarity model. Good results without regularization have been obtained for comparably low Reynolds-number isotropic turbulence [276]. A regularization proposed for ADM by Stolz et al. [278] is outlined in Sect. 5.3. The discrete filter operators used for ADM as given in Sect. 2.3 usually have a five-points stencil, which requires special considerations near computational-domain boundaries or computational-block boundaries [277].

5.2.3 Tensor-Diffusivity Model

An adaptation of the tensor-diffusivity model for compressible flow has been proposed by Vreman et al. [306–308]. For details on the tensor-diffusivity model and its variants for incompressible flow the reader can consult Ref. [244, Sect. 6.2]. As explained above, the tensor-diffusivity model can be interpreted as a truncation of ADM for a Gaussian filter. The resulting model for the SGS-stress is given by

$$\tau_{ij} = \frac{1}{12}\Delta_k^2 \bar{\rho} \left(\frac{\partial \tilde{u}_i}{\partial x_k} \frac{\partial \tilde{u}_j}{\partial x_k} \right), \tag{5.21}$$

where Δ_k is the filter width in x_k direction. According to the conservative form of the evolution equations, refer to Chap. 2.2, the model is formulated based on the momentum densities ρu_i. In terms of filter width the tensor-diffusivity model approximates the exact SGS stresses with order $\mathcal{O}(\Delta^4)$. The prefactor $1/12$ on the right-hand side of (5.21) holds for a Gaussian filter. More generally this factor corresponds to half the second moment of the filter. Since commonly the tensor-diffusivity model is not used with explicit filtering and the filter implied by the numerical discretization scheme is ignored, this factor could also be considered as a model parameter. As with the scale-similarity model the nonlinearity of the viscous stresses due to the temperature-dependency of the viscosity is commonly neglected. Unlike with the scale-similarity model or with ADM a straight-forward estimate by using the approximately deconvolved solution is awkward when Taylor-series expansions with the tensor-diffusivity model are required and has not been used so far.

5.3 Regularization Techniques

Whereas the previously mentioned models operate on the range of represented scales, the models of this section address the effect of the missing interaction with non-represented scales. Models which have been developed for this purpose can be classified as:

1. Physically motivated functional models, whose interaction with the main structural model has to be taken into account.
2. Mathematically motivated regularizations of the filtered discretized equations.

For reproducing the proper inertial-range energy transfer, structural models require a model for the effect of the non-represented scales. Structural models generally give a better approximation of the subgrid-scale residuals for sufficiently resolved simulations since a large fraction of the energy transfer and redistribution is taken care of by the represented scales. The missing, often small fraction due to the non-existent interaction with the non-represented scales, however, is essential for the sensible long-term evolution of turbulent flows. For this remaining energy transfer physically motivated and mathematically motivated models have been proposed. Since the main purpose of this part of the full model is to maintain some kind of nonlinear stability we will call it regularization in the following. For more precise mathematical notions of stability for SGS models the reader is referred to the book of Berselli et al. [19].

5.3.1 Eddy-Viscosity Regularization

For both, the scale-similarity model and the tensor-diffusivity model, applications show that in most cases computations become unstable. With the original version of the scale-similarity model by Bardina [13] a regularization by an eddy-viscosity extension was included. A similar observation holds for the tensor-diffusivity model, introduced by Leonard [163], which is very unstable if no regularization is applied. While there is no proof that these models necessarily lead to unstable discrete approximations there is proof that without further correction they cannot lead to energy-bounded discrete approximations. Berselli et al. [19] explain that existence of a solution to

$$\frac{\partial \bar{u}_i}{\partial t} + \bar{u}_j \frac{\partial \bar{u}_i}{\partial x_j} + \frac{\partial \bar{p}}{\partial x_i} = \frac{1}{Re} \frac{\partial^2 \bar{u}_i}{\partial x_k^2} - \frac{\partial}{\partial x_k} \left(\frac{\Delta^2}{12} \frac{\partial \bar{u}_k}{\partial x_j} \frac{\partial \bar{u}_j}{\partial x_i} \right) \tag{5.22}$$

can be shown under conditions which make the solution physically irrelevant, such as very high regularity and smallness. It is shown as well, that if the tensor-diffusivity model in this equation is regularized by an additional eddy-viscosity term as in

$$\begin{aligned} \frac{\partial \bar{u}_i}{\partial t} + \bar{u}_j \frac{\partial \bar{u}_i}{\partial x_j} + \frac{\partial \bar{p}}{\partial x_i} = {} & \frac{1}{Re} \frac{\partial^2 \bar{u}_i}{\partial x_k^2} - \frac{\partial}{\partial x_k} \left(\frac{\Delta^2}{12} \frac{\partial \bar{u}_k}{\partial x_j} \frac{\partial \bar{u}_j}{\partial x_i} \right) \\ & + \frac{\partial}{\partial x_k} \left(C_G \sqrt{\frac{\partial \bar{u}_i}{\partial x_j} \frac{\partial \bar{u}_i}{\partial x_j} \frac{\partial \bar{u}_i}{\partial x_k}} \right), \end{aligned} \tag{5.23}$$

the requirement for existence and uniqueness of a solution is that the eddy-viscosity parameter C_G be sufficiently large.

To the authors' knowledge there are no applications of the tensor-diffusivity model, either with or without regularization to compressible flows at Mach numbers beyond the incompressible limit. A low-Mach-number application, where the energy SGS residuals have been neglected, to mixing-layer turbulence is due to Vreman et al. [309]. Since the tensor-diffusivity model was meant to increase the modeling accuracy, as shown above it allows for a deconvolution error of order $\mathcal{O}(\Delta^4)$ in filter width, the addition of an eddy-viscosity regularization introduces an artificial term of order $\mathcal{O}(\Delta^2)$. Thus, in the case of not very smooth solutions, the eddy-viscosity term dominates over the tensor-diffusivity term, the latter probably being of not much relevance.

These observations, made for incompressible flows, also hold for compressible flows, although there is no strict mathematical analysis available.

Berselli et al. [19] shown that for the incompressible scale-similarity model without regularization, which can be written in the form

$$\frac{\partial \bar{u}}{\partial t} + \frac{\partial \bar{u}_i \bar{u}_j}{\partial x_j} = -\frac{\partial \left(\overline{\bar{u}_i \bar{u}_j} - \bar{\bar{u}}_i \bar{\bar{u}}_j\right)}{\partial x_j} + \frac{1}{Re} \frac{\partial^2 \bar{u}_i}{\partial x_k^2}, \tag{5.24}$$

energy stability requires

$$\left\| \frac{\partial \bar{u}_i}{\partial x_j} \right\| \le \frac{\text{const}}{\sqrt{\Delta} Re}. \tag{5.25}$$

This requirement can only be satisfied if the velocity gradient is limited by the inverse Reynolds number for fixed filter width, which is clearly unphysical since the turbulent velocity field does not exhibit this type of regularity with increasing Reynolds number. The eddy-viscosity regularization of Bardina [13] amounts to a $\mathcal{O}(\Delta^2)$-regularization of (5.24), for which no energy-stability proof is available. A sufficient condition for energy-stability is obtained for an $\mathcal{O}(\sqrt{\Delta})$ regularization [19], which, however, is undesirable since the regularization would overwhelm the scale-similarity part of the model. Concerning long-time stability of the regularized scale-similarity model currently one has to rely heavily on numerical experimentation.

The eddy-viscosity regularization of the scale-similarity model for compressible flows was introduced by Speziale et al. [273]. The regularized expression for the SGS-stresses with this formulation according to Erlebacher et al. [75] is

$$\tau_{ij} = \bar{\rho}\left(\widetilde{\tilde{u}_i \tilde{u}_j} - \tilde{\tilde{u}}_i \tilde{\tilde{u}}_j\right) - 2C_R \bar{\rho} \Delta^2 \sqrt{\tilde{S}_{ij}\tilde{S}_{ij}}\left(\tilde{S}_{ij} - \delta_{ij}\tilde{S}_{kk}\right). \tag{5.26}$$

Speziale et al. [273] add an isotropic part,

$$-\frac{2}{3} C_I \bar{\rho} \Delta^2 \sqrt{\tilde{S}_{ij}\tilde{S}_{ij}}\delta_{ij}, \tag{5.27}$$

corresponding to the model of Yoshizawa [320], to the right-hand side. The regularized SGS heat flux becomes

$$Q_j = \bar{\rho}\left(\widetilde{\tilde{u}_j \tilde{T}} - \tilde{\tilde{u}}_j \tilde{\tilde{T}}\right) - \frac{C_R}{Pr_T} \Delta^2 \sqrt{\tilde{S}_{ij}\tilde{S}_{ij}} \frac{\partial \tilde{T}}{\partial x_k}, \tag{5.28}$$

where an analog to the turbulent Prandtl number has been introduced to scale the eddy-viscosity parameter.

Special considerations are necessary for the regularization of the energy equation. Again, we have to distinguish between the different formulations of the energy equation:

1. For the enthalpy formulation (2.95) we have to consider the pressure dilatation Π_{dil} and ϵ_v, for which no regularization is employed. Erlebacher

et al. [75] suggest to neglect Π_{dil} by analogy to statistical averaging, where in particular for wall-bounded turbulent shear flows up to the low hypersonic range the statistically averaged pressure dilatation has been found to be negligible.

2. For the temperature formulation (2.100) the same terms occur as for the enthalpy formulation, and the same procedure concerning regularization applies.
3. The only difference between the pressure formulation (2.105) and the enthalpy or temperature formulation is that the three unclosed terms for ideal gases have different constant factors but otherwise are the same. Again, also the same procedure for regularization applies.
4. The entropy formulation has not been used with scale-similarity modeling to the authors' knowledge. Whether regularization is required for the SGS terms arising in the energy equation is unknown.
5. For the energy equation (2.122) the first candidate term for regularization is SGS turbulent diffusion. No explicit regularization of this term is performed. If a formulation which explicitly contains the SGS stress tensor is used, such as $\mathcal{J}_j = -\tau_{ij}\tilde{u}_i$, regularization is introduced implicitly by regularization of the SGS stress tensor. The same holds for the SGS residuals D_3, D_4, D_5. When the formulation with resolved energy, filtered pressure and filtered temperature is used the SGS terms B_1 through B_7 require consideration. For regularization, following Vreman [307], terms B_1 and B_2 are combined as

$$B_1 + B_2 = \text{scale-similarity} - \frac{\partial}{\partial x_j}\left(\frac{C_{23}\Delta^2}{2(\gamma-1)Pr_T M^2}\bar{\rho}\check{S}_{ij}\check{S}_{ij}\frac{\partial \tilde{T}}{\partial x_j}\right), \quad (5.29)$$

where

$$\check{S}_{ij} = \frac{\partial \tilde{u}_i}{\partial x_j} + \frac{\partial \tilde{u}_j}{\partial x_i} - \frac{2}{3}\delta_{ij}\frac{\partial \tilde{u}_k}{\partial x_k}. \quad (5.30)$$

Terms B_3 and B_4 are regularized implicitly by the regularization of τ_{ij}. No regularization is performed for B_5 and B_6. The term B_7 corresponds to D_5.

The regularization parameters C_R and Pr_T in the above eddy-viscosity regularizations can also be computed dynamically. For this purpose one defines according to Vreman [307] a generalized Germano identity as shown in the following. A vector field $\mathbf{w}$, for which on application of a nonlinear functional a SGS residual occurs as

$$\mathbf{res} = \overline{\mathbf{f}(\mathbf{w})} - \mathbf{f}(\bar{\mathbf{w}}). \quad (5.31)$$

On the test-filter level the residual is

$$\mathbf{Res} = \widehat{\overline{\mathbf{f}(\mathbf{w})}} - \mathbf{f}(\hat{\bar{\mathbf{w}}}). \quad (5.32)$$

The generalized Leonard term follows as

$$\mathbf{Leo} = \mathbf{Res} - \widehat{\mathbf{res}}, \tag{5.33}$$

for which the generalized Germano identity gives

$$\mathbf{Leo} = \widehat{\mathbf{f}(\bar{\mathbf{w}})} - \mathbf{f}(\hat{\bar{\mathbf{w}}}). \tag{5.34}$$

For a regularized model in which the dynamic procedure applies only to the regularization, say with linear parameter C_{reg}, one has to separate the respective model contributions in (5.26), when the SGS model is substituted

$$\mathbf{Leo}(\hat{\bar{\mathbf{w}}}, \bar{\mathbf{w}}) = \mathbf{H}(\hat{\bar{\mathbf{w}}}, \bar{\mathbf{w}}) - C_{reg}\mathbf{M}(\hat{\bar{\mathbf{w}}}, \bar{\mathbf{w}}). \tag{5.35}$$

Here, $\mathbf{H}$ stands for the scale-similarity part and $\mathbf{M}$ for the regularization part. The dynamic regularization parameter is then obtained by

$$C_{reg} = \frac{\langle \mathbf{M} \cdot (\mathbf{Leo} - \mathbf{H}) \rangle}{\langle \mathbf{M} \cdot \mathbf{M} \rangle}, \tag{5.36}$$

where for tensorial functionals an appropriate inner product has to be defined.

5.3.2 Relaxation Regularization

For the explanation of the relaxation regularization we resort again to the generic one-dimensional conservation law equation (5.2). For regularization the right-hand side of (5.4a) is supplemented by a relaxation term

$$\frac{\partial \bar{u}_N}{\partial t} + \frac{\partial \overline{F_N(u_N^*)}}{\partial x} = -\chi(\mathrm{I} - Q * G) * \bar{u}_N, \tag{5.37}$$

with relaxation parameter χ. By the operator $(I - Q * G)$ the relaxation term is wavenumber dependent and affects mainly the wavenumber range $k_c < \|k\| < k_N$. The filtered solution $\bar{u}_N$ is driven towards a solution $\overline{u^*}_N$ which is smoother. The order of the secondary filter $Q * G$ arising in the relaxation term is $r(N+1)$, where r is the order of the primary filter G and N is the regularization parameter of the deconvolution. I is the identity operator. Adams and Stolz [4] show that by this kind of regularization term even shock-capturing properties for simple flows can be recovered. It is also worthwhile to note that this term resembles the drift term in Langevin stochastic turbulence models.

The regularization effect is not very sensitive to the particular choice of the relaxation parameter, once an appropriate magnitude has been identified, which of course is solution dependent. For statistically steady flows it is not necessary to update the regularization parameter at every time step, leading to some savings of computational effort. The regularized ADM model in deconvolution form, where the deconvolved solution is inserted directly into the flux function, results in the continuity equation

$$\frac{\partial \bar{\rho}}{\partial t} + \overline{\frac{\partial (\rho u)^*_j}{\partial x_j}} = -\chi_\rho(\bar{\rho} - \bar{\rho}^*), \tag{5.38a}$$

the momentum equations

$$\frac{\partial \overline{(\rho u)}_i}{\partial t} + \overline{\frac{\partial}{\partial x_j}\left(\frac{(\rho u)^*_i (\rho u)^*_j}{\rho^*} + \breve{p}^* \delta_{ij} - \breve{\tau}^*_{ij}\right)} = -\chi_{\rho u}(\overline{\rho u} - \overline{\rho u}^*), \tag{5.38b}$$

and the energy equation

$$\frac{\partial \bar{E}}{\partial t} + \overline{\frac{\partial}{\partial x_j}\left(\frac{(\rho u)^*_j}{\rho^*}(E^* + \breve{p}^*) - \breve{\tau}^*_{ji}\frac{(\rho u)^*_i}{\rho^*} + \breve{q}^*_j\right)} = -\chi_E(\bar{E} - \bar{E}^*). \tag{5.38c}$$

For the generalized scale-similarity form the relaxation is simply added to the SGS-residuals on the right hand side. The same parameter $\chi_{\rho u}$ is used for all three components of the momentum equation. The solution procedure for (5.38) is:

(1) Dynamic estimation of the relaxation parameters, as detailed below.
(2) Deconvolution of the current filtered solution.
(3) Computation of the filtered flux derivatives.
(4) Filtering of the deconvolved solution (as needed in the regularization term).
(5) Computation of the relaxation regularization.
(6) Summation of filtered flux derivative and relaxation regularization to obtain the time derivative of the filtered solution.
(7) Integration in time by an explicit integration scheme.

Stolz et al. [277, 278] have proposed a dynamic relaxation-parameter determination. The regularization parameter can be obtained from a dynamic procedure based on an estimate of the SGS kinetic energy. Objective of the regularization is to prevent an increase of kinetic energy, respectively a corresponding energy norm of quantities other than velocity or momentum, in the range of non-resolved represented scales, whereas an energy redistribution within this range of scales is permitted. E.g., for density the considered range of scales is extracted by $\Phi_\rho = (\mathrm{I} - Q * G) * \bar{\rho}$, similar expressions hold for $\overline{\rho u_i}$ and $\bar{E}$. An estimate of the energy norm of Φ_ρ can be obtained locally by the second-order structure function $F_2(\mathbf{x}, t)$ as

$$F_2(\mathbf{x}, t) = \|\Phi_\rho(\mathbf{x} + \mathbf{r}, t) - \Phi_\rho(\mathbf{x}, t)\|^2_{\|\mathbf{r}=h\|}, \tag{5.39}$$

where the norm stands for integration over a ball with radius of the local grid spacing h at location $\mathbf{x}$. For grid data this integration is estimated from a trapezoidal rule involving the center point and its 6 next neighbors. For determining the regularization parameter χ_ρ, similarly for the other variables, first the increase of energy in the considered scale range is estimated by advancing the solution without regularization and obtaining the energy estimate after one time step as $F_2(\mathbf{x}, t+\tau)_{\chi_\rho=0}$. Also an energy estimate $F_2(\mathbf{x}, t+\tau)_{\chi_{\rho 0}}$ can be obtained based on an initial guess for χ_ρ. With the energy estimate at the current time step $F_2(\mathbf{x}, t)$ and due to the fact that the regularization is linear in χ_ρ, an updated estimate for the regularization parameter is obtained from

$$\chi_\rho = \chi_{\rho 0} \frac{F_2(\mathbf{x}, t+\tau)_{\chi_\rho=0} - F_2(\mathbf{x}, t)}{F_2(\mathbf{x}, t+\tau)_{\chi_\rho=0} - F_2(\mathbf{x}, t+\tau)_{\chi_{\rho 0}}}. \tag{5.40}$$

During time advancement χ_{ρ_0} can be estimated from χ_ρ at the previous time step.

An alternative version of ADM has been proposed by Adams and Stolz [4] for shock capturing. If the deconvolution error is neglected, i.e. $Q * G = I$, (5.4a) is equivalent to

$$\frac{\partial u_N^*}{\partial t} + \frac{\partial F_N(u_N^*)}{\partial x} = 0, \tag{5.41}$$

and with relaxation regularization one obtains

$$\frac{\partial u_N^*}{\partial t} + \frac{\partial F_N(u_N^*)}{\partial x} = -\chi(I - Q * G) u_N^*, \tag{5.42}$$

which is equivalent to (5.2) under the condition $Q*G = I$. From this equation, the filtered solution $\bar{u}_N$ can be obtained by postprocessing filtering, since

$$G * \left(\frac{\partial u_N^*}{\partial t} + \frac{\partial F_N(u_N^*)}{\partial x} \right) = \frac{\partial \bar{u}_N^*}{\partial t} + G * \frac{\partial F_N(u_N^*)}{\partial x}.$$

Relaxation regularization in this case amounts to adding the relaxation term to the right hand side of the discretized conservation law, as in (5.42), where for shock solutions an *a priori* estimate of a constant relaxation parameter is possible [4]. For transitional or turbulent flows, the relaxation term has been applied successfully within this framework [250].

5.3.3 Regularization by Explicit Filtering

The relaxation regularization can be implemented also as explicit filtering [276]. This can be seen from a fractional-time-step discretization of (5.37). First we predict an intermediate solution at $\tau/2$ without the regularization as

$$\bar{u}_N\left(t + \frac{\tau}{2}\right) = \bar{u}_N(t) + \tau \frac{\partial \overline{F_N(u_N^*)}}{\partial x}.$$

This solution is corrected by adding the contribution of the regularization, and one obtains

$$\begin{aligned} \bar{u}_N(t+\tau) &= \bar{u}_N\left(t + \frac{\tau}{2}\right) - \tau\chi\left(\mathrm{I} - Q * G\right) * \bar{u}_N\left(t + \frac{\tau}{2}\right) \\ &= \tau\chi Q * G * \bar{u}_N\left(t + \frac{\tau}{2}\right). \end{aligned}$$

For $\chi = 1/\tau$ this implies that the relaxation regularization for this one-step time integration is identical, up to temporal discretization error, to a post-processing filtering with $Q * G$ at each time step.

Filtering as regularization can be so effective that no additional SGS model is needed. Visbal et al. [303, 304] use a high-order compact filters of up to 10th-order and high-order compact finite differences (4th- to 6th-order) as spatial discretizations [166].

Based on the equivalence between relaxation regularization and explicit filtering and on assuming that the relaxation term is a model for the error $\partial \overline{u^*}_N/\partial t - \partial \bar{u}_N/\partial t$ an alternative version of ADM has been proposed by Mathew et al. [194]. In the first step, (5.4a) is rewritten, assuming that $\overline{u^*}_N = \bar{u}_N$, as

$$G * \left(\frac{\partial u_N^*}{\partial t} + \frac{\partial F_N(u_N^*)}{\partial x} \right) = G * \frac{\partial u_N^*}{\partial t} - \frac{\partial \overline{u^*}_N}{\partial t}. \tag{5.43}$$

Combined with the reformulation of the relaxation regularization as explicit filtering the solution is advanced in time, and after each time step the $(2N+1)$ order filter $Q * G$ is applied.

A modification of (5.41) has been proposed by Mathew et al. [195] as

$$\frac{\partial u_N^*}{\partial t} + \frac{\partial F_N(u_N^*)}{\partial x} = \frac{\partial F(Q * G * u_N^*)}{\partial x} - \frac{\partial F(u_N^*)}{\partial x}. \tag{5.44}$$

The time advancement of this equation is regularized by applying the above explicit filter at each time step. It is interesting to note that (5.44) can be interpreted as a modification of the generalized scale-similarity form of ADM equation (5.4b). Following the analysis of Mathew et al. [195] and by the property $F(u) = F'(u)u$ of the flux function one can estimate

$$-F(Q * G * u_N^*) + 2F(u_N^*) \approx -F' u_N^* - F' \cdot (Q * G * u_N^* - u_N^*) + 2F' u_N^*,$$

where the flux derivative $F' = dF/du$ is to be taken at u_N^*. The left-hand side of that equation corresponds to the effective flux term of (5.44), rewritten such that the right-hand side is zero. We can re-arrange now

$$\begin{aligned} &-F' u_N^* - F' \cdot (Q * G * u_N^* - u_N^*) + 2F' u_N^* \\ &= F' u_N^* + F' \cdot (u_N^* - Q * G * u_N^*) = F'(u_N^* + u_N^* - Q * G * u_N^*) \\ &= F' \sum_{m=0}^{1} (I - Q * G)^m * u_N^* \approx F(u_N^{**}), \end{aligned}$$

where u_N^{**} is an ADM approximation, truncated at $M = 1$, of u_N from u_N^* with respect to the filter $Q * G$. In terms of the estimation we obtain a re-formulation of (5.44), which is

$$\frac{\partial u_N^*}{\partial t} + \frac{\partial F_N(u_N^{**})}{\partial x} = 0.$$

This equation corresponds to ADM with a change of the primary filter from G to $Q{*}G$. The difference between this equation and (5.44) is that F' is evaluated as u_N^* whereas for (5.44) it would need to be evaluated at $Q * G * u_N^*$. This approximation is, however, consistent with the derivation of the model [195].

5.4 Multi-Scale Modeling of Subgrid-Scales

5.4.1 Multi-Level Approaches

The fundamental idea of multi-level approaches is that a part of the subgrid-scales is advanced along with the resolved field. It is straight-forward to distinguish between resolved and represented scales, where in the case of multi-level approaches the non-resolved represented scales again can be split into different sub-scale ranges. The benefit of these approaches is that the range of non-resolved represented scales is much smaller than the full range of physically relevant scales, and that a simplified transport can be applied to that range of scales.

A multilevel SGS closure was proposed by Terracol et al. [287]. The underlying concept is evident from Fig. 5.2. A set of $N+1$ filter levels is introduced, ranging from level 0, i.e. no filtering, corresponding to DNS resolution, up to level N the coarse-mesh resolution, to be considered as equivalent to the resolved range of the LES. Filtering can be performed by projective or non-projective kernels. A straight-forward multi-level scheme is given by dyadic

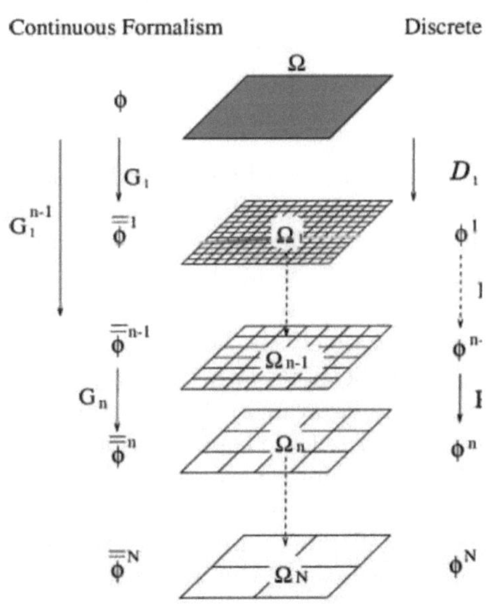

Fig. 5.2. Formalism of the multi-level-scheme of [287, 288], reproduced with permission

embedded meshes, where for transforming variables between the filter levels can be handled by multigrid prolongation and restriction operators, as indicated in Fig. 5.2.

The resolved-scale evolution is based on the conservative form of the compressible Navier-Stokes equations

$$\frac{\partial \bar{\rho}}{\partial t} + \frac{\partial \bar{\rho}\tilde{u}_i}{\partial x_i} = 0 \tag{5.45a}$$

$$\frac{\partial \bar{\rho}\tilde{u}_i}{\partial t} + \frac{\partial \bar{\rho}\tilde{u}_i\tilde{u}_j}{\partial x_j} = -\frac{\partial \bar{p}}{\partial x_i} + \frac{\partial \breve{\sigma}_{ij}}{\partial x_j} - \frac{\partial \tau_{ij}}{\partial x_j} \tag{5.45b}$$

$$\frac{\partial \check{E}}{\partial t} + \frac{\partial (\check{E} + \bar{p})\tilde{u}_j}{\partial x_j} = \frac{\partial (\breve{\sigma}_{ij} + \tau_{ij})\tilde{u}_j}{\partial x_j} - \frac{\partial \bar{Q}_j + q_j}{\partial x_j}. \tag{5.45c}$$

Filtering is to be taken at the level N and nonlinearities in the viscous terms are neglected, i.e. their SGS contributions are suppressed. Considering the velocity field, an approximation of the unfiltered field is obtained by adding the multi-level increments to that at level N

$$u_j^* = \tilde{u}_j + \sum_{n=1}^{N-1} \delta u_j^{(n)}. \tag{5.46}$$

Based on this approximation a generalized scale-similarity ansatz can be invoked, for the SGS stress tensor leading to

$$\tau_{ij} = \bar{\rho}\big(\widetilde{u_i^* u_j^*} - \tilde{u}_i\tilde{u}_j\big). \tag{5.47}$$

A version regularized by an eddy viscosity would be

$$\tau_{ij} = \bar{\rho}\big(\widetilde{u_i^* u_j^*} - \tilde{u}_i\tilde{u}_j\big) - 2C_d\bar{\rho}\Delta^2\sqrt{\check{S}_{ij}\check{S}_{ij}}\check{S}_{ij}. \tag{5.48}$$

The regularization parameter C_d can be obtained dynamically by the generalized Germano identity as outlined in Sect. 5.3, where one more coarse-grid level $N+1$ is introduced for test filtering [288]. Analogous procedures apply to the SGS heat flux, other SGS terms being neglected in (5.45). During a computational time step on the resolved level N a V-sweep is performed on the finest level $n = 0$, successively moving up to $n = N$ and subsequently back to $n = 0$, where to each level its own time-step size Δt^n is assigned, e.g. $\Delta t^{n+1} = 2\Delta t^n$. During the integration of the evolution equations at level n the smaller scales at levels $0, \ldots, n-1$ are kept constant, based on the frozen-flow or quasi-static hypothesis of small scales, see e.g. Ref. [60], and the solution at the finer resolution levels is not updated in time. That assumption is the main source of computational-time savings. Note that the concept of approximate inertial manifolds [83] rather leads to an instantaneous adaptation of the small scales to the larger ones.

In a further development of the multi-level method Terracol et al. [288] propose a method to estimate the time range during which the quasi-static hypothesis is valid. For this purpose the time-scale on which the evolution of the smallest represented scales, i.e. that on the resolution level $n = 0$, can be neglected is estimated from the condition that the relative increase of total energy remains below a chosen threshold

$$t_{QS} = \varepsilon_{max} \frac{\|\breve{E}^{(0)}\|}{\|\frac{\partial \delta \breve{E}^{(0)}}{\partial t}\|}, \tag{5.49}$$

where ε is typically between 10^{-3} and 10^{-4}. The increment in total energy $\delta \breve{E}^{(n)}$ is obtained by subtracting the total energy of the next coarser level $n+1$ and the time derivative is approximated by a backward difference for which available data from a previous time step can be used. The maximum number of resolution levels and their assigned maximum time-step sizes which can be considered are obtained by summing up the time-step sizes from each level and stop summing when t_{QS} is reached. The admissible time-step sizes at each resolution level $\Delta t^{(n)}$ are determined from a numerical stability (Courant-Friedrichs-Lewy) criterion. In a subsequent step the physical admissible time-step sizes for each level are computed by solving the following system of N equations

$$\sum_{m=n}^{N} \Delta t_m = \varepsilon_{max} \frac{\|\breve{E}^{(n-1)}\|}{\|\frac{\partial \delta \breve{E}^{(n-1)}}{\partial t}\|}, \quad 1 \leq n \leq N \tag{5.50}$$

for Δt_m. The admissible number of time steps at each level is then obtained from the integer of the ratio $\Delta t_n / \Delta t^{(n)}$.

A two-level-scheme is the SGS-estimation model of Domaradzki and Saiki [54], which has been formulated for compressible flows by Dubois et al. [61]. An $\bar{E}$, $\bar{p}$, $\tilde{T}$ formulation of the energy equation is used and nonlinearities in the viscous and heat-conduction terms due to the material-parameter dependence on the temperature are neglected, also neglected is SGS turbulent diffusion. With the SGS-estimation model, a two-level representation of the solution is employed, where $\mathbf{U}^{(2)}$ corresponds to the resolved scales and $\mathbf{U}^{(1)}$ to the represented scales. Commonly with the SGS-estimation model the resolution on the resolved level is half of that on the represented level, i.e. $\Delta^{(2)} = 2\Delta^{(1)}$, and orthogonal meshes are used. This choice is due to the fact that the dominant convective nonlinearity of scales up to $\Delta^{(2)}$ extends to a range of up to $\Delta^{(1)}$. One operation step is the restriction from the $n = 1$ level to the $n = 2$ level which is performed by a Simpson integration rule. The restriction can be inverted by solving a tridiagonal system of equations, resulting in a prolongation from the coarse to the fine mesh. This step can also be considered as regularized deconvolution. Since the prolongation results in a solution with the same smoothness as that on the coarse level, some fine level information needs to be generated, which is done by cubic-spline interpolation at the intermediate grid points, i.e. such grid points which the fine level does not share with the coarse level, resulting in a seed of the fine-scale field $\mathbf{u}^{(1)}$.

The characteristic time scale of the kinetic energy increase of this field can be estimated by considering the nonlinear convection of $\mathbf{u}^{(1)}$ by the small-scale field $\delta \mathbf{u}^{(1)}$ only, i.e. for the velocity field one obtains

$$\mathcal{N}_i^{(1)} = \delta u_j^{(1)} \frac{\partial u_i^{(1)}}{\partial x_j}. \tag{5.51}$$

The characteristic time by which a given energy in the non-resolved represented range is reached follows from dimensional arguments as

$$t_{let} = \text{const} \times \sqrt{\frac{(\delta \mathbf{u}^1)^2}{(\mathcal{N}^{(1)} - \bar{\mathcal{N}}^{(1)})^2}}, \tag{5.52}$$

where the proportionality factor has been found to be about $1/2$. t_{let} can be interpreted as a large-eddy turn-over time. The fine-scale information is then reconstructed as $\delta \mathbf{u}^{(1)} = t_{let}(\mathcal{N}^{(1)} - \bar{\mathcal{N}}^{(1)})$ leading to an estimated velocity field $\mathbf{u}^{(2)} + \delta \mathbf{u}^{(1)}$ on the fine level in time. The estimated field can then be inserted into a generalized scale-similarity form of the SGS-stress-tensor model, and the resolved field can be advanced in time. The analogous procedure is applied to the temperature for obtaining an estimate of the SGS heat flux. The SGS-estimation model has been further developed for the incompressible Navier-Stokes equations, resulting in the model of truncated Navier-Stokes dynamics [56]. In this model the fine-scale field $\mathbf{U}^{(1)}$ is advanced in time, but the energy in the non-resolved represented range is annihilated periodically, about every t_{ref} time period, by removing $\delta \mathbf{U}^{(1)}$.

5.4.2 Stretched-Vortex Model

The stretched-vortex model of Misra and Pullin [199] has been formulated for compressible flows by Kosovic et al. [147]. Although it leads to a functional expression of the SGS-stress tensor it can be considered as structural model since the derivation of the modeled SGS-stress tensor is based on an explicit construction of the SGS field. Fundamental assumption is a vortical nature of the SGS field so that it can be composed out of generic vortices [184]. Essential to the model is the preferred alignment of the subgrid vortices $\mathbf{e}$ which is linearly weighted between the direction of the principal extensional eigenvector $\mathbf{e}_3$ of the local resolved strain-rate tensor $\tilde{S}_{ij}$ and the direction of the resolved local vorticity $\mathbf{e}_{\tilde{\omega}}$. The weight is estimated from the eigenvalue λ_3 corresponding to $\mathbf{e}_3$ and from the vorticity magnitude $\|\tilde{\omega}\|$ as

$$\mu = \frac{\lambda_3}{\lambda_3 + \|\tilde{\omega}\|}.$$

The orientation of the subgrid vortices is then obtained from

$$\mathbf{e} = \mu \mathbf{e}_3 + (1 - \mu)\mathbf{e}_{\tilde{\omega}}.$$

The SGS stress tensor is constructed as

$$\tau_{ij} = \bar{\rho} K (\delta_{ij} - e_i e_j). \tag{5.53}$$

The SGS kinetic energy is estimated from

$$K = \int_{k_N}^{\infty} C\epsilon^{2/3} k^{-5/3} e^{-2k^2/(\nu|a|)},$$

where $|a|$ is the magnitude of the strain $\tilde{S}_{ij} e_i e_j$ along the vortex axis, and $C\varepsilon^{2/3}$ is estimated from the SGS kinetic energy, which again can be estimated by the second-order structure function of the resolved velocity field. The SGS heat flux is closed by invoking the model for scalar transport [225], resulting in

$$q_i = \frac{1}{2} \Delta \sqrt{K} (\delta_{ij} - e_i e_j) \frac{\partial \tilde{T}}{\partial x_j}. \tag{5.54}$$

The underlying equations follow the $\tilde{E}$, $\bar{p}$, $\tilde{T}$ formulation, so that besides τ_{ij} and Q_j also the SGS viscous diffusion $\mathcal{D}_j$ and the SGS turbulent diffusion $\mathcal{J}_j$ require modeling. Nonlinearities in the friction and heat flux due to the temperature dependency of viscosity and heat conductivity are neglected. Kosovic et al. [147] assume $\mathcal{J}_j$ and $\mathcal{D}_j$ to be small and therefore do not propose a special model for these terms.

5.4.3 Variational Multi-Scale Model

The variational multi-scale model (VMS) has been proposed for incompressible flow by Hughes et al. [128] based on a more fundamental exposure [127]. It can be considered as multi-scale model, or as another version of models which act on a high-pass filtered field [241]. VMS has been proposed originally with the finite-element discretization of the flow equations in mind, where a scale separation on each computational element is possible by distinguishing the different polynomial degrees of the trial functions. Hughes et al. [129] analyze the VMS method in the framework of a Fourier-spectral discretization by which the relation to the other approaches is more evident. For compressible flows VMS has been adapted by Koobus and Farhat [145] for unstructured meshes and by van der Bos et al. [298] investigated with respect to general properties of the VMS formulation. The most straight-forward way to outline the basic idea of VMS is to start with a 2-level formulation, where we split the dependent variables in a resolved part $\mathbf{U}^{(2)}$, corresponding to level 2 and a non-resolved represented part $\delta\mathbf{U}^{(1)} = \mathbf{U}^{(1)} - \mathbf{U}^{(2)}$ from level 1. With VMS it is assumed that

$$\mathbf{U} \approx \mathbf{U}^{(2)} + \delta\mathbf{U}^{(1)}, \tag{5.55}$$

i.e. that $\mathbf{U}^{(1)}$ approximates the unfiltered field $\mathbf{U}$. The particular way how to split $\mathbf{U}^{(2)}$ and $\delta\mathbf{U}^{(1)}$ depends on the numerical discretization and is not elaborated here, for details the reader can refer, e.g., to van der Bos et al. [298].

On inserting the split solution into the conservation equations, one can separate the nonlinear terms in a similar fashion as standard practice in LES modeling to obtain residuals corresponding to the Leonard (i.e. resolved) stresses, the cross (i.e. resolved / non-resolved) stresses and the subgrid (non-resolved) stresses. In terms of a Galerkin projection all these contributions need to be evaluated on the range of represented scales only, nevertheless cross- and subgrid-stress residuals need to be modeled. More than two-level versions of VMS have been proposed. As pointed out by Collis [44] VMS requires the standard assumptions invoked in functional SGS modeling, namely that (i) there is negligible direct effect of the non-represented scales on the evolution of the resolved scales, and (ii) the action of the non-represented scales on the resolved scales is mainly dissipative. The VMS belongs to the multi-level models and can be cast into the same formalism. Depending on the underlying discretization scheme, VMS usually employs restriction and prolongation operators which are consistent with the weak formulation of the discrete equations, e.g. the characteristic function as test function corresponding to finite-volume methods, a polynomial trial function as test function corresponding to Galerkin finite-element methods. Bearing this relation in mind a fully detailed exposure of VMS is possible for certain classes of discretizations schemes, whereas a general exposure essentially follows that of the multi-level approach above.

6

Relation Between SGS Model and Numerical Discretization

6.1 Systematic Procedures for Nonlinear Error Analysis

6.1.1 Error Sources

For simplicity we consider the initial-value problem for a generic scalar nonlinear transport equation for the variable v

$$\frac{\partial v}{\partial t} + \frac{\partial F(v)}{\partial x} = 0. \tag{6.1}$$

On a mesh $x_j = jh$ with equidistant spacing h and $j = \ldots, -1, 0, 1, \ldots$ the grid function $v_N = \{v_j \mid j = \ldots, -1, -0-1, \ldots\} = \{v_j\}$ represents a discrete approximation of $v(x)$ by $v_j \doteq v(x_j)$. A spectrally accurate interpolant of the grid function with the same Fourier transform can be constructed using the Whittaker cardinal function [301]. For finite h the representation of the continuous solution $v(x)$ by the grid function v_N results in a subgrid-scale error or residual

$$\mathcal{G}_{SGS} = \frac{\partial F_N(v_N)}{\partial x} - \frac{\partial F_N(v)}{\partial x} \tag{6.2}$$

which arises from the nonlinearity of $F(v)$. The modified differential equation (MDE) for v_N is

$$\frac{\partial v_N}{\partial t} + \frac{\partial F_N(v_N)}{\partial x} = \mathcal{G}_{SGS}. \tag{6.3}$$

Since for LES the ratio between characteristic flow scale and grid size h never can be considered as asymptotically small $\mathcal{G}_{SGS}$ cannot be neglected for proper evolution of v_N but requires approximation by modeling closures.

A similar MDE is obtained from a finite-volume discretization of (6.1) which corresponds to a convolution with the top-hat filter

$$G(x - x_j; h) = \begin{cases} 1/h, & |x - x_j| \leq h/2, \\ 0, & \text{else} \end{cases} \tag{6.4}$$

E. Garnier et al., *Large Eddy Simulation for Compressible Flows*,
Scientific Computation,

on the grid $x_N = \{x_j\}$. An application of the filter operation (6.4) to a function $u(x)$ returns the filtered solution in terms of a grid function u_j at x_j

$$\bar{u}_j = G * u = \frac{1}{h} \int_{x_{j-1/2}}^{x_{j+1/2}} u(x')dx'.$$

The resulting finite-volume approximation of (6.1) is given by

$$\frac{\partial \bar{u}_N}{\partial t} + G * \frac{\partial \tilde{F}_N(u_N^*)}{\partial x} = 0, \tag{6.5}$$

where $u_N^* \doteq u_N$ results from an approximate inversion $u_N^* = Q * \bar{u}_N$ of the filtering $\bar{u}_N = G * u$ and $\tilde{F}_N$ is a consistent numerical flux function. This equation approximates the exact filtered equation

$$\frac{\partial \bar{v}}{\partial t} + G * \frac{\partial F(v)}{\partial x} = 0. \tag{6.6}$$

Similar considerations hold for filter kernels other than the top-hat filter. We recall that although the inverse-filtering operation is ill-posed, an approximation u_N^* of u on the grid x_N can be obtained by regularized deconvolution, refer to Chap. 5.

Once deconvolution operation and numerical flux function are determined, the modified-differential-equation analysis of (6.5) leads to an evolution equation of $\bar{u}_N$ in the form of

$$\frac{\partial \bar{u}_N}{\partial t} + G * \frac{\partial F_N(u_N)}{\partial x} = \mathcal{G}_N, \tag{6.7}$$

where

$$\mathcal{G}_N = G * \frac{\partial F_N(u_N)}{\partial x} - G * \frac{\partial \tilde{F}_N(u_N^*)}{\partial x} \tag{6.8}$$

is the truncation error of the discretization scheme. If $\mathcal{G}_N$ approximates $\bar{\mathcal{G}}_{SGS}$ in some sense for finite h we obtain an implicit subgrid-scale model implied by the discretization scheme. Note that this requirement is similar to that for $\mathcal{G}_{SGS}$ and different from classical asymptotic truncation-error analysis, where $\mathcal{G}_N$ approximates $\bar{\mathcal{G}}_{SGS}$ for $h \to 0$.

A study of the relation between SGS model and truncation error reveals that due to the interference between SGS model and truncation error a grid-resolution dependent optimum choice for the model parameter exists [198] for different-order numerical discretization schemes. The dimensionality of the problem would increase further if the discretization scheme would contain itself parameters or would be nonlinear. The investigation of the truncation error, and in particular how the truncation error can be exploited for SGS modeling, is the main objective of this chapter. We follow the common interpretation of LES as result of an evolution equation with spatially reduced resolution, and thus we focus on the analysis of errors due to spatial discretizations and spatial filtering. This implies that the time-step size chosen for time integration is always sufficiently small so that error contributions from temporal discretization and temporal filtering (if applied) are negligible as compared to spatial error contributions. From now on this will be tacitly assumed.

The above SGS error or SGS residual subsumes other errors which arise from the particular choice of filter. For explicit models such contributions possibly are important, e.g. the error due to the lack of commutativity of filtering and derivative operators. The commutation error has been investigated by Ghosal and Moin [96] and has sparked a series of investigations on commutative or moment-preserving filters [300] which reduce the commutation-error terms to higher order in filter width. The issue of time-variable filter width, e.g. due to dynamic grid refinement, has been analyzed by Leonard et al. [164]. Fureby and Tabor [86] have given a comprehensive derivation of the commutation-errors for the compressible flow equations, see also Sect. 2.3.5. For the continuity equation one obtains

$$\frac{\partial \bar{\rho}}{\partial t} + \frac{\partial \bar{\rho}\tilde{u}_i}{\partial x_i} = -\left(\frac{\partial G}{\partial \Delta} * (\rho u_i)\right)\frac{\partial \Delta}{\partial x_i} - \left(G * (\rho u_i) n_i\right)\big|_{\mathrm{b}}\,, \tag{6.9}$$

with similar additional terms in the momentum and energy equations, where the index b indicates boundary terms. For commonly used second-order filter kernels the commutation error is $\mathcal{O}(\Delta^2)$. Special higher-order filter kernels can be constructed with respect to computational space [299] or with respect to physical space [277], reducing the effect of the truncation error. Near domain boundaries one-sided filters of higher-order can be constructed. A more detailed discussion can be found in Ref. [244, Sect. 2.2]. Another error source with explicit SGS models arises from aliasing errors [97, 150]. As such errors are inevitable with non-spectral discretizations and even for spectral discretizations are difficult to suppress [38] we do not further discuss aliasing errors here.

6.1.2 Modified Differential Equation Analysis

The modified differential equation analysis (MDEA) was essentially introduced by Warming and Hyett [311] following previous work by Hirt [119] and Yanenko and Shokin [317], see also Ref. [120]. Considering the initial-value problem of the generic one dimensional conservation law of (6.1) with the generic semi-discretization equation (6.5) we obtain as solution a grid function $u_N = \{u_j\}$, respectively a filtered grid function, on an equidistant mesh x_j, spanning the x-axis. Assuming that this grid function decays sufficiently fast with $x \longrightarrow \pm\infty$ it possesses a discrete Fourier transform

$$\hat{u}_k = h \sum_{j=-\infty}^{\infty} u(x_j) e^{-ikx_j}\,. \tag{6.10}$$

As explained by Vichnevetsky and Bowles [301] the same Fourier transform applies to the continuous function

$$u(x) = \sum_n \frac{\sin(\pi \frac{x-x_j}{h})}{\pi \frac{x-x_j}{h}} u_j \tag{6.11}$$

which is the continuous interpolation of the grid function u_N by virtue of the Whittaker cardinal function

$$\Psi_n = \frac{\sin(\pi \frac{x-x_n}{h})}{\pi \frac{x-x_n}{h}}.$$

The interpolated function is smooth $u(x) \in C^\infty$ and can be expanded as a Taylor series at each x_j. Note that $u(x)$ represents jumps in u_N, which are approximations of discontinuities in terms of the grid function, by a smooth approximation. Also, any filtering applied to u_N can be formally inverted by application of the inverse filter kernel to $\hat{u}_N$ since as a grid function u_N has compact support in Fourier space, and accordingly does $u(x)$.

Given $u(x)$ as continuous approximation of u_N, and given that this function is sufficiently smooth, the question is what differential equation is solved exactly by $u(x)$. Since $\bar{u}(x_j) = \bar{u}_j$, and $\bar{u}_j \doteq \bar{v}(x_j)$ is only an approximation of the solution of (6.6) certainly $\bar{v}(x)$ and $\bar{u}(x)$ differ in general. Therefore, also the differential equation whose exact solution is $\bar{u}(x)$ differs from (6.6). Given that $\bar{u}_N$ is the solution of (6.5) obtained by a consistent discretization, the differential equation for $\bar{u}(x)$ should differ from (6.6) for $\bar{v}(x)$ by the truncation error, which for consistent schemes should become small with small grid- and time-step sizes. As we consider here only spatial semi-discretizations it is the spatial truncation error, which is relevant. Now it is clear that (6.7) can be rewritten in terms of $u(x,t)$ as

$$\frac{\partial \bar{u}}{\partial t} + G * \frac{\partial F(u)}{\partial x} = \mathcal{G}, \tag{6.12}$$

where the continuous spatial truncation error $\mathcal{G}$ can be obtained by inserting the Taylor expansion of $u(x)$, after $u(x)$ has been reconstructed from $\bar{u}(x)$ by inverse filtering or deconvolution.

The cell-averaging, or top-hat filtering over one computational cell, can be inverted as follows. First, we use that $u(x_j) = u_j$, then $\bar{u}_j = G * u(x)|_j$ where G is the kernel from (6.4). Due to the equivalence between the Fourier-transforms of $u(x)$ and u_N the unfiltered solution can be reconstructed in Fourier space by using the Fourier transform of the filter kernel $\hat{G}(k)$ as

$$\hat{u}_k = \frac{\hat{\bar{u}}_k}{\hat{G}(k)}. \tag{6.13}$$

This inversion operator certainly is useful only for equidistant meshes.

For the purpose of MDEA we have to distinguish between deconvolution for SGS reconstruction, leading to u_N^*, and deconvolution for Taylor expansion of the truncation error, leading to $w(x)$. A deconvolution reconstruction which operates on local interpolants and can be extended to non-equidistant meshes has been proposed by Harten et al. [109]. For local deconvolution we do not rely on Fourier transforms of $u(x)$ and u_N. Instead we create an approximation $w(x)$ to $u(x)$ in a neighborhood of x_j by polynomial interpolation up to degree M. Since the polynomial representation is unique one can define the polynomial coefficients by comparison with a Taylor expansion of $u(x)$ at x_j

$$w(x) = \sum_{m=0}^{M} \frac{w^{(m)}}{m!} \xi^m \doteq \sum_{m=0}^{\infty} \frac{u^{(m)}}{m!} \xi^m = u(x), \tag{6.14}$$

where $\xi = x - x_j$ and $w^{(m)} = d^m w/dx^m$. Given $w(x)$ one can create an approximant $\bar{w}(x)$ for $\bar{u}_N$ from

$$\bar{w}(x) = \sum_{m=0}^{M} \frac{w^{(m)}}{m!} G * \xi^m = \sum_{m=0}^{M} w^{(m)} \alpha_m h^m, \tag{6.15}$$

where

$$\alpha_m = \begin{cases} 0, & m \text{ odd}, \\ \frac{1}{2^m (m+1)!}, & m \text{ even}. \end{cases}$$

Both sides of (6.15) can be differentiated $M-1$ times, resulting in a set of linear equations which can be solved for $w(x_j) \doteq u_j$

$$\begin{bmatrix} \bar{w}(x_j) \\ h\bar{w}'(x_j) \\ h^2\bar{w}''(x_j) \\ \vdots \\ h^{M-1}\bar{w}^{(M-1)}(x_j) \end{bmatrix} = \begin{bmatrix} 1 & 0 & \alpha_2 & 0 & \alpha_4 & 0 & \cdots & \alpha_{M-1} \\ & \ddots & \ddots & \ddots & \ddots & \ddots & \ddots & \\ & & \ddots & \ddots & \ddots & \ddots & \ddots & 0 \\ & & & \ddots & \ddots & \ddots & \ddots & \alpha_4 \\ & & & & \ddots & \ddots & \ddots & 0 \\ & & & & & \ddots & \ddots & \alpha_2 \\ & & & & & & \ddots & 0 \\ & & & & & & & 1 \end{bmatrix} \times \begin{bmatrix} w(x_j) \\ hw'(x_j) \\ h^2 w''(x_j) \\ \vdots \\ h^{M-1} w^{(M-1)}(x_j) \end{bmatrix}. \tag{6.16}$$

With $w(x)$ approximating $u(x)$ to desired order in h the Taylor expansion of the truncation error can be determined also to desired order in h. Note that the deconvolution to obtain $w(x)$ as approximation of $u(x)$ from $\bar{u}_N$ here serves the purpose of deriving the MDE and not the purpose of using $w(x)$ as reconstructed solution in the flux function during numerical computation instead of u_N^*. It is clear that M should be sufficiently larger than the polynomial approximation degree involved in the deconvolution procedure for u_N^*.

As simple example we can consider a finite-volume semi-discretization of the initial-value problem for the advection equation

$$\frac{\partial v}{\partial t} + \frac{\partial v}{\partial x} = 0.$$

The finite-volume discretization on an equidistant mesh is given by

$$\frac{\partial \bar{u}_j}{\partial t} + \frac{u_{j+1/2} - u_{j-1/2}}{h} = 0. \tag{6.17}$$

The cell-face values are approximated by linear interpolation from the cell-averaged solution $u_{j+1/2} \doteq u^*_{j+1/2} = (u^*_{j+1} + u^*_j)/2$ and $u_{j-1/2} \doteq u^*_{j-1/2} = (u^*_j + u^*_{j-1})/2$. Deconvolution based on linear interpolation gives $u^*_j = \bar{u}_j$, and one obtains the discretization scheme

$$\frac{\partial \bar{u}_j}{\partial t} + \frac{\bar{u}_{j+1} - \bar{u}_{j-1}}{2h} = 0. \tag{6.18}$$

The modified equation satisfied by $\bar{w}(x)$ which is obtained by the above procedure is

$$\frac{\partial \bar{w}}{\partial t} + G * \frac{\partial w}{\partial x} \doteq -\frac{1}{12} h^2 \bar{w}_j'''. \tag{6.19}$$

For more general transport equations and more complex, e.g. nonlinear, discretization schemes the derivation of the modified differential equation becomes significantly more complex. Also, when incompressible flows are considered, the spatial truncation error involves contributions of the divergence constraint. If zero divergence is satisfied by projection these analytical contributions require the solution of a Poisson equation by suitable Green functions. Overall, the real space modified-differential-equation analysis is hardly tractable for more complex discretizations and more general transport equations. Even if one can derive the truncation error for the considered equations and discretization scheme it arises as an infinite or truncated Taylor expansion in terms of grid size. Except for terms which carry second-order (dissipation) or third-order (dispersion) derivatives, or even higher-order derivatives (hyper-dissipation), it is hard to assign physical meaning to the individual terms. Modeling based on such a real-space representation of the truncation error can resort to term-by-term comparisons with other results, such as the SGS residual of explicit models (Taylor expanded). Since SGS energy transfer always involves the entire represented range of scales, a truncated Taylor expansion, however, is a bad approximation of the Taylor series. This issue is well known and leads, e.g. to the ill-posedness of the tensor-diffusivity model. An analogy exists in the kinetic theory of gases, where a higher-order term lifting the Chapman-Enskog expansion from the Navier-Stokes equations to the Burnett equations results in ill-posedness [236]. Therefore, for the development of implicit SGS models the preferred option is to consider a form of the truncation error which provides not a truncated but rather a direct approximation to the Taylor series form. With MDEA for implicit SGS modeling it always has to be kept in mind that an asymptotic truncation error analysis is irrelevant since the truncation error cannot be considered as small. Therefore, implicit SGS modeling amounts to a fully nonlinear optimization of the truncation error.

In the following we consider the momentum-conservation equations in the following generic form

$$\frac{\partial \boldsymbol{u}}{\partial t} + \boldsymbol{\nabla} \cdot \boldsymbol{F} + \nabla p = \boldsymbol{0}, \tag{6.20}$$

where $\boldsymbol{u} = [u, v, w]$ is the velocity, $\boldsymbol{F} = \boldsymbol{uu}$ is the nonlinear flux tensor, p is the pressure, and friction forces are suppressed for simplicity. The differential equations for the resolved scales are obtained by applying the filter to (6.20)

$$\frac{\partial \bar{\boldsymbol{u}}_N}{\partial t} + G * \boldsymbol{\nabla} \cdot \boldsymbol{N}_N(\mathbf{u}_N^*, p_N^*) = -G * \boldsymbol{\nabla} \cdot \boldsymbol{\tau}_{SGS}, \tag{6.21}$$

where $\boldsymbol{N} = \boldsymbol{F} + p$.

In (6.21) the represented-scale part of the unfiltered field is reconstructed for computing the nonlinear term by a regularized inverse-filter operation $\mathbf{u}_N^* = Q * \bar{\boldsymbol{u}}_N$ applied to represented scales. Since non-represented scales cannot be recovered it is $\mathbf{u}_N^* \neq \boldsymbol{u}$, which results in the subgrid-stress tensor

$$\boldsymbol{\tau}_{SGS} = \boldsymbol{F}(\boldsymbol{u}, p) - \boldsymbol{F}_N(\mathbf{u}_N^*, p_N^*). \tag{6.22}$$

Due to numerical approximations the exact solution of the discretization of (6.20) does not satisfy (6.21) with $\boldsymbol{\tau}_{SGS} = \boldsymbol{0}$, but rather a MDE which can be generally written as

$$\frac{\partial \bar{\boldsymbol{u}}_N}{\partial t} + \tilde{G} * \tilde{\boldsymbol{\nabla}} \cdot \tilde{\boldsymbol{N}}_N(\mathbf{u}_N^*, p_N^*) = \boldsymbol{0}. \tag{6.23}$$

Now we consider a three-dimensional top-hat filter-kernel G, which can be generated from the one-dimensional kernel by repeated application in the individual coordinate directions. For this filter kernel we recall that the LES equations are equivalent to a finite-volume discretization. $\widetilde{\boldsymbol{u}}_N$ denotes an approximant of the velocity $\mathbf{u}_N^*$. The local Riemann problem at the faces of computational cells can be handled by a consistent numerical flux function $\tilde{\boldsymbol{F}}_N$ in $\tilde{\boldsymbol{N}}_N = \tilde{\boldsymbol{F}}_N + \tilde{p}_N$. The symbols $\tilde{G} * \tilde{\boldsymbol{\nabla}}$ indicate that G and $\boldsymbol{\nabla}$ are replaced by their respective numerical approximations. Note that in particular $\tilde{G} * \tilde{\boldsymbol{\nabla}}$ can be a nonlinear operator. The truncation error due to the discretization of the flux divergence is

$$\mathcal{G}_N = G * \boldsymbol{\nabla} \cdot \boldsymbol{F}_N(\boldsymbol{u}_N, p_N) - \tilde{G} * \tilde{\boldsymbol{\nabla}} \cdot \tilde{\boldsymbol{F}}_N(\mathbf{u}_N^*, p_N^*). \tag{6.24}$$

6.1.3 Modified Differential Equation Analysis in Spectral Space

For implicit SGS modeling the question arises how physical modeling concepts can be imposed onto discretization schemes so that the truncation error has desired energy-transfer properties. As mentioned in the previous section, a term-by-term comparison of the modified differential equation analysis (MDEA) in real space with existing explicit models may be one option. In that case one would expect that at best the performance of the reference explicit model can be recovered. More promising is to design a discretization scheme in such a way that the total truncation error functions as physical subgrid scale model. One of the essential requirements for a SGS model is that for isotropic turbulence it has to recover the proper energy transfer between represented and non-represented scales. Only then one can expect that the kinetic-energy spectrum develops a Kolmogorov inertial range. Most suitable for such analyses is a representation of the truncation error in spectral space, leading the MDEA in spectral space.

Before considering MDEA in spectral space we begin with a related method, which is applicable to linear or linearized evolution equations, the modified-wavenumber analysis. The modified-wavenumber analysis was introduced essentially by Vichnevetsky and Bowles [301] for the analysis of discretizations of the advection equation. Considering the advection equation as example for a linear conservation law

$$\frac{\partial v}{\partial t} + \frac{\partial v}{\partial x} = 0, \tag{6.25}$$

and on assuming that v is approximated by the grid function u_N, a modified differential equation for its discretization can be derived as

$$\frac{\partial \bar{u}_N}{\partial t} + G * \frac{\partial u_N}{\partial x} = \left(G * \frac{\partial u_N}{\partial x} - \tilde{G} * \frac{\partial u_N^*}{\partial x} \right). \tag{6.26}$$

By proper definition of the regularized deconvolution $u_N^* = Q * \bar{u}_N$ any finite-difference scheme can be written in the above finite-volume form. Noting that $u_N = \sum_{k=-N/2}^{k=N/2-1} \hat{u}_k e^{\mathrm{i}kjh}$ the orthogonality of the Fourier modes can be exploited, and from (6.26) its spectral-space equivalent is obtained

$$\frac{d\hat{\bar{u}}_k}{dt} + \mathrm{i}k\hat{G}_k\hat{u}_k = \left(\mathrm{i}k\hat{G}_k\hat{u}_k - \mathrm{i}k\hat{\tilde{G}}_k\hat{\bar{u}}_k \right) = \mathrm{i}k \left(\hat{G}_k\hat{u}_k - \hat{\tilde{G}}_k\hat{Q}_k\hat{u}_k \right). \tag{6.27}$$

In spectral space the relative truncation error thus can be expressed as

$$\hat{\mathcal{G}}(k) = \left(1 - \frac{\hat{\hat{G}}_k \hat{Q}_k}{\hat{G}_k}\right). \tag{6.28}$$

The last term within the brackets on the right-hand side is commonly referred to as the modified wavenumber of the discretization, see also Ref. [166] To nonlinear equations a modified-wavenumber analysis was applied by Garnier et al. [91]. For a generic one-dimensional conservation law (6.1) their definition of a spectral distribution of numerical errors is

$$\mathcal{T}_k = \frac{|\widehat{(\frac{\partial \tilde{F}}{\partial x})}^2_k|}{|k^2 \hat{F}^2_k|}, \tag{6.29}$$

where $\tilde{F}$ is a numerical flux function consistent with F. For more than one spatial dimension the terms in the nominator and denominator each are averaged over shells with $\|\underline{k}\| = k$ in spectral space, and for vectorial transport equations the spectral error distribution can be evaluated and analyzed for each component.

As an extension of the MDE analysis in real space or as an extension of the modified-wavenumber analysis to nonlinear equations we consider now MDE analysis in spectral space. The truncation error, as given in (6.24) for a generic three-dimensional conservation law is responsible for deviations of the spectral energy transfer from purely truncated Navier-Stokes dynamics, see e.g. Ref. [56]. Implicit LES modeling (ILES) is based on requiring that this deviation from truncated Navier-Stokes dynamics represents the correct SGS energy transfer. Since theoretically well established results exist for isotropic, homogeneous turbulence, the most suitable way to analyze the energy transfer implied by the truncation error is in Fourier spectral space which has been proposed by Domaradzki et al. [57] and Hickel et al. [112]. In the following, we analyze the MDE in Fourier space in order to develop a theoretical framework for the numerical evaluation of the subgrid dissipation and the spectral numerical viscosity. For consistency with a truncated representation of isotropic turbulence, where wavenumbers up to a certain cut-off wavenumber k_C are considered, we define a truncated filtered velocity field as

$$\hat{\bar{\boldsymbol{u}}}_C(\mathbf{k}) = \begin{cases} \hat{\bar{\boldsymbol{u}}}_N(\boldsymbol{k}), & |\mathbf{k}| \leq k_C, \\ \mathbf{0}, & \text{otherwise}, \end{cases} \tag{6.30}$$

where the hat denotes the Fourier transform. The corresponding spectral-space MDE is

$$\frac{\partial \hat{\bar{\boldsymbol{u}}}_C}{\partial t} + \hat{G} \mathrm{i} \boldsymbol{k} \cdot \hat{\boldsymbol{F}}_C(\boldsymbol{u}_C) = \hat{\boldsymbol{\mathcal{G}}}_C. \tag{6.31}$$

On the represented wave-number range, the kinetic energy of the unfiltered field

$$\hat{\boldsymbol{u}}_C(\boldsymbol{k}) = \hat{G}^{-1}(\boldsymbol{k})\hat{\bar{\boldsymbol{u}}}_C(\boldsymbol{k}), \quad \text{with } |\boldsymbol{k}| \leq k_C \tag{6.32}$$

is

$$\hat{E}(\boldsymbol{k}) = \frac{1}{2}\hat{\boldsymbol{u}}_C(\boldsymbol{k}) \cdot \hat{\boldsymbol{u}}_C^*(\boldsymbol{k}). \tag{6.33}$$

Multiplying (6.31) by the complex-conjugate Fourier coefficient $\hat{\boldsymbol{u}}_C^*$, we obtain

$$\hat{G}(\mathbf{k})\frac{\partial \hat{E}(\boldsymbol{k})}{\partial t} - \hat{G}(\boldsymbol{k})\hat{T}_C(\boldsymbol{k}) = \hat{\boldsymbol{u}}_C^*(\boldsymbol{k}) \cdot \hat{\boldsymbol{\mathcal{G}}}_C(\boldsymbol{k}). \tag{6.34}$$

The nonlinear energy transfer $\hat{T}_C(\boldsymbol{k})$ is the Fourier transform of the flux-divergence and has to be computed from the deconvolved velocity $\hat{\boldsymbol{u}}_C$. Finally, we deconvolve (6.34) by multiplication with the inverse filter coefficient $\hat{G}^{-1}(\boldsymbol{k})$, which is exactly known on the range of represented scales $|\boldsymbol{k}| \leqq k_C$,

$$\frac{\partial \hat{E}(\boldsymbol{k})}{\partial t} - \hat{T}_C(\boldsymbol{k}) = \hat{G}^{-1}(\boldsymbol{k})\hat{\boldsymbol{u}}_C^*(\boldsymbol{k}) \cdot \hat{\boldsymbol{\mathcal{G}}}_C(\boldsymbol{k}). \tag{6.35}$$

The right-hand side of this equation gives by definition of the numerical dissipation

$$\varepsilon_{num}(\boldsymbol{k}) = \hat{G}^{-1}(\boldsymbol{k})\hat{\boldsymbol{u}}_C^*(\boldsymbol{k}) \cdot \hat{\boldsymbol{\mathcal{G}}}_C(\boldsymbol{k}), \tag{6.36}$$

which for ILES is expected to represent the physical subgrid dissipation $\varepsilon_{SGS}(\boldsymbol{k})$.

For isotropic turbulence equation (6.36) can be integrated over wavenumber shells with radius $\|\mathbf{k}\| = \boldsymbol{k}$, resulting in the evolution equation of the 3D energy spectrum of the numerical solution

$$\frac{\partial \hat{E}(k)}{\partial t} - \hat{T}_C(k) = \varepsilon_{num}(k). \tag{6.37}$$

Invoking a spectral-eddy-viscosity assumption the spectral truncation error $\varepsilon_{\text{num}}(k)$ can be expressed as numerical spectral viscosity

$$\nu_{\text{num}} = \frac{\varepsilon_{num}(k)}{2k^2\hat{E}(k)}, \tag{6.38}$$

which relates to the SGS spectral eddy viscosity

$$\nu_{SGS} = \frac{\varepsilon_{SGS}(k)}{2k^2\hat{E}(k)}. \tag{6.39}$$

For a given numerical scheme the spectral numerical viscosity is

$$\nu_{num}(k) = -\frac{1}{2k^2\hat{E}(k)}\int_{|\boldsymbol{k}|=k} \hat{G}^{-1}(\boldsymbol{k})\hat{\boldsymbol{u}}_N^*(\boldsymbol{k}) \cdot \hat{\boldsymbol{\mathcal{G}}}_N(\boldsymbol{k})d\boldsymbol{k}. \tag{6.40}$$

Usually it is non-dimensionalized by

$$\nu_{num}^{+}(k^{+}) = \nu_{num}\left(\frac{k}{k_C}\right)\sqrt{\frac{k_C}{\hat{E}(k_C)}}, \quad \text{with } k^{+} = \frac{k}{k_C}. \tag{6.41}$$

Modern analytical statistical theories of low-Mach number turbulence employ a wavenumber-dependent eddy viscosity, as first proposed by Heisenberg [111]. The Eddy-Damped Quasi-Normal Markovian (EDQNM) theory defines the eddy viscosity by

$$\nu_{SGS}(k^{+}) = \nu_{SGS}^{\infty} X(k^{+}), \tag{6.42}$$

in spectral space [173]. Here, ν_{SGS}^{∞} is the asymptotic eddy viscosity and $X(k^{+})$ is a non-dimensional function with a plateau equal to 1 at small wavenumbers $k^{+} \lesssim 1/3$ and a sharply rising cusp in the vicinity of the cut-off wave number $k^{+} = 1$. The plateau and the corresponding asymptotic eddy viscosity can be calculated analytically from a given subgrid energy spectrum by means of EDQNM. A parameter fit gives

$$\nu_{Chollet}^{+}(k^{+}) = 0.441 C_K^{-3/2}\sqrt{\frac{E(k_C)}{k_C}}\left(1 + 34.47 e^{0.33 k^{+}}\right) \tag{6.43}$$

as sufficient approximation of the exact result [41] where C_K is the Kolmogorov constant [173]. This expression can serve as reference for optimizing ν_{num} [112].

6.2 Implicit LES Approaches Based on Linear and Nonlinear Discretization Schemes

6.2.1 The Volume Balance Procedure of Schumamm

With the Volume-Balance Procedure [253] the incompressible Navier-Stokes equations are averaged over finite volumes. As mentioned already this is equivalent to top-hat filtering, where the filter kernel is centered at discrete points in space. Although the method as such has not been transferred to the compressible Navier-Stokes equations, we outline the method here since it can be considered as one of the first physically motivated implicit LES models. The flow-evolution equations are averaged over finite volumes, using a staggered arrangement of the grid for the different components of the momentum equation, see Fig. 6.1. Volume averaging gives rise to a SGS stress at the cell faces, resulting from area averaging over the cell faces. Due to the latter property the volume-balance approach can naturally deal with grid anisotropies, unlike the filtering approach. Schumann [253] did not further consider the truncation error of the numerical discretization due to interpolation of off-grid values. The main focus is placed on physically motivated modeling of the SGS stresses. For this purpose isotropic and inhomogeneous contributions are considered separately. A particular truncation-error term arising from the discretization, or volume-averaging, respectively, is modeled, whereas not the full truncation error is considered. In this respect the volume-balance procedure rather could be considered as explicit LES model, but the notion that the discrete equations should be the starting point of modeling is shared with implicit LES modeling.

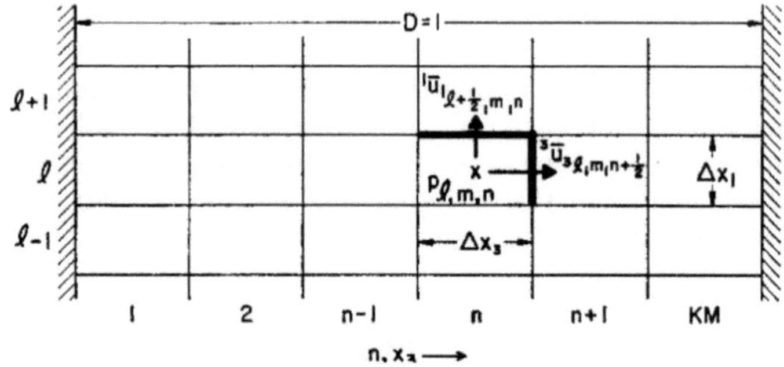

Fig. 6.1. Averaging volume and grid arrangement for the volume-balance procedure, reproduced with permission from Ref. [253]

6.2.2 The Kawamura-Kuwahara Scheme

A linear implicit LES method has been proposed by Kawamura and Kuwahara [137] for incompressible flow. The method is based on a standard pressure-correction scheme with non-staggered arrangement of grid points for the discretization of momentum and pressure-Poisson equations. While for the discrete derivative operators in the Poisson equation and in the friction term of the momentum equations second-order central differences are used, the essence of this simple implicit model is in the discretization of the convective terms in the momentum equations. The discretization of the convective terms such as

$$u\frac{\partial v}{\partial x}$$

by the Kawamura-Kuwahara scheme amounts to

$$\begin{aligned} u\frac{\partial v}{\partial x} &\doteq u_j \frac{-v_{j+2} + 8v_{j+1} - 8v_{j-1} + v_{j-2}}{12h} \\ &\quad + |u_j| \frac{v_{j+2} - 4v_{j+1} + 6v_j - 4v_{j-1} + v_{j-2}}{4h} . \end{aligned} \tag{6.44}$$

This discretization constitutes a third-order upwind scheme for the convective flux. For more than one dimension the discretization operator is applied in a tensorial fashion in each coordinate direction. Furthermore, in two dimensions an additional grid is introduced by rotating the original grid at each grid point by 45° and using the data at the diagonal grid points, which are spaced by $\sqrt{2}h$ with respect to the original grid. On the rotated grid the above upwind discretization operator is computed as well and at each grid point the results are superimposed with a weighting of 2 : 1 in favor of the original grid, see Fig. 6.2. In three dimensions three additional grid systems are introduced by rotating the grid at each grid point by 45° around one of the original coordinate axes. Time discretization is performed by a Crank-Nicolson scheme. We are not aware of an assessment of the Kawamura-Kuwahara scheme with respect to the SGS energy transfer in high-Reynolds number isotropic turbulence. A multi-directional scheme for compressible flow, following the concept of the Kawamura-Kuwahara scheme was proposed by Lee et al. [161], without an emphasis on turbulent-flow computations, however. The linear upwind scheme in that case has been replaced by a MUSCL TVD scheme [290].

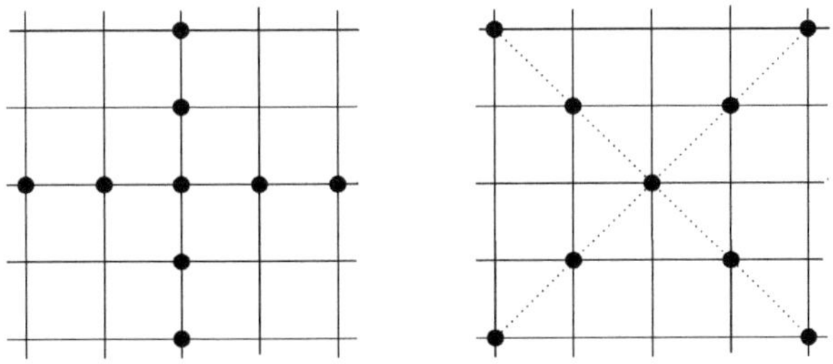

Fig. 6.2. Grid arrangement for the Kawamura-Kuwahara scheme. *Left*: scheme aligned with coordinate axes. *Right*: diagonal scheme

6.2.3 The Piecewise-Parabolic Method

The Piecewise-Parabolic Method (PPM) was proposed by Woodward and Colella [315] for the numerical solution of gas-dynamics problems, see also Ref. [106, Chap. 4b], and applied to turbulence flows by Porter et al. [224]. Basis of PPM is a MUSCL (Monotone UPstream-centered Scheme for Conservation Laws) [162] reconstruction of the unfiltered solution by quadratic local approximation polynomials. If we consider cell j as in Fig. 6.3 a quadratic local approximation polynomial for the deconvolved (reconstructed) solution p_j can be obtained. For more general local polynomial reconstructions the reader can refer to Harten et al. [109] or Shu [263]. The same can be done for the neighboring cells $j \pm 1$. From each of these approximants an estimate of the volume-averaged solution $\bar{u}_j$ in cell j can be obtained, where the estimates from the left and right neighbor cells differ in general from that of cell j itself. From these differences a so-called fractional error ϵ_j at cell j can be computed which serves as a measure of the local smoothness of the solution. Based on this reconstruction procedure and the accompanying smoothness measure a

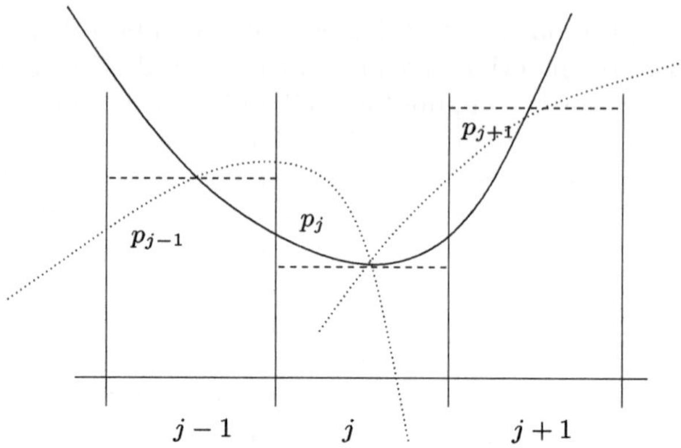

Fig. 6.3. Sketch of the PPM interpolation

MUSCL-type, formally third-order accurate spatial discretization can be constructed, where the interpolation procedure for compressible flow problems is commonly applied to the Riemann variables (i.e. to a local characteristic projection of the transport equations). To maintain nonlinear stability a slope limiting is introduced based on the above-mentioned smoothness measure. For contact discontinuities a steepening operation is constructed. With this framework PPM introduces a method for the reconstruction of subgrid information which is motivated by its function as approximate Riemann solver for gas dynamics.

6.2.4 The Flux-Corrected-Transport Method

For implicit SGS modeling the discretization scheme has to be specifically designed so that the truncation error $\mathcal{G}_N$ has physical significance. In a numerical analysis of Garnier et al. [91] several approaches to implicit LES are investigated and difficulties in satisfying this requirement are demonstrated. It was found that artificial dissipation introduced by common nonlinearly-stable discretizations indeed stabilizes under-resolved turbulence simulations. For the investigated schemes, however, small scales suffer from excessive numerical damping such that the probability-density functions of velocity increments and pressure exhibit the typical behavior of low Reynolds-number flows rather than that of high Reynolds-number turbulence. Thus it appears that for these schemes the prediction accuracy of subgrid effects is poor, although some general trends were reproduced.

With Flux-Corrected Transport Method (FCT), however, good results for a wide range of complex flows have been reported [103]. One of the initial attempts to use FCT for regularization of the effect of non-resolved flow scales other than shocks and contact-lines are the applications to two-dimensional laminar mixing layers [105]. The further development of FCT lead to the version proposed by Fureby and Grinstein [88, 104] which for the spatial semi-discretization becomes

$$\frac{\partial \rho_P}{\partial t} + \frac{1}{\delta V_P} \sum_f (\rho u_i)_f (\delta A_i)_f = 0, \tag{6.45a}$$

$$\frac{\partial(\rho u_i)}{\partial t} + \frac{1}{\delta V_P} \sum_f \left((\rho u_i u_j)_f - \left(\mu \frac{\partial u_i}{\partial x_j} + \mu \frac{\partial u_j}{\partial x_i} \right)_f \right.$$
$$\left. + \frac{1}{3} \left(\mu \frac{\partial u_k}{\partial x_k} \delta_{ij} \right)_f \right) (\delta A_j)_f + \left(\frac{\partial p}{\partial x_i} \right)_P = 0, \tag{6.45b}$$

$$\frac{\partial (\rho E)_P}{\partial t} + \frac{1}{\delta V_P} \sum_f \left((\rho E u_i)_f - \left(\frac{\kappa}{C_V} \frac{\partial E}{\partial x_i} \right)_f + (p u_i)_f \right.$$
$$\left. - \left(\mu \frac{\partial u_i}{\partial x_j} u_i + \mu \frac{\partial u_j}{\partial x_i} u_i \right)_f + \frac{1}{3} \left(\mu \frac{\partial u_j}{\partial x_j} u_i \right)_f \right) (\delta A_i)_f = 0. \tag{6.45c}$$

Here, the index P stands for an average over the cell volume V_P, the index f indicates the cell face with vector-valued area $\delta \mathbf{A}_f$. Convective fluxes are discretized by blending a component based on a high-order reconstruction, superscript H, and a component based on a low-order reconstruction, superscript L

$$(F_{ij})_f^C = (F_{ij})_f^{C,H} - (1 - \Psi_f(u_i)) \left((F_{ij})_f^{C,H} - (F_{ij})_f^{C,L} \right). \tag{6.46}$$

For non-convective fluxes linear central reconstructions are used. Essential to the FCT scheme is that a flux limiter Ψ as functional of the respective component of the velocity $\mathbf{u}$ is used to blend between a high-order velocity reconstruction $\mathbf{u}^H$ and a low-order velocity reconstruction $\mathbf{u}^L$. The high-order reconstruction is commonly obtained from linear or cubic polynomial approximants, leading to a second- or fourth-order approximation of the flux difference, whereas the low-order reconstruction results from a zero-order (piecewise constant) upwind approximation, resulting in a first-order approximation of the flux difference. The latter involves the upwind-indicators $\beta^{\pm} = 1, 0, -1$ (see below) which select out the upwind contribution of the velocity, depending on the projection of the cell-face velocity on the face-normal.

A MDE of the resulting FCT scheme gives when truncated after the second-order term in terms of grid spacing [106, Chap. 4]), see also Ref. [170]

$$\frac{\partial \rho}{\partial t} + \frac{\partial \rho u_i}{\partial x_i} = \frac{\partial}{\partial x_i} \left(\rho \chi \frac{\partial u_i}{\partial x_j} h_j + \frac{\rho}{8} \frac{\partial^2 u_i}{\partial x_j \partial x_k} h_j h_k \right) + \cdots, \tag{6.47a}$$

$$\frac{\partial \rho u_i}{\partial t} + \frac{\partial \rho u_i u_j}{\partial x_j} + \frac{\partial p}{\partial x_i} + \frac{\partial \tau_{ij}}{\partial x_j}$$
$$= \frac{\partial}{\partial x_k} \left(\rho \chi u_i h_j \frac{\partial u_k}{\partial x_j} + \rho \chi u_j h_i \frac{\partial u_j}{\partial x_k} + \rho \chi^2 \frac{\partial u_i}{\partial x_j} \frac{\partial u_k}{\partial x_l} h_j h_l \right)$$

$$+ \frac{\partial}{\partial x_k}\left(\frac{\rho}{8}u_i\frac{\partial^2 u_j}{\partial x_l^2}h_k h_j + \frac{\rho}{8}u_k\frac{\partial^2 u_j}{\partial x_l^2}h_i h_j\right) + \cdots , \tag{6.47b}$$

$$\begin{aligned} &\frac{\partial \rho E}{\partial t} + \frac{\partial \rho u_i E}{\partial x_i} + \frac{\partial p u_i}{\partial x_i} - \frac{\partial q_i}{\partial x_i} - \frac{\partial \tau_{ij} v_i}{\partial x_j} \\ &= \frac{\partial}{\partial x_i}\left(\rho\chi u_i h_j \frac{\partial E}{\partial x_j} + \chi E \frac{\partial u_i}{\partial x_j}h_j + \chi^2 \frac{\partial E}{\partial x_j}h_j\frac{\partial u_i}{\partial x_k}h_k\right) \\ &+ \frac{\partial}{\partial x_i}\left(\frac{\rho}{8}\frac{\partial^2 E}{\partial x_j^2}h_i h_k v_k + \frac{\rho}{8}E\frac{\partial^2 v_j}{\partial x_k^2}h_j h_i\right) \\ &+ \frac{\partial}{\partial x_i}\left(\frac{\partial^2 q_j}{\partial x_k^2}h_i h_j\right) + \cdots . \end{aligned} \tag{6.47c}$$

Here, $\chi = \pm(1-\Psi)/2$ occurs as potential model parameter which depends on the flux-limiting function Ψ. For Ψ different choices are possible, such as the van Leer limiter $\Psi(r) = 2r/(1+r)$, where $r = (\rho_{i,j,k}-\rho_{i-1,j,k})/(\rho_{i+1,j,k}-\rho_{i,j,k})$ for the i direction, see also Ref. [106]. The leading order terms of the spatial truncation error in (6.47) function as implicit subgrid scale model, whose particular SGS energy transfer properties depend on the choice of the flux-limiter Ψ.

With the above FCT scheme several investigations have been made, demonstrating the suitability for a wide range of turbulent flows, e.g. isotropic turbulence, channel flow, jet flow, backward-facing step flow [102]. However, no further attempt has been made of a systematic model optimization based on theoretical or physical knowledge. Results for post-transition stages of the three-dimensional Taylor-Green vortex show a developed inertial range, Fig. 6.4. The channel flow computations at $Re_\tau = 395$ show a clearly improved prediction of the near-wall region as compared with the dynamic Smagorinsky model, whereas some small deviations occur in the channel center, Fig. 6.5.

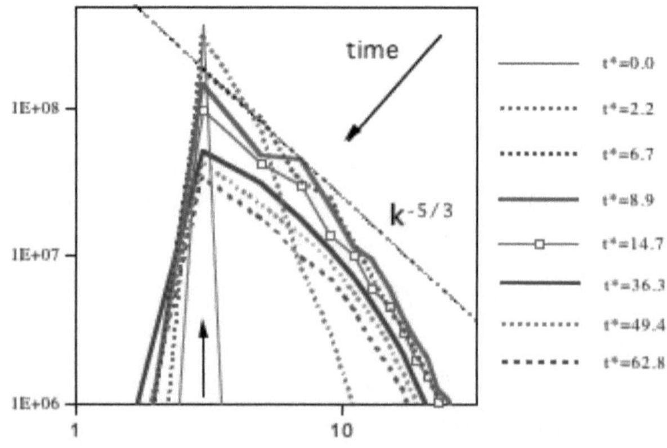

Fig. 6.4. Turbulent kinetic energy spectrum decay of the 3D Taylor-Green vortex after transition with FCT, reproduced from Ref. [104] with permission

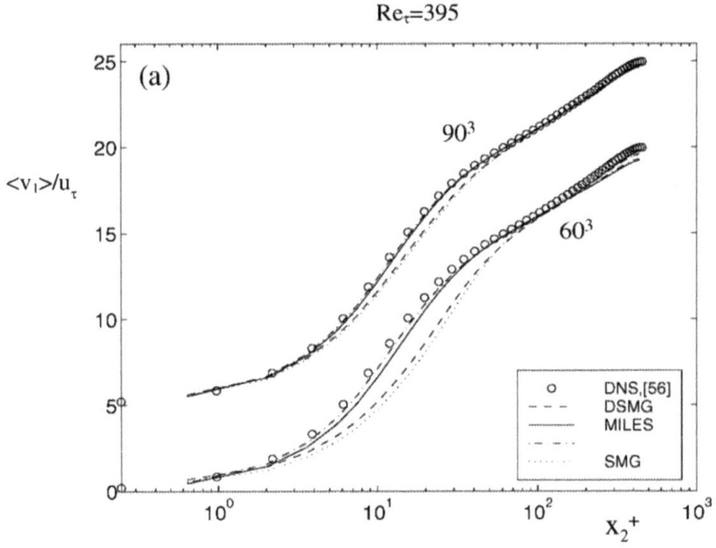

Fig. 6.5. Mean velocity for the turbulent channel flow at $Re_\tau = 395$, reproduced from Ref. [88] with permission

A sensitivity study of the modeling parameter of a simple FCT scheme has been performed by Honein and Moin [122] for quasi-incompressible isotropic turbulence and compressible isotropic turbulence with different initial ratios of dilatational to total kinetic energy. For this particular FCT scheme a first-order upwind scheme based on Steger-Warming flux-vector splitting [290] is used for the low-order flux formulation, and a 6-th order compact scheme [166] for the high-order flux formulation. A flux limiter is used which blends linearly between the two proposed limiters of Zalesak [321]. On the solution at the new time step the following limits can be imposed on the admissible maximum or minimum values of the solution in cell j, here shown for simplicity in one spatial dimension,

$$\begin{aligned} a^{max} &= \max\left(u_{j-1}^{td}, u_j^{td}, u_{j+1}^{td}\right), \\ a^{min} &= \min\left(u_{j-1}^{td}, u_j^{td}, u_{j+1}^{td}\right), \end{aligned} \tag{6.48a}$$

$$\begin{aligned} b^{max} &= \max\left(u_{j-1}^{td}, u_j^{td}, u_{j+1}^{td}, u_{j-1}^{n}, u_j^{n}, u_{j+1}^{n}\right), \\ b^{min} &= \min\left(u_{j-1}^{td}, u_j^{td}, u_{j+1}^{td}, u_{j-1}^{n}, u_j^{n}, u_{j+1}^{n}\right), \end{aligned} \tag{6.48b}$$

where the superscript td denotes the estimate of the solution at time step $n+1$ as obtained by the low-order flux formulation. The two limiters are blended to obtain the maximum admissible value for u_j^{n+1}

$$u_j^{max} = a^{max} + (1-f)\left(b^{max} - a^{max}\right), \tag{6.49}$$

and similarly for the minimum, where the linear-blending parameter f serves as model parameter.

For this simple FCT scheme a strong dependence of the computed results on the particular value of the model parameter has been observed. Figure 6.6 shows the results for incompressible isotropic turbulence, where the parameter dependence is evident. For compressible turbulence the solution quality for a fixed parameter f becomes dependent on the initial data. In Fig. 6.7 results for the initial turbulent Mach number $M_t = 0.4$ and micro-scale Reynolds number $Re_\lambda = 60$ are shown for different initial ratios χ of dilatational turbulent kinetic energy and total turbulent kinetic energy. Apparently, for this particular FCT scheme there is no consistent behavior of the solution with fixed f for different χ.

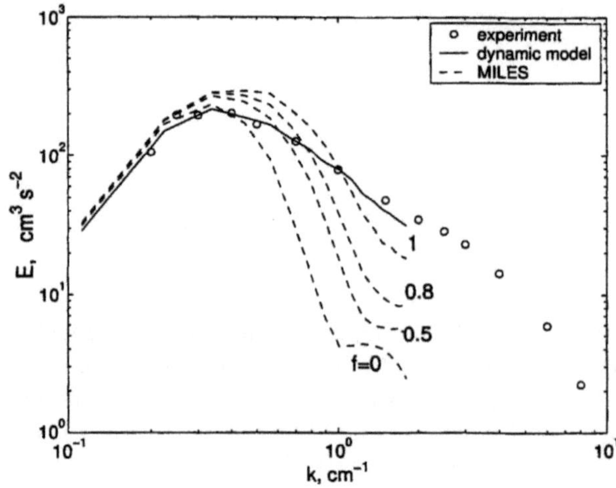

Fig. 6.6. Effect of the parameter f on MILES computation with 32^3 grid points of the Comte-Bellot-Corrsin experiments [45]. Energy spectra $E(k)$ are shown at the non-dimensional experimental time of $U_0 t/M = 98$, reproduced from Ref. [122] with permission

6.2.5 The MPDATA Method

The Multidimensional Positive Definite Advection Transport Algorithm (MPDATA) has been introduced by Smolarkiewicz [268, 269] for meteorological applications. Reviews of MPDATA can be found in Refs. [270, 271]. Essential to MPDATA is upwinding of the numerical error, which is achieved in a two-step scheme, where a spatially first-order upwind step is followed by an anti-diffusive step, involving upwinding of the truncation error. Overall second-order in space is achieved. Commonly, the scheme is developed based on considering a generic scalar advection equation [106, 232]

$$\frac{\partial u}{\partial t} + \frac{\partial a u}{\partial x} = 0 \tag{6.50}$$

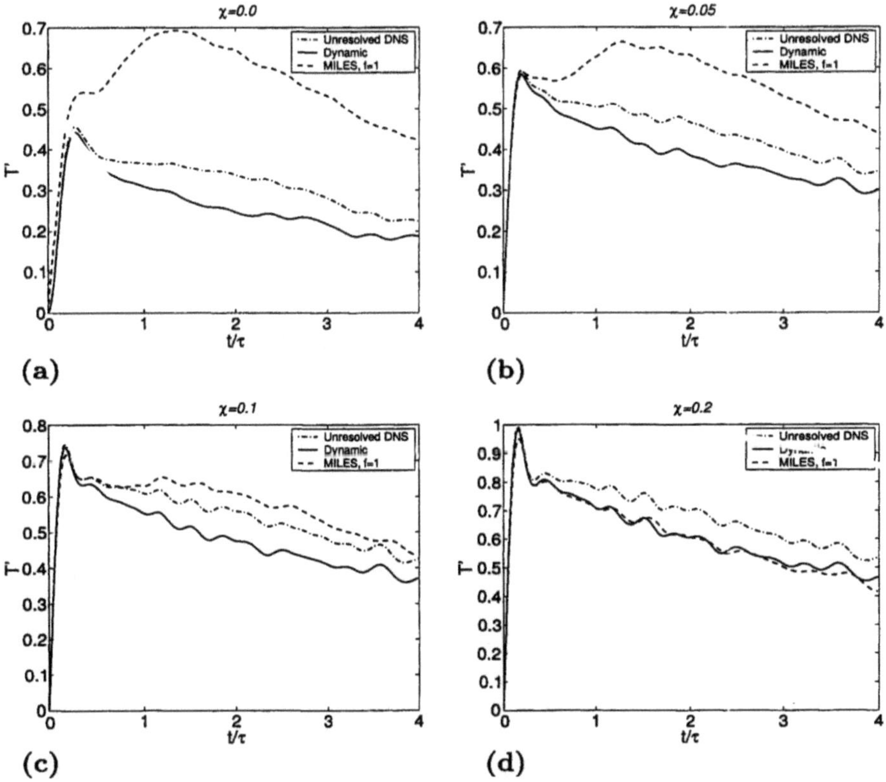

Fig. 6.7. Time evolution of RMS density fluctuations ρ' in compressible isotropic turbulence with 32^3 grid points, initial turbulent Mach number $M_t = 0.4$, maximum-energy wavenumber $k_0 = 4$, and micro-scale Reynolds number $Re_\lambda = 60$: (**a**) $\chi = 0$, (**b**) $\chi = 0.05$, (**c**) $\chi = 0.1$, (**d**) $\chi = 0.2$. Reproduced from Ref. [122] with permission

which for the first step is discretized by an upwind scheme in space

$$u'_j = u^n_j + F_{j+1/2} - F_{j-1/2}, \tag{6.51}$$

where

$$F_{j+1/2} = \frac{1}{2}\left(1 + \frac{a_{j+1/2}}{|a_{j+1/2}|}\right)u_j + \frac{1}{2}\left(1 - \frac{a_{j+1/2}}{|a_{j+1/2}|}\right)u_{j+1},$$

and accordingly for $F_{j-1/2}$, and an Euler-forward scheme in time. The leading-order truncation error of the upwind step is

$$\mathcal{G}_N = \frac{\partial}{\partial x}\left(\frac{\Delta x^2}{2\Delta t}\left(|A| - A^2\right)\frac{\partial u}{\partial x}\right),$$

where $A = a\Delta t/\Delta x$. The upwind step is followed by an anti-diffusive step involving an upwind transport of the truncation error

$$u_j^{n+1} = u'_j + F'_{j+1/2} - F'_{j-1/2}, \tag{6.52}$$

where

$$F'_{j+1/2} = \frac{1}{2}\left(1 + \frac{b_{j+1/2}}{|b_{j+1/2}|}\right)u_j + \frac{1}{2}\left(1 - \frac{b_{j+1/2}}{|b_{j+1/2}|}\right)u_{j+1}.$$

The diffusive pseudo velocity herein is

$$b_{j+1/2} = \left(|a| - a^2\frac{\Delta t}{\Delta x}\right)\frac{u'_{j+1} - u'_j}{u'_{j+1} + u'_j}.$$

The definition of $F'_{j+1/2}$ follows accordingly. As analyzed by Rider [232] for a nonlinear scalar conservation law in one dimension with flux $f(u)$ the MPDATA scheme amounts to

$$\mathcal{G}_N = \frac{\partial}{\partial x}\left(\frac{\Delta x^2}{6}\frac{\partial f}{\partial u}\left(\frac{\partial u}{\partial x}\right)^2 + \frac{\Delta x^2}{6}\left|\frac{\partial f}{\partial u}\right|\frac{\partial^2 u}{\partial x^2} + \frac{\Delta x^2}{4|u|}\left|\frac{\partial u}{\partial x}\right|\left(\frac{\partial u}{\partial x}\right)\right) + \cdots,$$

which contains implicitly a Smagorinsky term. In multiple dimensions, the diffusive pseudo velocity contains second-degree cross derivatives of the different coordinate directions [106].

Applied to incompressible Navier-Stokes turbulence the MPDATA method was analyzed by Domaradzki et al. [57]. Simulations for isotropic incompressible turbulence at formally infinite Reynolds number for fixed numerical parameters (resolution and time-step size). It was found that the compensated inertial-range spectra approximate the value for the Kolmogorov constant in a quality comparable to classical explicit LES models. A spectral-space MDEA reveals that the spectral numerical viscosity is in the range of the theoretically predicted spectral eddy viscosity, it does not, however return the plateau values and does not reproduce the cusp near the cut-off wavenumber, see Fig. 6.8

An extension of MPDATA for compressible flows has been developed by Smolarkiewicz and Szmelter [272]. They propose to apply the MPDATA algorithm to the convective terms of the conservation equations, with the exception that for the total-energy transport the pressure-work per time is considered as source term on the right-hand side, along with the viscous and heat-conduction terms. The source terms are treated implicitly in time and require iterative solution in general.

6.2.6 The Optimum Finite-Volume Scheme

Starting point of the Optimum Finite-Volume Scheme of Zandonade et al. [322] is the finite-volume discretization of the incompressible Navier-Stokes equations. As with the Volume-Balance Method the problem of modeling the cell-face fluxes is considered directly

$$\Delta x^3 \frac{d\bar{u}_j^{\mathrm{i}}}{dt} + \int_{\partial V_j^{\mathrm{i}}} u_j u_k n_k dS = -\int_{\partial V_j^{\mathrm{i}}} p n_j dS + \frac{1}{Re}\int_{\partial V_j^{\mathrm{i}}} \frac{\partial u_j}{\partial x_k} n_k dS. \tag{6.53}$$

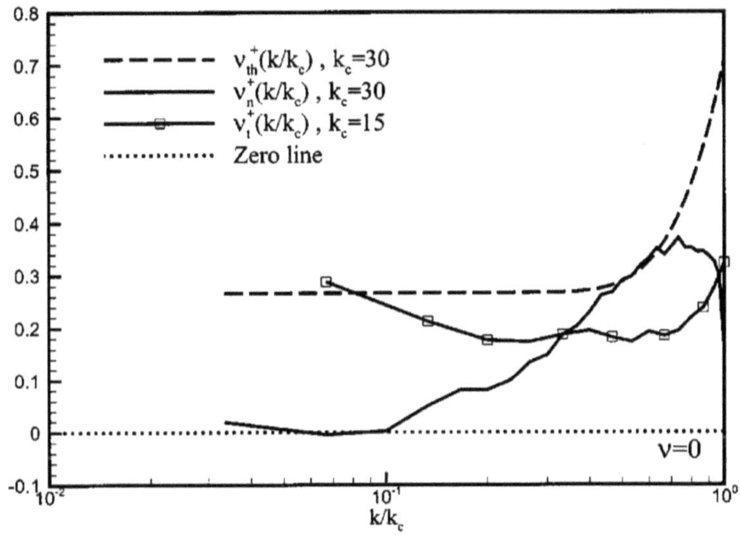

Fig. 6.8. Eddy viscosities for isotropic turbulence at vanishing viscosity. *Broken line*: theoretical eddy viscosity; *solid line with symbols*: intrinsic eddy viscosity for $k_C = 15$; *solid line*: numerical eddy viscosity. Reproduced from Ref. [57] with permission

The convective cell-face fluxes

$$M^{\mathbf{i}}_{jk} = \int_{\partial V^{\mathbf{i}}_j} u_j u_k n_k dS$$

require modeling. Due to isotropy two different cell-face flux operators (1 diagonal and 1 off-diagonal term) can occur and need to be considered. Zandonade et al. [322] also model viscous cell-face fluxes, since they contain a derivative operation whose exact evaluation requires knowledge of subgrid-scales. For very large Reynolds numbers the contribution should be small. For the cell-face fluxes the following ansatz is made

$$m^{\mathbf{i}}_{jk} = A + \sum_{\mathbf{m}} B^{\mathbf{m}}_j \bar{u}^{\mathbf{i}+\mathbf{m}}_j + \sum_{\mathbf{m}} \sum_{\mathbf{n}} C^{\mathbf{mn}}_{jk} \bar{u}^{\mathbf{i}+\mathbf{m}}_j \bar{u}^{\mathbf{i}+\mathbf{n}}_k \tag{6.54}$$

which is then subjected to stochastic estimation. In effect, the above ansatz leaves all finite-volume stencil weights A, $B^{\mathbf{m}}_j$ and $C^{\mathbf{mn}}_{jk}$ free for optimization. Numerical requirements, such as consistency or stability, are not enforced by constraining the optimization and have to be recovered by the estimation procedure. Also, as the ansatz is linear, it cannot be expected that the resulting model can adapt locally to the flow character, whether turbulent or laminar. An optimum parameter set was identified for isotropic turbulence at a Taylor-micro-scale Reynolds number of $Re_\lambda = 164$. For optimization the stencil was restricted to different staggered (denoted with a leading "S") and collocated configurations (denoted with a leading "C"). *A posteriori* results for isotropic turbulence are shown in Fig. 6.9. No further applications of the optimum finite-volume method have been reported, also no extension to compressible flows can be found in literature.

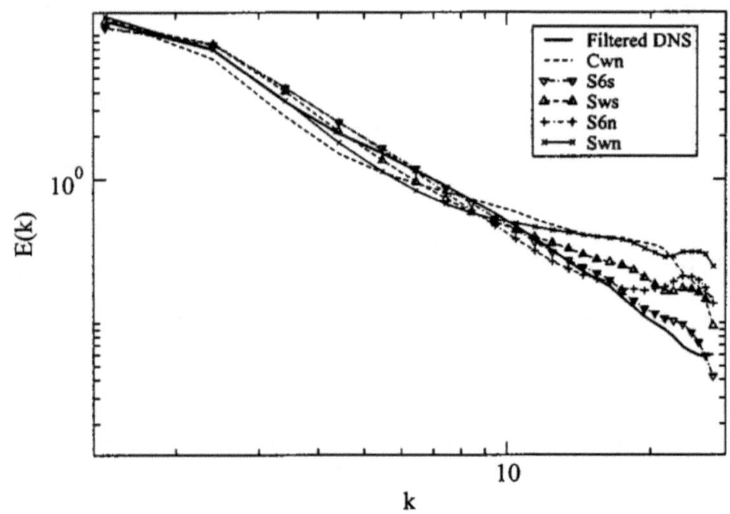

Fig. 6.9. Three-dimensional energy spectrum $E(k)$, filtered DNS data compared with optimal LES results. Reproduced from Ref. [322] with permission

6.3 Implicit LES by Adaptive Local Deconvolution

The Adaptive Local Deconvolution Method (ALDM) constitutes a general framework for ILES which is inherently solution adaptive and involves only few discretization parameters which can be identified by incorporating theoretical knowledge about turbulence physics. For any discretization parameters ALDM results in a consistent spatial discretization scheme which is at least second-order accurate. The focus here is on a one-dimensional development, following Ref. [5], of the fundamental concept of ALDM. The extension to the incompressible and the compressible Navier-Stokes equations is described. ALDM also has been formulated for scalar transport [114].

6.3.1 Fundamental Concept of ALDM

The generic discretized conservation law equation (6.7) is considered, with G being a real-space top-hat filter, centered at the grid nodes x_j. For the filtered flux term one obtains accordingly

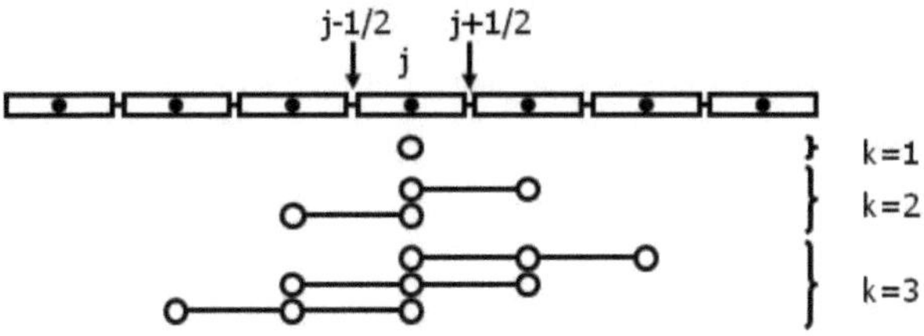

Fig. 6.10. Admissible stencils for polynomial order $k = 1, 2, 3, 4$

$$G * \frac{\partial F(u)}{\partial x} = \frac{F(u_{j+1/2}) - F(u_{j-1/2})}{h}, \tag{6.55}$$

where $x_{j\pm1/2}$ denotes the faces of cell j. Obviously an approximation of the unfiltered solution $u(x)$ at the left and right faces of each cell j which are called $u^+_{j-1/2}$ and $u^-_{j+1/2}$, respectively, is required. The top-hat filtering being equivalent to a finite-volume discretization allows for a primitive-function reconstruction of $u(x)$ from $\bar{u}_N$ at $x_{j\pm1/2}$ as proposed by Harten et al. [109] as basis of an essentially non-oscillatory (ENO) discretization scheme with stencil selection. We introduce a set of local interpolation polynomials of order $k = 1, \ldots, K$, for each k with shift $r = 0, \ldots, k-1$ of the left-most stencil point which also identifies the respective stencil. Admissible stencils, i.e. stencils which include the interpolation points $x_{j\pm1/2}$ in their support, range from $j - r$ to $j - r + k - 1$, expressed by (k, r). The stencil setup is illustrated in Fig. 6.10. Right-face interpolants at $x_{j+1/2}$ and left-face interpolants at $x_{j-1/2}$ of order k can be found e.g. in Ref. [263] as

$$\begin{aligned} p^-_{k,r}(x_{j+1/2}) &= \sum_{l=0}^{k-1} c^{(k)}_{r,l}(j)\bar{u}_{j-r+l}, \\ p^+_{k,r}(x_{j-1/2}) &= \sum_{l=0}^{k-1} c^{(k)}_{r-1,l}(j)\bar{u}_{j-r+l}, \end{aligned} \tag{6.56}$$

respectively. For each k these expressions represent the information contained in admissible polynomials, where deconvolution and interpolation are expressed by the coefficients $c^{(k)}_{r,l}(j)$ as given by Shu [263]

$$c^{(k)}_{r,l}(j) = h_{j-r+l} \sum_{\mu=l+1}^{k} \frac{\sum_{\substack{p=0 \\ p\neq\mu}}^{k} \prod_{\substack{\nu=0 \\ \nu\neq\mu,p}}^{k} x_{j+1/2} - x_{j-r+\nu-1/2}}{\prod_{\substack{\nu=0 \\ \nu\neq\mu}}^{k} x_{j-r+\mu-1/2} - x_{j-r+\nu-1/2}}. \tag{6.57}$$

As indicated, this rule holds for variable mesh spacing. If $h_j = h = \text{const}$ it can be simplified accordingly. The index range of $c^{(k)}_{r,l}(j)$ is $r = 0, \ldots, k-1$ and $l = 0, \ldots, k-1$ for each $k = 1, \ldots, K$.

Whereas for ENO or weighted-ENO (WENO) approaches a single interpolation-polynomial order is chosen, here a quasi-linear combination of all possible interpolation polynomials up to a certain order K is constructed

$$\tilde{u}^{\mp}_{j\pm 1/2} = \sum_{k=1}^{K} \sum_{r=0}^{k-1} \omega^{\mp}_{k,r}(j) p^{\mp}_{k,r}(x_{j\pm 1/2}). \tag{6.58}$$

As restriction it is imposed that the sum of all weights $\omega^{\pm}_{k,r}(j)$ over k and r is unity

$$\sum_{k=1}^{K} \sum_{r=0}^{k-1} \omega^{\pm}_{k,r} = 1. \tag{6.59}$$

Equation (6.58) gives the resulting approximants for the deconvolved solution at the left and right cell faces.

Finally, an appropriate numerical flux function $\tilde{F}_N$ needs to be devised which approximates the physical flux F. Among several possible choices is a modified Lax-Friedrichs flux function

$$\tilde{F}_N(x_{j+1/2}) = \frac{1}{2}\left(F(\tilde{u}^{-}_{j+1/2}) + F(\tilde{u}^{+}_{j+1/2})\right) - \sigma_{j+1/2}\left(\tilde{u}^{+}_{j+1/2} - \tilde{u}^{-}_{j+1/2}\right), \tag{6.60}$$

where $\sigma_{j+1/2}$ can be any shift-invariant functional of $\bar{u}_N$. In Ref. [5] it was found that the following numerical flux function leads to favorable error cancellations

$$\tilde{F}_N(x_{j+1/2}) = F\left(\frac{\tilde{u}^{-}_{j+1/2} + \tilde{u}^{+}_{j+1/2}}{2}\right) - \sigma_{j+1/2}\left(\tilde{u}^{+}_{j+1/2} - \tilde{u}^{-}_{j+1/2}\right). \tag{6.61}$$

The implicit SGS model is expressed in the above framework by the choices for the smoothness measure which enters the computation of $\omega^{\pm}_{k,r}$ and the dissipative weights $\sigma_{j+1/2}$. For the weights $\omega^{\pm}_{k,r}$ the WENO approach [171] provides weights according to

$$\omega^{\pm}_{k,r} = \frac{1}{K} \frac{\alpha^{\pm}_{k,r}}{\sum_{\mu=0}^{k-1} \alpha^{\pm}_{k,\mu}}. \tag{6.62}$$

In this definition the coefficients $\alpha^{\pm}_{k,r}$ are computed from

$$\alpha^{\pm}_{k,r} = \gamma^{\pm}_{k,r}(\varepsilon + \beta_{k,r})^{-2}, \tag{6.63}$$

where ε is a small number to prevent singularity. The smoothness measure $\beta_{k,r}$ can be computed as the total variation of $\bar{u}_N$ on the considered stencil

$$\beta_{k,r} = \sum_{\mu=-r}^{k-r-2} |\bar{u}_{j+\mu+1} - \bar{u}_{j+\mu}|, \tag{6.64}$$

which is used for the one-dimensional analysis below. Alternatively, other, e.g. WENO smoothness measures [263] can be used.

The dissipative weight in (6.61) can be chosen, e.g., as $\sigma_{j+1/2} = |\bar{u}_{j+1} - \bar{u}_j|$, by which the numerical flux function is Galilean invariant. Equation (6.63) introduces free parameters for the right-face interpolant $\gamma^{-}_{k,r}$ and the left-face interpolant $\gamma^{+}_{k,k-1-r}$, which by isotropy are symmetric with respect to the stencil center $\gamma^{-}_{k,r} = \gamma^{+}_{k,k-1-r}$. These free discretization parameters can be determined by incorporating theoretical knowledge about turbulent flows or following an MDE analysis.

The real-space MDE analysis is performed for the semi-discretization only as for LES the time-step size can be assumed to be sufficiently small so that the temporal truncation error is small compared to the spatial truncation error [5]. As result of MDE analysis a differential equation for the continuous extension of $\bar{u}_N$ is obtained. The implicit SGS model can be identified by computing $\mathcal{G}_N$ as defined in (6.8). As example we consider the viscous Burgers equation for which $F(v) = v^2/2 - \nu\partial v/\partial x$ is to be substituted in (6.6). For SGS modeling only the convective term is relevant, so that we will not consider the diffusion terms within MDE analysis. The exact expression for the convective part of the Burgers equation is

$$G * \frac{\partial F_N(u_N)}{\partial x} = \bar{u}_N \frac{\partial \bar{u}_N}{\partial x} + \frac{1}{12}\frac{\partial \bar{u}_N}{\partial x}\frac{\partial^2 \bar{u}_N}{\partial x^2} - \frac{1}{720}\frac{\partial \bar{u}_N}{\partial x}\frac{\partial^4 \bar{u}_N}{\partial x^4} h^4 + \frac{1}{30240}\frac{\partial \bar{u}_N}{\partial x}\frac{\partial^6 \bar{u}_N}{\partial x^6} h^6 - + \cdots , \tag{6.65}$$

where derivatives are to be taken at the cell centers x_N.

The Smagorinsky model for the Burgers equation substituted for $\mathcal{G}_{SGS}$ in (6.3) can be matched by $\mathcal{G}_N$ for a certain choice of parameters $\gamma^{\pm}_{k,r}$, so that the discretization implicitly reproduces the Smagorinsky model.

$$\bar{\mathcal{G}}_{SGS} = -\frac{\partial \tau_{Smag}}{\partial x} = 2C_S h^2 \left|\frac{\partial \bar{u}}{\partial x}\right| \frac{\partial^2 \bar{u}}{\partial x^2} . \tag{6.66}$$

With the implicit SGS approach, we can for instance identify model parameters in such a way that the resulting implicit formulation matches with the explicit model for $K = 3$ up to order $\mathcal{O}(h^3)$ as given in Table 6.1. Choosing $\sigma_{j+1/2} = 9C_S|\tilde{u}^{-}_{j+1/2} - \tilde{u}^{+}_{j+1/2}|$ in (6.61), the truncation error $\mathcal{G}_N$ follows as

$$\mathcal{G}_N = 2C_S \left|\frac{\partial \bar{u}}{\partial x}\right| \frac{\partial^2 \bar{u}}{\partial x^2} h^2 - \frac{1}{6} C_S \left|\frac{\partial \bar{u}}{\partial x}\right| \frac{\partial^4 \bar{u}}{\partial x^4} h^4 + \mathcal{O}\left(h^6\right) . \tag{6.67}$$

Computational results in Ref. [5] confirm that the implicit SGS reproduces the result for an explicit LES with the Smagorinsky model for Burgers turbulence. It is also shown that better results are obtained when the model parameters $\gamma^{\pm}_{k,r}$ are identified in such a way that an optimum representation of the SGS energy transfer is obtained.

6.3.2 ALDM for the Incompressible Navier-Stokes Equations

In three dimensions a top-hat filter kernel can be defined by factorization into three one-dimensional operators

Table 6.1. Result for the discretization parameters $\gamma^{\pm}_{k,r}$ to match the explicit Smagorinsky model

Parameter	Value
$\gamma^{+}_{1,0}$	1
$\gamma^{+}_{2,0}$	$\frac{2}{3}$
$\gamma^{+}_{2,1}$	$\frac{1}{3}$
$\gamma^{+}_{3,0}$	$\frac{3}{10}$
$\gamma^{+}_{3,1}$	$\frac{3}{10}$
$\gamma^{+}_{3,2}$	$\frac{4}{10}$

$$G(\boldsymbol{x}) = G_x(x) * G_y(y) * G_z(z). \tag{6.68}$$

A regularized inverse-filter operation can be defined as a convolution with the inverse kernel

$$Q(\boldsymbol{x}) = Q_x(x) * Q_y(y) * Q_z(z), \tag{6.69}$$

factorized into three one-dimensional operators which can be obtained by modifying and extending the one-dimensional formulations of the previous section. This operator, denoted by Q^{λ}_{x}, is defined on a one-dimensional grid $x_N = \{x_i\}$. Applied to the filtered grid function $\bar{\varphi}_N = \{\bar{\varphi}(x_i)\}$ it returns the approximately deconvolved grid function $\phi^{*\lambda}_{N} \doteq \{\varphi(x_{i+\lambda})\}$ on the shifted grid $x^{\lambda}_{N} = \{x_{i+\lambda}\}$

$$Q^{\lambda}_{x}\bar{\varphi}_N = \{\varphi\left(x_{i+\lambda}\right) + \mathcal{O}\left(\Delta x_i{}^{\kappa}\right)\} = \phi^{*\lambda}_{N}. \tag{6.70}$$

The filtered data are given at the cell centers $\{x_i\}$. Reconstruction at the left cell face $\{x_{i-\frac{1}{2}}\}$ is indicated by $\lambda = -1/2$ and at the right face by $\lambda = +1/2$. For obtaining a three-dimensional reconstruction by successive one-dimensional operations yet another approximation of the partially deconvolved solution is required at the cell centers. The respective operator is indicated by $\lambda = 0$.

Deconvolution and interpolation are done simultaneously by Lagrangian interpolation polynomials [109]. Given a generic k-point stencil ranging from x_{i-r} to $x_{i-r+k-1}$ the one-dimensional expression for the case of a top-hat kernel is

$$\varphi(x_{i+\lambda}) = \sum_{l=0}^{k-1} c^{\lambda}_{k,r,l}(x_i)\bar{\varphi}_N(x_{i-r+l}) + \mathcal{O}\left(\Delta x_i{}^{k}\right), \tag{6.71}$$

with $r \in \{0, \ldots, k\}$. The grid-dependent coefficients are

$$c^{\lambda}_{k,r,l}(x_i) = \left(x_{i-r+l+\frac{1}{2}} - x_{i-r+l-\frac{1}{2}}\right) \sum_{m=l+1}^{k} \frac{\sum_{\substack{p=0\\p\neq m}}^{k} \prod_{\substack{n=0\\n\neq p,m}}^{k} x_{i+\lambda} - x_{i-r+n-\frac{1}{2}}}{\prod_{\substack{n=0\\n\neq m}}^{k} x_{j-r+m-\frac{1}{2}} - x_{i-r+n-\frac{1}{2}}}, \tag{6.72}$$

as given by Shu [263] apply to grids with variable mesh width and for arbitrary target positions $x_{i+\lambda}$. In case of a staggered grid the values of x_i are different for each velocity component, and the coefficients $c^{\lambda}_{k,r,l}$ have to be specified accordingly.

In three dimensions the ALDM reconstruction becomes

$$\phi^{*\lambda}_{N}(x_{i+\lambda}) = \sum_{k=1}^{K}\sum_{r=0}^{k-1} \omega^{\lambda}_{k,r}(\bar{\varphi}_N, x_i) \sum_{l=0}^{k-1} c^{\lambda}_{k,r,l}(x_i)\ \bar{\varphi}_N(x_{i-r+l}). \tag{6.73}$$

The sum of all weights satisfies

$$\sum_{r=0}^{k-1} \omega^{\lambda}_{k,r} = \frac{1}{K} \tag{6.74}$$

with $k = 1, \ldots, K$, where each weight is given by

$$\omega^{\lambda}_{k,r}\left(\bar{\varphi}_N, x_i\right) = \frac{1}{K}\frac{\gamma^{\lambda}_{k,r}\beta_{k,r}\left(\bar{\varphi}_N, x_i\right)}{\sum_{s=0}^{k-1}\gamma^{\lambda}_{k,s}\beta_{k,s}\left(\bar{\varphi}_N, x_i\right)}, \tag{6.75}$$

with $r = 0, \ldots, k-1$ for each $k = 1, \ldots, K$. For the incompressible Navier-Stokes equations a different definition for the smoothness measure than in (6.64) is used, namely

$$\beta_{k,r}\left(\bar{\varphi}_N, x_i\right) = \left(\varepsilon_\beta + \sum_{l=-r}^{k-r-2}\left(\bar{\varphi}_{i+m+1} - \bar{\varphi}_{i+m}\right)^2\right)^{-2}, \tag{6.76}$$

where ε_β is a small number to prevent division by zero.

The parameters $\gamma^{+1/2}_{k,r}$, $\gamma^{-1/2}_{k,r}$, and $\gamma^{0}_{k,r}$ represent a stencil-selection preference that would become effective in the statistically homogeneous case. The requirement of an isotropic discretization for this case implies the following symmetries

$$\gamma^{-1/2}_{k,r} = \gamma^{+1/2}_{k,k-1-r} \quad \text{and} \quad \gamma^{0}_{k,r} = \gamma^{0}_{k,k-1-r}. \tag{6.77}$$

As consequence of (6.74) the number of independent parameters is further reduced by

$$\sum_{r=0}^{k-1}\gamma^{+1/2}_{k,r} = 1 \quad \text{and} \quad \sum_{r=0}^{k-1}\gamma^{0}_{k,r} = 1. \tag{6.78}$$

For $K = 3$ four parameters $\{\gamma^{+1/2}_{2,0}, \gamma^{+1/2}_{3,0}, \gamma^{+1/2}_{3,1}, \gamma^{0}_{3,1}\}$ can be used for implicit modeling. The three-dimensional adaptive local deconvolution operator $Q^{\boldsymbol{\lambda}}$ is constructed from one-dimensional operators of the kind Q^{λ}_{x}. Application to the filtered solution results in the reconstructed solution

$$\phi^{*\boldsymbol{\lambda}}_{N} = Q^{\boldsymbol{\lambda}}\bar{\varphi}_N = Q^{\lambda_3}_{z} * Q^{\lambda_2}_{y} * Q^{\lambda_1}_{x}\bar{\varphi}_N. \tag{6.79}$$

The vector $\boldsymbol{\lambda} = \{\lambda_1, \lambda_2, \lambda_3\}$ indicates the relative reconstruction position. Required operations are summarized in Table 6.2. The sequence of one-dimensional operators should satisfy certain constraints to maintain rotational invariance of the resulting three-dimensional operator [112].

Table 6.2. Interpolation directions for three-dimensional reconstruction

Direction	Relative target index λ	Example
(R) rightward	$[+\frac{1}{2}, 0, 0]$	$\widetilde{\boldsymbol{u}}^R_{i,j,k} \approx \boldsymbol{u}(x_{i+\frac{1}{2}}, y_j, z_k)$
(L) leftward	$[-\frac{1}{2}, 0, 0]$	$\widetilde{\boldsymbol{u}}^L_{i,j,k} \approx \boldsymbol{u}(x_{i-\frac{1}{2}}, y_j, z_k)$
(F) forward	$[0, +\frac{1}{2}, 0]$	$\widetilde{\boldsymbol{u}}^F_{i,j,k} \approx \boldsymbol{u}(x_i, y_{j+\frac{1}{2}}, z_k)$
(B) backward	$[0, -\frac{1}{2}, 0]$	$\widetilde{\boldsymbol{u}}^B_{i,j,k} \approx \boldsymbol{u}(x_i, y_{j-\frac{1}{2}}, z_k)$
(U) upward	$[0, 0, +\frac{1}{2}]$	$\widetilde{\boldsymbol{u}}^U_{i,j,k} \approx \boldsymbol{u}(x_i, y_j, z_{k+\frac{1}{2}})$
(D) downward	$[0, 0, -\frac{1}{2}]$	$\widetilde{\boldsymbol{u}}^D_{i,j,k} \approx \boldsymbol{u}(x_i, y_j, z_{k-\frac{1}{2}})$

Once the reconstruction of the velocity solution at the cell faces is known, the numerical flux $\widetilde{\boldsymbol{F}}_N = \{\overset{1\sim}{\boldsymbol{f}}, \overset{2\sim}{\boldsymbol{f}}, \overset{3\sim}{\boldsymbol{f}}\}$ can be computed. A consistent numerical flux function approximates the physical flux function

$$\widetilde{\boldsymbol{F}}_N \approx \boldsymbol{F} = \boldsymbol{u}\boldsymbol{u} \quad \text{and} \quad \overset{l\sim}{\boldsymbol{f}} \approx \overset{l}{\boldsymbol{f}} = u_l \boldsymbol{u}. \tag{6.80}$$

In analogy to the flux function defined for one-dimension in (6.61) we formulate the following numerical flux function for the incompressible Navier-Stokes equations

$$\begin{aligned}
\overset{1\sim}{\boldsymbol{f}}_{i+\frac{1}{2},j,k} = {} & \frac{1}{4}\left(u^{*L}_{i+1,j,k} + u^{*R}_{i,j,k}\right)\left(\mathbf{u}^{*L}_{i+1,j,k} + \mathbf{u}^{*R}_{i,j,k}\right) \\
& - {}^1\sigma_{i,j,k}\begin{bmatrix} |\bar{u}_{i+1,j,k} - \bar{u}_{i,j,k}|(u^{*L}_{i+1,j,k} - u^{*R}_{i,j,k}) \\ |\bar{v}_{i+1,j,k} - \bar{v}_{i,j,k}|(v^{*L}_{i+1,j,k} - v^{*R}_{i,j,k}) \\ |\bar{w}_{i+1,j,k} - \bar{w}_{i,j,k}|(w^{*L}_{i+1,j,k} - w^{*R}_{i,j,k}) \end{bmatrix},
\end{aligned} \tag{6.81a}$$

$$\begin{aligned}
\overset{2\sim}{\boldsymbol{f}}_{i,j+\frac{1}{2},k} = {} & \frac{1}{4}\left(v^{*B}_{i,j+1,k} + v^{*F}_{i,j,k}\right)\left(\mathbf{u}^{*B}_{i,j+1,k} + \mathbf{u}^{*F}_{i,j,k}\right) \\
& - {}^2\sigma_{i,j,k}\begin{bmatrix} |\bar{u}_{i,j+1,k} - \bar{u}_{i,j,k}|(u^{*B}_{i,j+1,k} - u^{*F}_{i,j,k}) \\ |\bar{v}_{i,j+1,k} - \bar{v}_{i,j,k}|(v^{*B}_{i,j+1,k} - v^{*F}_{i,j,k}) \\ |\bar{w}_{i,j+1,k} - \bar{w}_{i,j,k}|(w^{*B}_{i,j+1,k} - w^{*F}_{i,j,k}) \end{bmatrix},
\end{aligned} \tag{6.81b}$$

$$\begin{aligned}
\overset{3\sim}{\boldsymbol{f}}_{i,j,k+\frac{1}{2}} = {} & \frac{1}{4}\left(w^{*D}_{i,j,k+1} + w^{*U}_{i,j,k}\right)\left(\mathbf{u}^{*D}_{i,j,k+1} + \mathbf{u}^{*U}_{i,j,k}\right) \\
& - {}^3\sigma_{i,j,k}\begin{bmatrix} |\bar{u}_{i,j,k+1} - \bar{u}_{i,j,k}|(u^{*D}_{i,j,k+1} - u^{*U}_{i,j,k}) \\ |\bar{v}_{i,j,k+1} - \bar{v}_{i,j,k}|(v^{*D}_{i,j,k+1} - v^{*U}_{i,j,k}) \\ |\bar{w}_{i,j,k+1} - \bar{w}_{i,j,k}|(w^{*D}_{i,j,k+1} - w^{*U}_{i,j,k}) \end{bmatrix}.
\end{aligned} \tag{6.81c}$$

The superscripts L, R, D, U, B, F indicate deconvolution at the left, right, down, up, back, front side, respectively, of cell (i, j, k). The first term on the right-hand side corresponds to the physical Navier-Stokes flux. For maximum order of consistency it is computed from the mean of both interpolants of the deconvolved velocity at the considered cell face. The difference between them serves as an estimate for the local reconstruction error. In the second term on the right-hand side it is multiplied with the magnitude of a filtered velocity increment which corresponds to the first-order structure function. For developed turbulence the Kolmogorov theory predicts a scaling with a 1/3 power of the two-point separation. The following compensation coefficients ensure a resolution independent numerical viscosity

$$ {}^{1}\sigma_{i,j,k} = \sigma \left(\frac{x_{i+1,j,k} - x_{i,j,k}}{\Delta_0} \right)^{-\frac{1}{3}}, \tag{6.82a} $$

$$ {}^{2}\sigma_{i,j,k} = \sigma \left(\frac{y_{i,j+1,k} - y_{i,j,k}}{\Delta_0} \right)^{-\frac{1}{3}}, \tag{6.82b} $$

$$ {}^{3}\sigma_{i,j,k} = \sigma \left(\frac{z_{i,j,k+1} - z_{i,j,k}}{\Delta_0} \right)^{-\frac{1}{3}}. \tag{6.82c} $$

Δ_0 is a reference length scale for which the scalar parameter σ, which arises as additional modeling parameter, has been identified, see below.

For model-parameter identification a spectral-space MDE analysis following Sect. 6.1.3 is performed. We consider freely decaying homogeneous isotropic turbulence in the limit of vanishing molecular viscosity. Reference is a numerical simulation in a $(2\pi)^3$-periodic box, discretized by $32 \times 32 \times 32$ uniform finite volumes. The computed velocity fields $\bar{\boldsymbol{u}}_N(t_n)$ are Fourier-transformed and truncated at $k_C = 15$

$$ \hat{\boldsymbol{u}}_C(\boldsymbol{k}, t_n) = \begin{cases} \mathcal{F}\{\bar{\boldsymbol{u}}_N\}(\boldsymbol{k}, t_n), & |\boldsymbol{k}| \leq k_C, \\ \boldsymbol{0}, & \text{otherwise.} \end{cases} \tag{6.83} $$

The spectral-energy decay is approximated by

$$ \frac{\partial \hat{E}(\boldsymbol{k}, t_{n-1/2})}{\partial t} \approx \frac{\hat{E}(\boldsymbol{k}, t_n) - \hat{E}(\boldsymbol{k}, t_{n-1})}{\Delta t} \tag{6.84} $$

at times $t_{n-1/2} = \frac{1}{2}(t_{n-1} + t_n)$. Energy spectrum and spectral transfer function are interpolated as

$$ \hat{E}(\boldsymbol{k}, t_{n-1/2}) = \frac{\hat{E}(\boldsymbol{k}, t_n) + \hat{E}(\boldsymbol{k}, t_{n-1})}{2}, \tag{6.85a} $$

$$ \hat{T}_C(\boldsymbol{k}, t_{n-1/2}) = \frac{\hat{T}_C(\boldsymbol{k}, t_n) + \hat{T}_C(\boldsymbol{k}, t_{n-1})}{2}. \tag{6.85b} $$

Following (6.48) the spectral numerical viscosity is

$$ \nu_{num}(\boldsymbol{k}, t_{n-1/2}) = \frac{1}{2\boldsymbol{k}^2 \hat{E}(\boldsymbol{k}, t_{n-1/2})} \left(\hat{T}_C(\boldsymbol{k}, t_{n-1/2}) - \frac{\partial \hat{E}(\boldsymbol{k}, t_{n-1/2})}{\partial t} \right) - \nu. \tag{6.86} $$

The three-dimensional numerical-viscosity spectrum is obtained by summation over integer-wavenumber shells $k - \frac{1}{2} \leq |\boldsymbol{k}| \leq k + \frac{1}{2}$

$$\nu_{num}(k, t_{n-1/2}) = \frac{4\pi k^2}{M(k)} \sum_{\boldsymbol{k}} \nu_{num}(\boldsymbol{k}, t_{n-1/2}), \tag{6.87}$$

where $M(k)$ is the number of integer wavenumbers on each shell with radius k. A subsequent normalization gives

$$\nu^+_{num}(k^+, t_{n-1/2}) = \nu_{num}(k_C k^+, t_{n-1/2}) \sqrt{\frac{k_C}{\hat{E}(k_C)}}. \tag{6.88}$$

Isotropic decaying turbulence does not lose memory of the initial data. An evaluation of ν^+_{num} for one data set only does not necessarily represent the statistical average. To cope with this problem the spectral numerical viscosity from a number of uncorrelated realizations needs to be evaluated and averaged. Model parameters can be found by automatic optimization. A suitable cost function is define the root-mean-square difference between the spectral numerical viscosity $\nu^+_{num}(k^+)$ and the spectral eddy viscosity $\nu^+_{Chollet}(k^+)$ of EDQNM [112]. The resulting set of model parameters for the incompressible Navier-Stokes equations is given in Table 6.3. With these parameters a good match of the theoretical spectral eddy viscosity and the spectral numerical viscosity is found, see Fig. 6.11.

Table 6.3. Result obtained by evolutionary optimization for the discretization parameters of ALDM

Parameter	Optimal value
$\gamma^0_{3,1}$	0.0500300
$\gamma^{+1/2}_{2,0}$	1.0000000
$\gamma^{+1/2}_{3,0}$	0.0190200
$\gamma^{+1/2}_{3,1}$	0.0855000
σ	0.0689100

ALDM for the incompressible Navier-Stokes equations has been applied to a wide range of flows. A representative result for homogeneous turbulence is the transition of the three-dimensional Taylor-Green vortex, whose dissipation evolution is very well captured by ALDM, see Fig. 6.12. Inhomogeneous turbulent flows also are well reproduced by ALDM, this holds e.g. for turbulent channel flow [113], see Fig. 6.13, and for massively separated boundary layer flows [115, 116].

6.3.3 ALDM for the Compressible Navier-Stokes Equations

ALDM as given in the previous section for the incompressible Navier-Stokes equations can be extended in a straight-forward fashion to the compressible

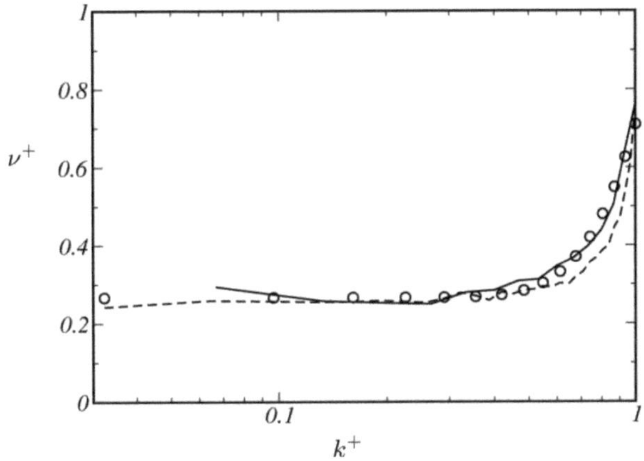

Fig. 6.11. Numerical viscosity of ALDM with optimized parameters compared to the prediction of turbulence theory. —— LES with $N = 32$, ---- LES with $N = 64$, ∘ EDQNM theory. Reproduced from Ref. [112] with permission

Navier-Stokes equations [117]. For the extension, the conservative formulation of the compressible Navier-Stokes equations with a transport equation for the total energy is used. The viscous stress tensor employs a Newtonian law and heat conduction follows a Fourier law. The conservation laws are complemented by an ideal-gas equation of state. As in Sect. 6.3.2 a staggered arrangement of finite volumes is the basis of the discretization, Fig. 6.14. With a tilde the approximately reconstructed cell-face values for the respective variables are denoted, $H = E - p$ is the enthalpy. The reconstruction follows that in the previous Sect. 6.3.2, in a simplified version with reduced operations [113]. The reconstructed solution is inserted in the numerical flux function which is a extension of the incompressible definition. The numerical flux function for the density is

$$\tilde{F}_\rho\left(x_{j-1/2}\right) = \frac{1}{2}\begin{bmatrix} u^{*\circ}_{i-1,j,k}(\rho^{*+}_{i-1/2,j,k} + \rho^{*-}_{i-1/2,j,k}) \\ v^{*\circ}_{i,j-1,k}(\rho^{*+}_{i,j-1/2,k} + \rho^{*-}_{i,j-1/2,k}) \\ w^{*\circ}_{i,j,k-1}(\rho^{*+}_{i,j,k-1/2} + \rho^{*-}_{i,j,k-1/2}) \end{bmatrix}$$
$$- \sigma_\rho \begin{bmatrix} |u^{*+}_{i-3/2,j,k} - u^{*-}_{i-1/2,j,k}|(\rho^{*+}_{i-1/2,j,k} - \rho^{*-}_{i-1/2,j,k}) \\ |v^{*+}_{i,j-3/2,k} - v^{*-}_{i,j-1/2,k}|(\rho^{*+}_{i,j-1/2,k} - \rho^{*-}_{i,j-1/2,k}) \\ |w^{*+}_{i,j,k-3/2} - w^{*-}_{i,j,k-3/2}|(\rho^{*+}_{i,j,k-1/2} - \rho^{*-}_{i,j,k-1/2}) \end{bmatrix}, \tag{6.89}$$

where we use u, v, w for the velocity components in the coordinate directions x_1, x_2, x_3. The convection velocities need to be computed from the momentum fluxes $\overline{\rho u}_{i,j,k}$, $\overline{\rho v}_{i,j,k}$, $\overline{\rho w}_{i,j,k}$, where it was found that for proper scaling of the

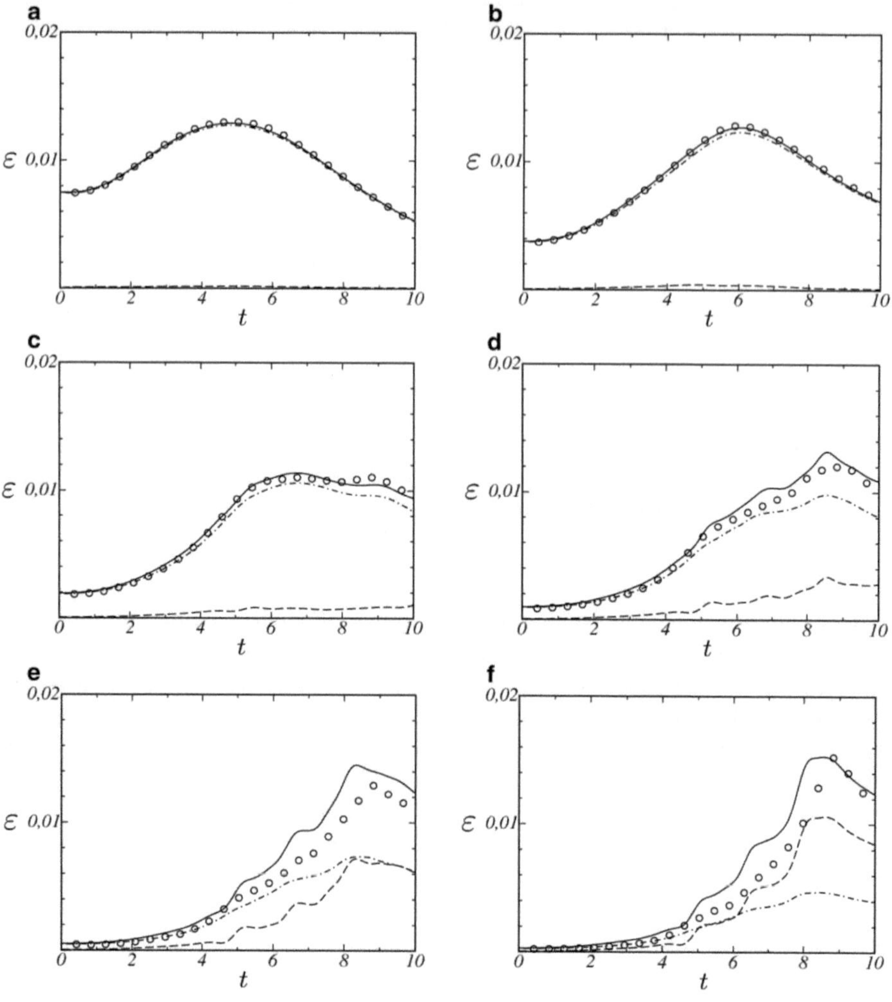

Fig. 6.12. Contributions to energy dissipation in ALDM for LES of the Taylor Green vortex at Re = (**a**) 100, (**b**) 200, (**c**) 400, (**d**) 800, (**e**) 1600, (**f**) 3000; —·— molecular dissipation, - - - - implicit SGS dissipation, —— total dissipation, *symbols* DNS data. Reproduced from Ref. [112] with permission

dilatation these convective velocities are given by

$$
\begin{aligned}
u^{*\circ}_{i,j,k} &= \frac{2\overline{\rho u}_{i,j,k}}{\rho^{*+}_{i+1/2,j,k} + \rho^{*-}_{i+1/2,j,k}}, \\
v^{*\circ}_{i,j,k} &= \frac{2\overline{\rho v}_{i,j,k}}{\rho^{*+}_{i,j+1/2,k} + \rho^{*-}_{i,j+1/2,k}},
\end{aligned}
\tag{6.90}
$$

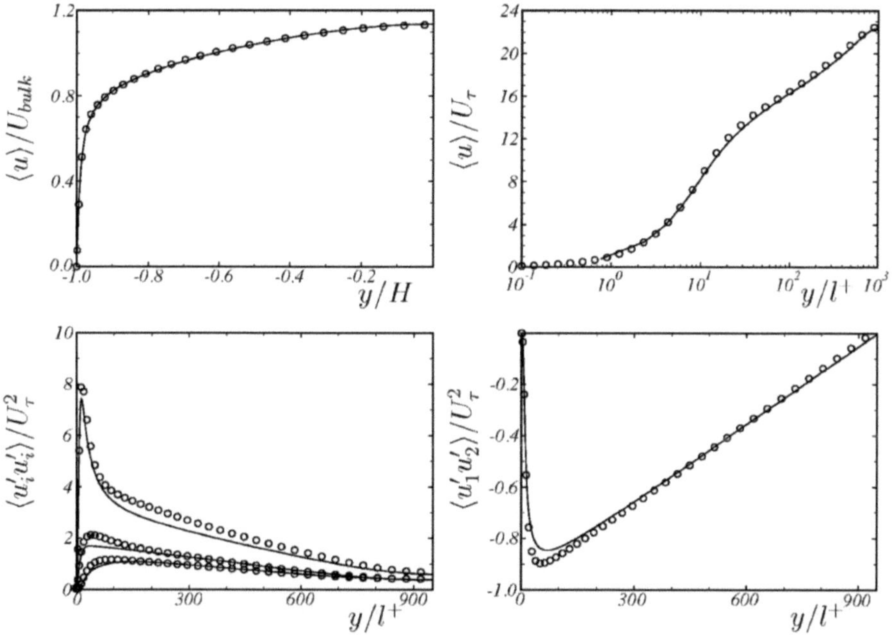

Fig. 6.13. Mean velocity profile and resolved Reynolds stresses for LES of turbulent channel flow at $Re_\tau = 950$. *Lines*: ALDM, *symbols*: DNS. Reproduced from Ref. [113] with permission

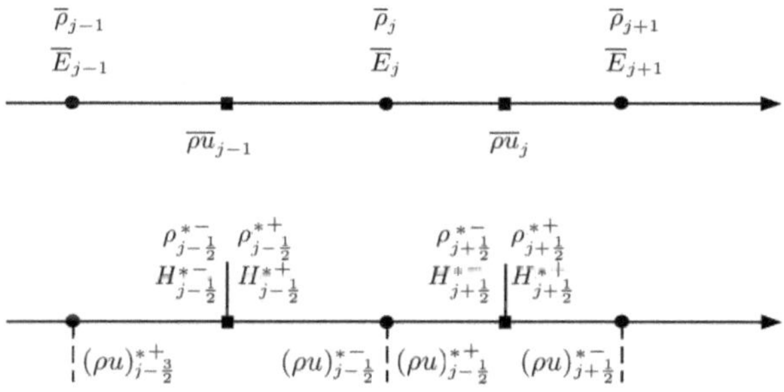

Fig. 6.14. Staggered grid arrangement. Filtered solution (*top*) and reconstructions of the unfiltered solution (*bottom*)

$$w^{*\circ}{}_{i,j,k} = \frac{2\overline{\rho w}_{i,j,k}}{\rho^{*+}_{i,j,k+1/2} + \rho^{*-}_{i,j,k+1/2}} .$$

For the enthalpy H the same definition is used by just replacing ρ by H. The numerical flux function for the momentum equation in x_1 direction is

$$
\begin{aligned}
&\tilde{F}_{\rho u}\left(x_{j-1/2}\right) \\
&= \frac{1}{4}\begin{bmatrix} (u^{*+}_{i-1/2,j,k} + u^{*-}_{i-1/2,j,k})((\rho u)^{*+}_{i-1/2,j,k} + (\rho u)^{*-}_{i-1/2,j,k}) + p_{i,j,k} \\ (v^{*+}_{i+1/2,j-1,k} + v^{*-}_{i+1/2,j-1,k})((\rho u)^{*+}_{i,j-1/2,k} + (\rho u)^{*-}_{i,j-1/2,k}) \\ (w^{*+}_{i+1/2,j,k-1} + w^{*-}_{i+1/2,j,k-1})((\rho u)^{*+}_{i,j,k-1/2} + (\rho u)^{*-}_{i,j,k-1/2}) \end{bmatrix} \\
&\quad - \begin{bmatrix} \sigma_{\rho u}|u^{*+}_{i-1/2,j,k} - u^{*-}_{i-1/2,j,k}|((\rho u)^{*+}_{i-1/2,j,k} - (\rho u)^{*-}_{i-1/2,j,k}) \\ \sigma'_{\rho u}|v^{*+}_{i+1/2,j-1,k} - v^{*-}_{i+1/2,j-1,k}|((\rho u)^{*+}_{i,j-1/2,k} - (\rho u)^{*-}_{i,j-1/2,k}) \\ \sigma'_{\rho u}|w^{*+}_{i+1/2,j,k-1} - w^{*-}_{i+1/2,j,k-1}|((\rho u)^{*+}_{i,j,k-1/2} - (\rho u)^{*-}_{i,j,k-1/2}) \end{bmatrix} .
\end{aligned}
\tag{6.91}
$$

Similar expressions apply to the momentum fluxes in x_2 and x_3 directions. By using a different parameter for the numerical diffusion in the numerical flux function for the diagonal terms $\sigma_{\rho u}$ and for the off-diagonal terms $\sigma'_{\rho u}$, aside of the reconstruction weights in (6.78) four additional model parameters are introduced σ_ρ, $\sigma_{\rho u}$, $\sigma'_{\rho u}$, σ_E. Note that reflectional invariance requires that the same parameters $\sigma_{\rho u}$, $\sigma'_{\rho u}$ are used for the momentum fluxes in x_2 and x_3 directions.

Whereas the reconstruction parameters of Table 6.3 are used, the above numerical-diffusion parameters are subjected to optimization. Optimization is performed for low-Mach number isotropic turbulence following the same procedure as described in the previous section. The identified parameters are given in Table 6.4.

Table 6.4. Result obtained by evolutionary optimization for the discretization parameters of ALDM for compressible turbulence

Parameter	Optimal value
σ_ρ	0.6699
$\sigma'_{\rho u}$	0.6302
$\sigma_{\rho u}$	0.0638
σ_E	0.8093

The resulting implicit model well reproduces incompressible isotropic turbulence. For a case of decaying isotropic turbulence with significant compressibility at an initial turbulence Mach number of $M_t = 0.3$ and an initial microscale Reynolds number of $Re_\lambda = 100$ a comparison with DNS data shows that the decay of kinetic energy, density fluctuations, pressure fluctuations and temperature fluctuations is well predicted, see Fig. 6.15. In Ref. [117] it is also shown that the spectra of these fluctuating quantities agree well with DNS.

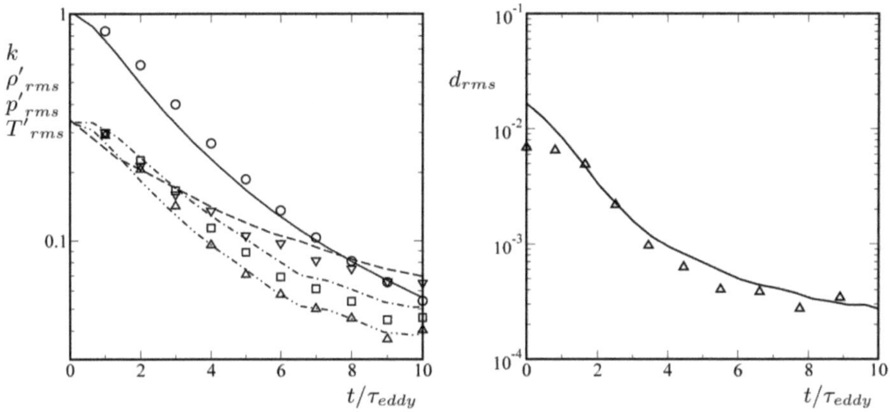

Fig. 6.15. Decay of kinetic energy k, density ρ'_{rms}, pressure p'_{rms}, temperature T'_{rms}, and dilatation fluctuations d_{rms} for implicit LES (*lines*) and filtered DNS (*symbols*) of compressible isotropic turbulence. The time is normalized by the eddy-turnover time

7

Boundary Conditions for Large-Eddy Simulation of Compressible Flows

7.1 Introduction

The definition of relevant boundary conditions for compressible large-eddy simulation is another very difficult issue, much more complex than its counterpart in LES for incompressible flows. The additional difficulty originates in that fact that boundary conditions must provide an information on all resolved scales of motion taking into account that the three Kovasznay modes exist at all scales, and that different conditions should be applied on each of the modes. This dual decomposition in terms of scales and Kovasznay modes renders the problem so difficult that only partially relevant, case-specific solutions have been proposed up to now.

The issue of non-reflecting boundary conditions for large-eddy simulation will not be discussed in this chapter, since LES does not introduce new elements here. The problem is therefore the same as for direct numerical simulation or unsteady Reynolds-averaged simulations. Two specific topics are addressed below, namely unsteady turbulent inflow conditions and wall models for compressible LES. In these two cases, LES has some specific features which deserve a detailed discussion.

The very definition of boundary conditions is also a mathematical problem: if one assumes that the scale separation is achieved by a suitable mathematical operator (e.g. a convolution filter), then the boundary conditions must also be defined using the same mathematical operator. A major difficulty arises here, since most of the mathematical models for LES rely on non-local operators, such as a convolution filter whose kernel has a finite support, making the boundary conditions non-local. This point is very often neglected, and almost all published boundary conditions for compressible LES are not consistent from this point of view. This issue will not be discussed further on, since no theoretical framework has been developed to address it up to now.

E. Garnier et al., *Large Eddy Simulation for Compressible Flows*,
Scientific Computation,

7.2 Wall Modeling for Compressible LES

7.2.1 Statement of the Problem

The wall modeling problem is nothing but a particular subgrid modeling problem, which arises when the first computational cell at the wall is too coarse to allow for a direct capture of the large turbulent scales (i.e. the production mechanisms) that are present in the inner part of the turbulent boundary layer. In the wall modeling problem some turbulence production has to be modeled, whereas in the usual subgrid closure problem the production mechanisms are assumed to be directly represented on the computational grid, and only kinetic energy cascade and dissipation are modeled. Since the first grid cell at the wall is too coarse, the usual no-slip boundary condition must be replaced by another condition which accounts for the details of the subcell motion. The new condition is to be provided by a wall model. The required output depends on the numerical method (finite volume, finite difference, finite element), the data location (staggered or non-staggered grids, cell-vertex or cell-centered methods, ...).

Let us emphasize here that in the incompressible case most of the published wall models have been designed for finite volume/difference schemes on staggered grids, in which the main output of the wall model is the skin friction at the wall. Only very few published works address the compressible case, and a vast majority of them rely on a cell-centered finite volume approach. Here again, the main output is the skin friction at the wall.

As it will be discussed at the end of this section, only very little attention has been paid to the development of a really compressible wall model. The main reason is that, according to the celebrated Morkovin's hypothesis, the compressible boundary layer dynamics is very similar to those of the strictly incompressible case for values of the Mach number based of the external velocity up to 5. In a first step, we will recall the key elements of the turbulent boundary layer dynamics, the emphasis being put on the velocity and the temperature field (i.e. the vorticity and the entropy modes in the parlance of Kovasznay). In a second step, the acoustic boundary layer will be discussed. In a third and last step, the problem of developing a fully compressible wall model will be analyzed. To close this chapter, the rare existing extensions of incompressible wall models to the compressible case will be presented.

7.2.2 Wall Boundary Conditions in the Kovasznay Decomposition Framework: an Insight

We present here a general discussion about the boundary condition at an impenetrable solid surface, the field being split into contributions of the three Kovasznay modes. The purpose is not to discuss the turbulent boundary layer case (this will be done later on), but to show the couplings between the modes induced by the boundary condition. The interested reader is referred to Chapter 10 of Ref. [215] for a more detailed discussion.

The conditions to be enforced are the usual conditions: the velocity is zero at the wall according to the no-slip boundary condition (it is recalled that subscripts Ω, e and p denote contributions of the vorticity mode, the entropy mode and the acoustic mode, respectively)

$$\boldsymbol{u}_{wall} = \boldsymbol{u}_\Omega + \boldsymbol{u}_e + \boldsymbol{u}_p, \qquad z = 0, \tag{7.1}$$

and both the temperature T and the normal component of the heat flux $\boldsymbol{q} \cdot \boldsymbol{n}$ must be continuous at the solid-fluid interface, where $\boldsymbol{q}$ and $\boldsymbol{n}$ are the heat flux and the surface normal unit vector, respectively. If the considered solid is a much better conductor than the fluid, these two conditions can be replaced by the assumption that solid's surface is at ambient temperature, i.e. $T' = 0$ at the surface

$$T' = T_e + T_p = 0, \qquad z = 0. \tag{7.2}$$

Considering the case in which the three Kovasznay modes have the same frequency ω, the boundary conditions for the three complex modes in a perfect gas are [215] (where $\imath^2 = -1$):

$$\nabla_\| \cdot \boldsymbol{u}_\Omega^\| - \frac{(1-\imath)}{\delta_\Omega} u_\Omega^\perp = 0, \tag{7.3}$$

$$\boldsymbol{u}_e^\| = \frac{\beta\chi}{\rho c_p} \nabla_\| T_e \simeq 0, \tag{7.4}$$

$$u_e^\perp = -\frac{\beta\chi}{\rho c_p} \frac{(1-\imath)}{\delta_e} T_e, \tag{7.5}$$

$$\nabla_\| \cdot \boldsymbol{u}_p^\| = -\frac{(1-\imath)}{\delta_\Omega} u_\Omega^\perp, \tag{7.6}$$

$$u_p^\perp - (1+\imath)\frac{\delta_\Omega}{2} \nabla_\| \cdot \boldsymbol{u}_p^\| + (1-\imath)(\gamma-1)\frac{\omega}{c}\frac{\delta_e}{2}\frac{p_p}{\rho c} = 0, \tag{7.7}$$

where any vector $\boldsymbol{u}$ is decomposed as $\boldsymbol{u} = \boldsymbol{u}^\| + u^\perp \boldsymbol{n}$ and $\nabla_\|$ denotes the nabla operator restricted in directions tangential to the solid surface. The boundary layer thicknesses (for the considered monochromatic fluctuating field) are defined as follows:

$$\delta_\Omega = \sqrt{\frac{2\mu}{\omega\rho}}, \qquad \delta_e = \sqrt{\frac{2\chi}{\omega\rho c_p}} = \frac{\delta_\Omega}{\sqrt{Pr}}. \tag{7.8}$$

It can seen that these thicknesses are much smaller than the acoustic wavelength λ_p at frequencies of practical interest

$$2\pi\frac{\delta_\Omega}{\lambda_p} = \frac{\omega}{c}\delta_\Omega = \sqrt{\frac{2\omega\mu}{\rho c^2}} \ll 1. \tag{7.9}$$

This set of boundary conditions allows for the analysis of the reflection of acoustic waves on the solid boundary. Equation (7.7) yields a general definition of the acoustic apparent specific admittance Y, which is defined as the reciprocal of the specific impedance Z

$$Y(\omega) \equiv \frac{1}{Z(\omega)} \equiv -\frac{u_p^{\perp}}{p_p} = \frac{(1-\imath)}{2}\frac{\omega}{\rho c^2}(\delta_\Omega \sin^2\theta_i + (\gamma - 1)\delta_e), \qquad (7.10)$$

where θ_i is the angle of incidence of the acoustic plane wave (see Fig. 7.1). It is observed that the vorticity and the entropy modes modify the reflection angle and yield an attenuation of the reflected acoustic wave, since Y is an

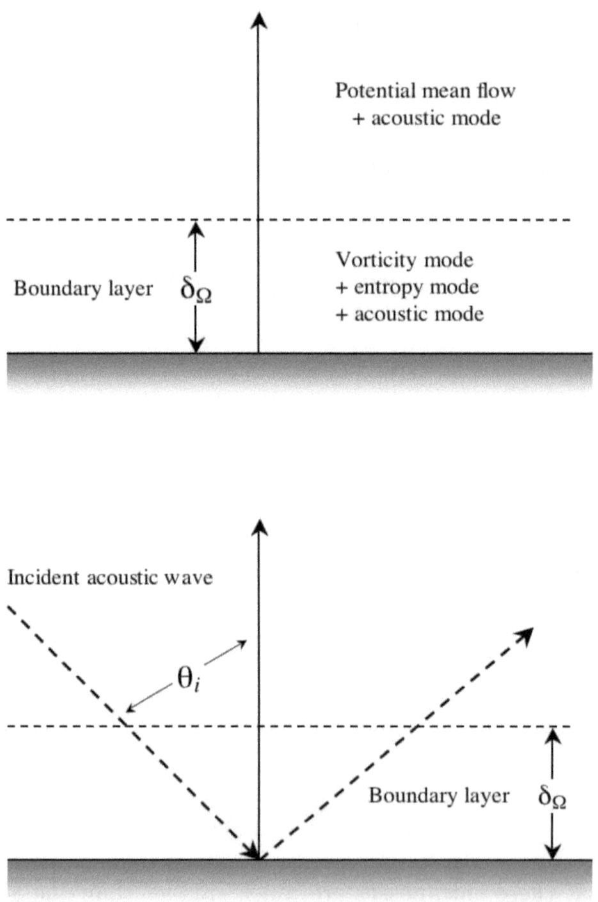

Fig. 7.1. Schematic view of (*top*) the boundary layer problem including the three Kovasznay modes and (*bottom*) the diffraction/refraction of an incident plane acoustic wave by a turbulent boundary layer

explicit function of δ_Ω and δ_e. In a more general way, the total admittance is split as the sum of a thermal admittance Y_χ and a shear stress admittance Y_μ, i.e. $Y(\omega) = Y_\chi(\omega) + Y_\mu(\omega)$, which are tied to the entropy and the vorticity modes, respectively, and whose expressions are generalized in the case of a turbulent boundary layer, as discussed below.

7.2.3 Turbulent Boundary Layer: Vorticity and Temperature Fields

Turbulent Boundary Layer Vortical Dynamics: a Brief Reminder

The aim of this section is to recall the main features of the vortical structure dynamics in a canonical turbulent boundary layer. The reader is referred to the book by Smits and Dussauge [266] for a detailed description of compressible boundary layer dynamics. Let us first recall that, according to Morkovin's hypothesis, compressibility effects are small in turbulent boundary layers for Mach numbers (based on the external velocity) up to 5. This stems from the fact that compressibility effects become significant when the turbulent velocity divergence $\nabla \cdot \boldsymbol{u}'$ becomes large with respect to the turbulent velocity gradient. In adiabatic supersonic boundary layers with $M < 5$, the following estimate for small scale behavior can be found

$$\frac{\nabla \cdot \boldsymbol{u}'}{u'/L} \sim 10(\gamma - 1)M_t^2\left(1 + \frac{M^2}{2\gamma}\right), \tag{7.11}$$

where u'/L is related to the magnitude of the large-scale gradient. The companion relation for large scales (i.e. energy-containing scales) based on the mean dissipation is

$$\frac{\nabla \cdot \boldsymbol{u}'}{u'/L} \sim (\gamma - 1)^2 M_t^3 M\left(1 + \frac{M^2}{2\gamma}\right). \tag{7.12}$$

The last term in the right-hand side accounts for both Mach and heating effects. Since maximum values of the turbulent Mach number M_t are at most 0.2 in boundary layers, it is seen that compressibility effects should remain small. Therefore, the emphasis is put here on the incompressible or nearly incompressible case, since it is the most common one in practical applications. The focus is also put on the inner part of the boundary layer, since this is the region of the flow whose dynamics must be accounted for by the wall model.

In the inner layer, which is a very thin region adjacent to the wall, several phenomena take place, which are responsible for the main part of the production of turbulent kinetic energy and turbulent drag. Elongated sinuous arrays of alternating streamwise jets superimposed on the mean shear, referred to as high/low-speed streamwise velocity streaks are observed very close to the wall. The wall shear is higher than the average at locations where the jets point forward (resp. backward) for high speed (resp. low-speed) streaks. The typical streamwise extent L_x of the streaks is about $L_x \sim 1000\nu/u_\tau$, where u_τ denotes the skin friction. The high-speed streaks are observed to be shorter than low-speed ones, but they have a larger width in the spanwise direction: the typical width of a high-speed streak is $L_y \sim 100\nu/u_\tau$, while the width of low-speed streaks is about $50\nu/u_\tau$. On the lateral sides of the streaks, streamwise vortices are found. The quasi-streamwise vortices are slightly tilted from the wall, and stay near the wall in a region of streamwise extent roughly equal to $200\nu/u_\tau$. Several vortices are associated with each streak, with a preferential longitudinal spacing $400\nu/u_\tau$. They are advected at speed $u_v \simeq 10u_\tau$.

Longitudinal vortices and streaks are the key elements of the near wall dynamics, and interact to generate an autonomous turbulent dynamical system, i.e. a self-sustained nonlinear turbulent process which is responsible for a non-negligible part of the turbulent kinetic energy production and turbulent drag. It is known that some longitudinal vortices are connected to legs of hairpin-like vortices which are present in the logarithmic layer at the top of the inner layer, but most of them merge in incoherent vorticity away from the wall. These vortices are independent from the wall, since they are also present in homogeneous shear flows (see [245]). Streaky velocity structures are wakes left by individual vortices which moves in the mean-shear in the buffer layer, and they are involved in the nonlinear mutual regeneration cycle with streamwise vortices. Other structures are found in the upper part of the inner layer. First, there are very long streaky structures generated by viscous coalescence of shorter ones. These very long streaks are not involved in the nonlinear regeneration process, they do not cascade energy into shorter wavelengths, and carry almost half of the Reynolds stresses. Second, self-similar vortex clusters are also present. That category encompasses nearly isotropic small vortex packets, which are detached from the wall, and tall clusters, which originate from the near-wall region.

Turbulent Boundary Layer: Mean Flow Features

We now discuss the statistical features of the turbulent boundary layer. Details of the algebra will be omitted and are available in specific books [251, 266]. The emphasis is put here on the temperature and velocity fields, i.e. on quantities that are mainly governed by the vorticity and entropy modes according to Kovasznay's decomposition. The very reason for this is that, since these modes have the same characteristic advection speed, the associated correlations are significantly correlated. It appears that in boundary layers on adiabatic walls the Strong Reynolds Analogy proposed by Morkovin holds. This hypothesis is usually written as:

$$\frac{\sqrt{\overline{T'^2}}}{\tilde{T}} = (\gamma - 1)M^2 \frac{\sqrt{\overline{u'^2}}}{\tilde{u}}, \tag{7.13}$$

$$R_{uT} \equiv -\frac{\overline{u'T'}}{\sqrt{\overline{T'^2}}\sqrt{\overline{u'^2}}} = \text{constant}. \tag{7.14}$$

In supersonic boundary layers over adiabatic walls one observes that $R_{uT} \simeq$ 0.6–0.8 throughout the layer, showing that passive advection of temperature by velocity fluctuations is the dominant mechanism in temperature dynamics in that case.

The mean velocity profile can be computed using the four hypotheses given below:

- Morkovin's hypothesis holds.
- Convective terms are small compared to viscous terms in the momentum and energy equations.
- The total stress $\mu(T)\frac{\partial \bar{u}}{\partial z} - \bar{\rho}\overline{u'w'}$ is constant across the inner layer and is equal to τ_w.
- The turbulent shear stress is reliably modeled using Prandtl's mixing length approach.

The viscosity is a temperature-dependent quantity, which is assumed to obey the simple law:

$$\bar{\mu}(T) = \bar{\mu}_w \left(\frac{\bar{T}}{T_w}\right)^{\omega}, \tag{7.15}$$

where the subscript w denotes mean values taken at the solid surface. The peak of dissipation in the inner region of the boundary layer is observed to yield a significant local heating, and an increase of the local viscosity.

In the viscous sublayer, i.e. neglecting the turbulent shear stress, the analytic solution for the mean velocity profile is

$$\frac{u^s}{u_\tau} = \frac{z u_\tau}{\nu_w}, \tag{7.16}$$

where u^s is obtained using the van Driest transform

$$u^s = \int_0^{\bar{u}} \left(\frac{\bar{T}}{T_w}\right)^{\omega} d\bar{u}. \tag{7.17}$$

The main advantage of the solution (7.16) is that it is formally identical to the incompressible solution if u^s is replaced by u (i.e. taking $\omega = 0$ in (7.15)). Taking $\omega = 1$ (instead of $\omega = 0.76$ for air), this equation has the following exact solution

$$u^s = \bar{u}\left(1 + \frac{a}{2}\frac{\bar{u}}{u_e} - \frac{b^2}{2}\left(\frac{\bar{u}}{u_e}\right)^2\right) \tag{7.18}$$

along with

$$\frac{\bar{T}}{T_w} = 1 - Pr\frac{T_\tau}{T_w}\bar{u}^+ - rM_\tau^2\frac{\gamma-1}{2}(\bar{u}^+)^2, \tag{7.19}$$

with

$$a = \left(1 + r\frac{\gamma-1}{2}M_e^2\right)\frac{T_e}{T_w} - 1, \tag{7.20}$$

$$b^2 = r\frac{\gamma-1}{2}M_e^2\frac{T_e}{T_w}, \tag{7.21}$$

where $\bar{u}^+ = \bar{u}/u_\tau$, $M_\tau = u_\tau/\sqrt{\gamma R T_w}$ is the friction Mach number, $T_\tau = q_w/\rho_w c_p u_\tau$ is the friction temperature and $u_\tau = \sqrt{\tau_w/\rho_w}$ is the friction velocity. The recovery factor r is close to $Pr^{1/3}$ for a boundary layer over an adiabatic flat plate. Subscript e denotes mean values taken at the outer edge of the boundary layer.

Using the same hypotheses, analytical solutions can be found in the logarithmic regions, which are formally identical to those found in incompressible boundary layer flows using the transformed velocity

$$u^s = \int_{u_1}^{\bar{u}} \sqrt{\frac{T_w}{\bar{T}}} d\bar{u}, \tag{7.22}$$

where subscript 1 denotes mean values at the bottom edge of the logarithmic layer

$$\frac{u^s}{u_\tau} = \frac{1}{\kappa} \ln \frac{z u_\tau}{\nu_w} + C, \tag{7.23}$$

where $\kappa \sim 0.41$ is the Karman's constant, and, using Fernholz's expression

$$u^s = \frac{u_e}{b} \sin^{-1}\left(\frac{2b^2(\bar{u}/u_e) - a}{\sqrt{a^2 + 4b^2}}\right), \tag{7.24}$$

$$C = \frac{1}{\kappa} \ln \frac{z_1 u_\tau}{\nu_w} + \frac{u_e}{b u_\tau}\left(\frac{2b^2(u_1/u_e) - a}{\sqrt{a^2 + 4b^2}}\right). \tag{7.25}$$

For an adiabatic wall, $a = 0$ and C is very close to its value for incompressible flows, i.e. $C \simeq 5.1$. The corresponding mean temperature profile is

$$\frac{\bar{T}}{T_w} = c_1 - \frac{\kappa}{\kappa_H} \bar{u}^+ \left(\frac{T_\tau}{T_w} - M_\tau^2 \frac{\gamma - 1}{2} \bar{u}^+\right), \tag{7.26}$$

where the von Karman's constant for the thermal boundary layer κ_H is such that κ/κ_H lies close to the recovery factor r or the Prandtl number Pr. The constant c_1 may be flow-dependent, but one expects that $c_1 \simeq 1$.

Since the aim of the present section is only to recall the main features of the mean profiles that are used to built wall models for large-eddy simulations, scaling laws—or the absence of universal scaling laws due to the inactive modes of motion—for the turbulent field statistical moments will not be discussed. We just point out that very recent high Reynolds number direct numerical simulations performed by Jimenez and coworkers for incompressible wall bounded flows [123] support the idea that the logarithmic law of the wall is nothing but an approximation (since $z(\partial\bar{u}/\partial z)$ is observed to vary with the distance to the wall) and that no universal scaling laws depending solely on the velocity scale u_τ for turbulent data can be found, since very large scale inactive fluctuations scale with u_e.

7.2.4 Turbulent Boundary Layer: Acoustic Field

All data presented above where obtained neglecting the acoustic field. This approximation is valid if one is interested in aerodynamic or aerothermal properties of the turbulent flow since most of the kinetic energy is contained in the vortical motion and that the acoustic field has a negligible feedback on both the vorticity and entropy dynamics in most cases. But it is not reliable if one is interested in the acoustic field or in fine details of the pressure and density field (e.g. as for aero-optics oriented studies). This is why a complete description of the compressible boundary layer must encompass the acoustic field. This section presents different issues associated to acoustic properties of the turbulent boundary layer. First, the scale separation between acoustic and hydrodynamic pressure fluctuations is introduced discussing the wall pressure spectrum. In a second step, the production of the acoustic field by the vortical motion is considered. The last subsection is devoted to the attenuation of sound across the boundary layer by vortical and temperature turbulent fluctuations.

A First Insight: Surface Pressure Fluctuations

The pressure fluctuations that develop on a solid surface beneath the turbulent boundary layer are conveniently described through their wavenumber-frequency spectrum (see [124] for a detailed discussion)

$$P(\boldsymbol{k},\omega) \equiv \frac{1}{(2\pi)^3} \int_{-\infty}^{+\infty} R_{pp}(y_1, y_2, \tau) e^{-\imath(\boldsymbol{k}\cdot\boldsymbol{y}-\omega\tau)} dy_1 dy_2 d\tau, \tag{7.27}$$

where $\boldsymbol{k} = (k_1, k_2, 0)$ is the wave vector restricted to the plane parallel to the wall and $R_{pp}(y_1, y_2, \tau) = \overline{p'(x_1, x_2, 0, t) p'(x_1 + y_1, x_2 + y_2, 0, t + \tau)}$ is the two-point two-time correlation factor of wall pressure fluctuations. The wall rms pressure is recovered as

$$\overline{p'^2} = R_{pp}(0,0,0) = \int_{-\infty}^{+\infty} P(\boldsymbol{k},\omega) dk_1 dk_2 d\omega. \tag{7.28}$$

Introducing the wall point pressure frequency spectrum $\Phi_{pp}(\omega)$

$$\Phi_{pp}(\omega) \equiv \int_{-\infty}^{+\infty} P(\boldsymbol{k},\omega) dk_1 dk_2, \tag{7.29}$$

one has

$$\overline{p'^2} = \int_{-\infty}^{+\infty} \Phi_{pp}(\omega) d\omega. \tag{7.30}$$

At low Mach number, $P(\boldsymbol{k},\omega)$ exhibits two different peaks (for a fixed frequency satisfying $\omega\delta/u_e > 1$, with δ being the boundary layer thickness), which are related to two different physical mechanisms (see Fig. 7.2). The

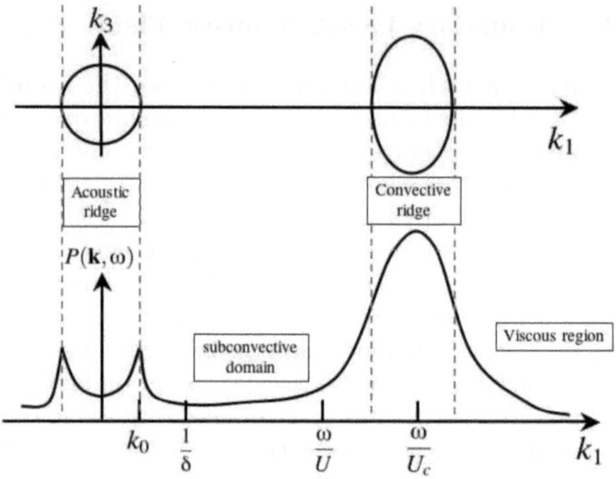

Fig. 7.2. Schematic view of the wavenumber-frequency wall pressure spectrum in a turbulent boundary layer. The frequency ω is assumed to be such that $\omega\delta/u_e > 1$

existence of the two peaks originates in the difference in the characteristic speed between the acoustic fluctuations and the hydrodynamic modes.

The strongest peak occurs within the so-called *convective ridge* ($k \gg |\omega|/c$) centered on $k_1 = \omega/U_c, k_2 = 0$, and is associated to the pressure fluctuations due to the hydrodynamic advection of turbulent structures at advection speed $U_c \simeq 0.5u_e - 0.7u_e$. Several models for $P(\boldsymbol{k},\omega)$ within that convective ridge have been proposed, among then Chase's model

$$\frac{P(\boldsymbol{k},\omega)}{\bar{\rho}\delta^3 u_\tau^3} = \frac{1}{((k_+\delta)^2 + 1/\beta^2)^{5/2}}\left(C_R(k_1\delta)^2 + C_S(k\delta)^2\frac{(k_+\delta)^2 + 1/\beta^2}{(k\delta)^2 + 1/\beta^2}\right), \quad (7.31)$$

where

$$k_+^2 = \left(\frac{\omega - U_c k_1}{3u_\tau}\right)^2 + k^2. \quad (7.32)$$

The coefficients C_R and C_M are associated to the rapid and slow pressure terms,[1] respectively. Recommended values of the constants are $\beta = 0.75$, $C_R = 0.1553$ and $C_S = 0.0047$.

Many models have been proposed for the wall point pressure frequency spectrum. Let us illustrate these models using the model proposed by Goody [101]

$$\frac{(\omega)u_e}{\tau_w\delta}\Phi_{pp} = \frac{3(\omega\delta/u_e)^2}{((\omega\delta/u_e)^{3/4} + 0.5)^{3.7} + (1.1R_T^{-0.57}(\omega\delta/u_e))^7}, \quad (7.33)$$

[1] We recall that the rapid pressure term originates in interactions between turbulent fluctuations and the mean flow, while the slow term comes from turbulence self-interactions.

where the non-dimensional parameter is defined as the ratio of the outer-layer-to-inner-layer timescale

$$R_T = \frac{(\delta/u_e)}{(\nu/u_\tau^2)}. \quad (7.34)$$

The second peak is located at smaller wave numbers and is associated to acoustic waves that propagate at speed c with respect to the fluid. It belongs to the region of the spectrum referred to as the *acoustic ridge*. Howe's model in this region is

$$\frac{P(\boldsymbol{k},\omega)}{\bar{\rho}\delta^3 u_\tau^3} = \frac{1}{(k_+\delta)^2 + 1/\beta^2)^{5/2}} \left(\frac{C_R (k_1\delta)^2 k^2}{|k^2 - k_0^2| + \epsilon^2 k_0^2} + C_S (k\delta)^2 \frac{(k_+\delta)^2 + 1/\beta^2}{(k\delta)^2 + 1/\beta^2} \left(\frac{2}{3} + \frac{|k^2 - k_0^2|}{6k^2} + \frac{k^2/6}{|k^2 - k_0^2| + \epsilon^2 k_0^2} \right) \right), \quad (7.35)$$

where ϵ is related to the attenuation and diffraction of surface propagating acoustic waves by the boundary layer. The wavenumber k_0 is related to the peak in the acoustic ridge.

The above expressions hold at low Mach numbers for which there is a clear distinction between the acoustic and the convective ridge. At high Mach numbers, the frequency gap vanishes and new models must be used. A general model valid for all Mach numbers can be defined using the Laganelli–Wolfe formula

$$\frac{u_e/\delta_\theta}{(\rho_e u_e^2)^2} \Phi_{pp}(\omega) = \frac{10^{-4}\lambda^{0.566}}{4\pi(1 + \lambda^{-2.868}(\omega\delta_\theta/u_e)^2)}, \quad (7.36)$$

where δ_θ refers to the boundary layer displacement thickness and

$$\frac{1}{\lambda} = \frac{1}{2}\left(1 + \frac{T_w}{T_e}\right) + 0.1(\gamma - 1)M_e^2. \quad (7.37)$$

This formula is consistent with the following approximation for the rms pressure:

$$\sqrt{\overline{p'^2}} = 0.01\lambda \frac{1}{2} \rho_e u_e^2. \quad (7.38)$$

Production of Pressure Fluctuations by the Vorticity Field

As said above, pressure fluctuations can be split into hydrodynamic fluctuations and acoustic fluctuations. Acoustic fluctuations originate from two different mechanisms: a volumic direct production of acoustic modes by the nonlinear mechanisms listed in Table 3.1, and surfacic conversion of vorticity/entropy modes into acoustic ones at the solid wall thanks to the boundary conditions. Both physical mechanisms are detailed below.

The hydrodynamic fluctuations are usually considered as the superposition of the rapid and the slow contributions, which are solution of

$$-\frac{\partial^2 p'}{\partial x_j \partial x_j} = -2\frac{\partial \bar{u}_i}{\partial x_j}\frac{\partial u'_j}{\partial x_i} \quad \text{(rapid pressure)}, \tag{7.39}$$

$$-\frac{\partial^2 p'}{\partial x_j \partial x_j} = -2\left(\frac{\partial u'_i}{\partial x_j}\frac{\partial u'_j}{\partial x_i} - \frac{\partial^2}{\partial x_i \partial x_j}\overline{u'_i u'_j}\right) \quad \text{(slow pressure)}. \tag{7.40}$$

The influence of each region of the turbulent boundary layer on the production of wall pressure fluctuations has been analyzed in [39] in an incompressible plane channel flow. Results are summarized in Table 7.1. It can be seen that the viscous sublayer and the buffer layer are important contributors to the slow pressure term at all wavenumbers, while low wavenumbers in the rapid pressure spectrum are not sensitive to the viscous sublayer. In both cases, the viscous sublayer and the buffer layer dominate the production of intermediate and high wavenumber fluctuations.

Table 7.1. Production of wall-pressure fluctuations in incompressible turbulent channel flow. Regions necessary to reconstitute the wave number ranges of the pressure spectra are indicated. Data from [39]

Pressure term	Wavenumber range			
	Lowest $k_x\delta < 1$	Low $1 < k_x\delta$ & $k_y\delta < 5$	Intermediate $5 < k_x\delta$ & $k_y\delta < 30$	High $30 < k_y\delta < 70$
Rapid	buffer layer + log layer	buffer layer + log layer	viscous sublayer + buffer layer	viscous sublayer + buffer layer
Slow	full channel	viscous sublayer + buffer layer + log layer	viscous sublayer + buffer layer	viscous sublayer + buffer layer

Let us now address the generation of acoustic fluctuations inside the turbulent boundary layer. The most useful mathematical model for that purpose is certainly the Ffowcs Williams–Hawkings acoustic analogy, which is an extension of Lighthill's analogy to wall-bounded flows (see Chap. 2 of [124] for a detailed description). All these approaches model the acoustic field as the results of the propagation of acoustic waves emitted by acoustic sources across a medium at rest. The propagation operator, which is derived manipulating the mass conservation and the momentum equations, is the d'Alembertian $\Box \equiv ((1/c_0^2)\partial^2/\partial t^2 - \nabla^2)$ of the fluid density fluctuations. The acoustic sources, which appear as source term on the right-hand-side of the d'Alembertian operator, are distinguished by Lighthill according to their mathematical structure in three classes

- *Monopole.* In the boundary layer case, monopoles are related to the production of acoustic waves by the viscous dissipation.
- *Dipole.* In usual turbulent wall-bounded flows, a dipole layer is associated to the wall-shear stress at the solid surface.

- *Quadrupole.* In the present case, the main quadrupolar term originates in the fluctuating Reynolds stresses.

A generic form of the Ffowcs Williams–Hawkings equation is:

$$\Box(pH_f) = \underbrace{\frac{\partial^2 (T_{ij} H_f)}{\partial x_i \partial x_j}}_{\text{Quadrupoles}} - \underbrace{\frac{\partial (G_i H_f + B_i \delta_w)}{\partial x_i}}_{\text{Dipoles}} + \underbrace{\frac{\partial (Q H_f + B \delta_w)}{\partial t}}_{\text{Monopoles}}, \tag{7.41}$$

where H_f and δ_w are the Heaviside and the Dirac functions associated to the volume filled by the turbulent fluid and the solid wall, respectively. G_i is the applied force per unit volume. The volumic sources (T_{ij}, Q) and the surfacic terms (B_i, B) are defined as follows

$$\begin{aligned} T_{ij} &= (1+\rho) u_i u_j - \sigma_{ij}, \qquad & Q &= -\frac{\partial}{\partial t}(\rho - M^2 p), \\ B_i &= T_{ij} n_j + p n_i, \qquad & B &= (1+\rho) u_i n_i, \end{aligned} \tag{7.42}$$

where σ_{ij} is the viscous stress and $\boldsymbol{n}$ is the outward unit vector associated to the solid surface. This fully general expression can be simplified in the case of description of the sound radiation from subsonic weakly compressible viscous flows that are adiabatic and of uniform composition. According to the no-slip boundary condition, $B\delta_w = 0$ at an impermeable solid surface at rest. In the absence of volumic force, the dipole distribution simplifies to

$$G_i H_f = -\frac{\partial (\rho_0 u_i H_f)}{\partial t}, \tag{7.43}$$

while other source terms can be rewritten as (the expression for Q holds for a perfect gas only)

$$Q = (\gamma - 1) M^2 \sigma_{ij} \frac{\partial u_i}{\partial x_j} + O(M^4), \qquad B_i = -\sigma_{ij} n_j + p n_i. \tag{7.44}$$

The relative importance of the three types of sources has been investigated by many authors. Recent DNS results by Hu, Morfey and Sandham [125, 126] prove that, at low Mach number, the far-field sound radiation from a turbulent channel flow is dominated by dipolar sources, i.e. by wall-shear stress fluctuations. Therefore, the sound power per unit channel area scales like M^2. At higher Mach number (typically for $M > 0.1$), the Reynolds-stress quadrupolar sources, whose generated sound varies as M^4, begin to dominate at low frequencies. The monopolar dissipation contribution is dominated in all cases. The generation of acoustic waves by the wall-shear stress fluctuations is easily understood by the coupling between the Kovasznay modes induced at the wall by the no slip-boundary condition (see (7.3)–(7.7)).

Attenuation of Acoustic Modes by Vorticity and Entropy Modes

The last physical mechanism that will be addressed in this section is the dissipation of acoustic wave energy by turbulent vorticity and entropy fluctuations. Let us first note that at very high and very low frequencies, acoustic waves are almost uncoupled with vortical and entropy fluctuations, since their characteristic time scales can differ by several orders of magnitude. In these two asymptotic cases, acoustic energy is not dissipated by turbulence. But, at intermediate frequencies, acoustic waves experience both thermal and vortical diffusion.

Turbulent diffusion for an acoustic wave with frequency ω at a distance z from the wall can be reliably approximated using an extended version of Prandtl's mixing length model (see Chap. 5 in Ref. [124])

$$\nu_{t,ac} = \begin{cases} 0 & z < z_c(\omega), \\ \kappa u_\tau(z - z_c(\omega)) & z \geq z_c(\omega), \end{cases} \tag{7.45}$$

where the frequency-dependent cutoff length is given by

$$\frac{u_\tau}{\nu} z_c(\omega) = 6.5\left(1 + \frac{1.7(\omega/\omega_c)^3}{1 + (\omega/\omega_c)^3}\right), \quad \omega_c \simeq 0.01\frac{u_\tau^2}{\nu}. \tag{7.46}$$

The acoustic turbulent thermal diffusivity is usually computed as $\chi_{t,ac} = \nu_{t,ac}/Pr_t$, where the turbulent Prandtl number is approximately equal to 0.7 in air.

The total effect experienced by an acoustic wave that crosses the full turbulent boundary layer can be represented extending the admittance concept introduced above. The shear stress and the thermal admittances can be written as

$$Y_\mu(\omega) = \frac{\imath}{p_p}\int_0^{+\infty}(u - u_p)dz, \qquad Y_\chi(\omega) = \frac{\imath\omega}{\rho_e p_p}\int_0^{+\infty}(\rho - \rho_p)dz. \tag{7.47}$$

Using the mixing length approximation for turbulent diffusivities, one obtains [124]

$$Y_\mu(\omega) \simeq \frac{1}{\rho_e\omega}\sqrt{\frac{\nu}{\imath\omega}}\mathcal{F}\left(\sqrt{\frac{\imath\omega\nu}{\kappa^2 u_\tau^2}}, z_c\sqrt{\frac{\imath\omega}{\nu}}\right), \tag{7.48}$$

with

$$\mathcal{F}(x,z) = \imath\frac{H_1^{(1)}(x)\cos z - H_0^{(1)}(x)\sin z}{H_0^{(1)}(x)\cos z + H_1^{(1)}(x)\sin z}. \tag{7.49}$$

The cylinder functions are given in the asymptotic limit $x \longrightarrow 0$ by

$$H_0^{(1)}(x) = 1 + \frac{2\imath}{\pi}(\gamma_E + \ln(x/2)) + O(x^2\ln x), \tag{7.50}$$

$$H_1^{(1)}(x) = -\frac{2\imath}{\pi x} + \frac{\imath x}{\pi}\left(\gamma_E - \frac{\imath\pi}{2} + \ln(x/2)\right) + O(x^3\ln x), \tag{7.51}$$

where $\gamma_E \simeq 0.57722$ is Euler's constant. The turbulent thermal admittance is given by

$$Y_\chi(\omega) \simeq \frac{\omega\beta}{\rho_e c_p}\sqrt{\frac{\chi}{\imath\omega}}\mathcal{F}\left(Pr_t\sqrt{\frac{\imath\omega\chi}{\kappa^2 u_\tau^2}}, z_c(\omega)\sqrt{\frac{\imath\omega}{\chi}}\right), \tag{7.52}$$

where β is the coefficient of expansion at constant pressure ($\beta = 1/T$ in an ideal gas of temperature T).

7.2.5 Consequences for the Development of Compressible Wall Models

Wall models in large-eddy simulation aim at representing the turbulence dynamics within the first computational cell off the wall, whose characteristic size is $O(100)$–$O(1000)$ wall units. Therefore, the full dynamics of both the viscous sublayer and the buffer layer must be parametrized. Looking at previous results, one can see that:

- A fully compressible wall model should take into account the existence of the three Kovasznay modes, since their physics is coupled and exhibit very different scales. More precisely, one can see that:
  - The wall model for the vorticity mode can be designed neglecting the acoustic mode, since there is almost no feedback of the acoustic mode on the vorticity dynamics in canonical turbulent boundary layer and that the kinetic energy of acoustic velocity fluctuations if very small compared with those of the vortical fluctuations within the turbulent region. Temperature fluctuations (i.e. the entropy mode) appear through the induced fluctuations of the molecular viscosity (buoyancy forces are neglected here). Therefore, they can be neglected if these viscosity fluctuations have little influence on the velocity dynamics.
  - The wall model for the temperature mode is enslaved to the description of the vorticity mode, but the acoustic mode can be neglected since thermoacoustic effects are negligible in a usual compressible turbulent boundary layer.
  - The wall model for the acoustic mode may be the most difficult part to develop (the authors are not aware of any attempt to develop such a model), since it should account for both production and diffraction/refraction effects. Even if acoustic production is dominated by vortical events, attenuation by temperature fluctuations (measured by the turbulent thermal admittance) must be taken into account to correctly capture the acoustic field.
- Efficient implementation of such a fully general wall model will certainly raise new numerical issues, since conditions to be imposed on computational unknowns will originate in the combination of different physical models (one for each Kovasznay mode), whose mathematical properties may be very different. Let us exemplify this noticing that dynamics of the acoustic mode is usually described in the frequency domain, while vorticity

and temperature dynamics are most often discussed in the time domain. Another problem, which is well known in the field of Computational Aeroacoustics, is that, since the hydrodynamic pressure fluctuations are more intense than the acoustic ones at the wall, the acoustic part of the boundary condition applied to a computational variable can be masked by the hydrodynamic one, leading to an erroneous prediction of the acoustic field.

- The wall modeling problem is much more difficult than the usual subgrid modeling issue discussed in previous chapters, since turbulence production effects must be parametrized and that the presence of the solid wall induces a coupling between the three Kovasznay modes, even in the linearized approximation. This point is clearly illustrated by the preceding sections, which are devoted to the simplest case, i.e. the canonical flat plate turbulent boundary layer without pressure gradient, wall curvature effects, roughness effects, hot/cold wall effects etc. The lack of universality of the physics to be incorporated in the wall model is certainly the main difficulty.

7.2.6 Extension of Existing Wall Models for Incompressible Flows

Since almost no results dealing with a fully compressible extension of incompressible LES wall models are available, this section displays the fundamentals of wall modeling and suggests some possible extension strategies. The reader is referred to Chap. 10 of Ref. [244] for an exhaustive description of existing incompressible wall models.

Algebraic Two-Layer Wall Models

Most LES wall models rely on an algebraic closure, which is based on the key assumption that the instantaneous subgrid motion within the first off-wall cell can be accurately described by the statistical mean flow features. Consequently, the instantaneous subgrid velocity profile is usually parametrized by the 1D zero pressure gradient turbulent boundary layer mean velocity profile. Assuming that the first point at which the LES field is computed lies within the boundary layer, and enforcing a matching condition between the instantaneous wall-parallel instantaneous resolved velocity component and the analytical mean velocity profile within the first cell, required quantities such as the instantaneous local skin friction can be estimated. Either explicit or implicit wall models can be derived, if power-law or logarithmic descriptions of the turbulent mean velocity profile are used, respectively. In the latter case, a Newton method is used to compute the skin friction. A two layer approach is required in both cases to account for the existence of the viscous sublayer, in which the velocity grows linearly versus the distance to the wall.

Many variants have been proposed since the pioneering works of Schumann, Grötzbach and Wengle in the 1970s and the early 1980s, but no significant and robust improvement has been reported up to now. An important limitation is that these models are not able to account for transition to turbulence, as expected since they rely on the key assumption that the flow is fully turbulent. The approach can be straightforwardly extended to compressible flows using the compressible mean flow profiles given in Sect. 7.2.3. The implementation can be further simplified thanks to van Driest's change of variables.

Thin-Boundary Layer Equations Based Models

All algebraic wall models suffer from the same weakness: they rely on the very strong assumption that the instantaneous subgrid motion within the first cell obeys an arbitrary ideal distribution. Obviously, this is wrong in most practical cases, and algebraic wall models are not able to cope with separation and strong pressure gradients. To alleviate this problem, Balaras, Benocci and Piomelli proposed in the mid 1990s to render the model more flexible by solving a simplified boundary layer equation within the first cell. An auxiliary grid is defined within the first off-wall LES cell, on which the boundary layer equation is solved. In the incompressible case, the resulting equation for the wall-parallel velocity components is

$$\frac{\partial}{\partial z}\left((\nu+\nu_t)\frac{\partial \bar{u}_i}{\partial z}\right) = \underbrace{\frac{\partial \bar{u}_i}{\partial t} + \frac{\partial \bar{u}_1\bar{u}_i}{\partial x} + \frac{\partial \bar{u}_3\bar{u}_i}{\partial z}}_{F_i}, \tag{7.53}$$

where z is the wall-normal direction and ν_t is an adequate eddy-viscosity model The eddy-viscosity model is most of the time a very simple RANS model, e.g. Prandtl's mixing length model, but some dynamic variants have been proposed. The use of more complex RANS models, such as the Spalart-Allmaras one-equation model has also been suggested. Most authors used $F_i = 0$, leading to a 1D nonlinear diffusion equation, which can be easily solved. Once the two tangential velocity component have been computed thanks to the boundary layer equation, the wall-normal velocity component is estimated using the incompressibility constraint

$$\bar{u}_3(z) = \int_0^z \frac{\partial \bar{u}_3}{\partial z}(\zeta)d\zeta = -\int_0^z \left(\frac{\partial \bar{u}_1}{\partial x}(\zeta) + \frac{\partial \bar{u}_2}{\partial y}(\zeta)\right)d\zeta. \tag{7.54}$$

These models are observed to perform better in separated flows and accelerated/decelerated flows than algebraic wall models. But, of course, they cannot account for transition to turbulence.

This approach may be extended to compressible flows using the compressible boundary layer equations (see Chap. 19 of [251] for a detailed derivation)

$$\frac{\partial \bar{\rho}\tilde{u}}{\partial x} + \frac{\partial \bar{\rho}\tilde{w}}{\partial z} = 0, \tag{7.55}$$

$$\frac{\partial}{\partial z}\left((\bar{\mu}+\mu_t)\frac{\partial \tilde{u}_i}{\partial z}\right) = \frac{dp_e}{dx} + \bar{\rho}\left(\tilde{u}\frac{\partial \tilde{u}_i}{\partial x} + \tilde{w}\frac{\partial \tilde{u}_i}{\partial z}\right), \tag{7.56}$$

$$\frac{\partial}{\partial z}\left(\bar{\lambda}\frac{\partial \bar{T}}{\partial z} - \overline{\rho w'' h''}\right) = -\tilde{u}\frac{dp_e}{dx} + \bar{\mu}\left(\frac{\partial \tilde{u}}{\partial z}\right)^2 + \bar{\rho}\tilde{\epsilon} - \bar{\rho}\left(\tilde{u}\frac{\partial \tilde{h}}{\partial x} + \tilde{w}\frac{\partial \tilde{h}}{\partial z}\right), \tag{7.57}$$

where $p_e = \bar{p} + \overline{\rho w''^2}$, $\tilde{h}$ and $\tilde{\epsilon}$ are the mean total pressure, the mean specific enthalpy and the mean turbulent dissipation, respectively. μ_t is the turbulent viscosity, which must be provided by a turbulence model.

7.3 Unsteady Turbulent Inflow Conditions for Compressible LES

7.3.1 Fundamentals

The problem discussed here is the following: *what conditions should be applied in LES at the inlet plane when the incoming flow is already turbulent?* Answering that question is a key step toward the use of LES as a practical tool for the analysis of most practical turbulent flows, since the computational domain cannot be taken large enough to encompass the full transition region.

The definition of turbulent inflow conditions is certainly one of the hardest open problem in the field of LES research, since it mixes several aspects:

- A mathematical problem: boundary conditions must be such that the full problem is well-posed. Despite the fact that the LES approach holds for viscous flows only, and that compressible Navier–Stokes equations are not strictly hyperbolic ones, the characteristic analysis is often used and results stemming from it, which rigorously holds for the Euler equations only, serve as guidelines to know how many unknowns can be prescribed at the inlet plane. The issue of turbulent inlet boundary conditions also raises the non-trivial problem of the definition of unsteady non-reflecting boundary conditions for Navier–Stokes conditions: while a part of the solution must be prescribed using Dirichlet-type conditions at the inlet, the complementary part, which corresponds to upstream traveling waves generated inside the computational domain, must be allowed to exit the computational domain without spurious reflections. This problem is even harder than the one of the definition of non-reflecting outflow boundary conditions. The reason is that adapted techniques, such that sponge regions in which the solution is strongly damped, are often used to prevent spurious reflections at the outlet plane. But such damping techniques cannot be used at the inlet, since the incoming signal would be also destroyed. This problem is still open.
- A physical problem: practical experience shows that if some basic features of turbulence are not taken into account, a relaxation region exists downstream the inlet plane, in which a physical solution is rebuilt (see Fig. 7.3). This relaxation region can be very large, and its existence may sometimes preclude the recovery of reliable results. It is known that both one-point (e.g. Reynolds stresses, spectra) and two-point/two-time correlations of turbulence must be adequately prescribed to minimize the length of the relaxation region. But almost all available knowledge deals with the incompressible case. In the compressible case, the problem is rendered more

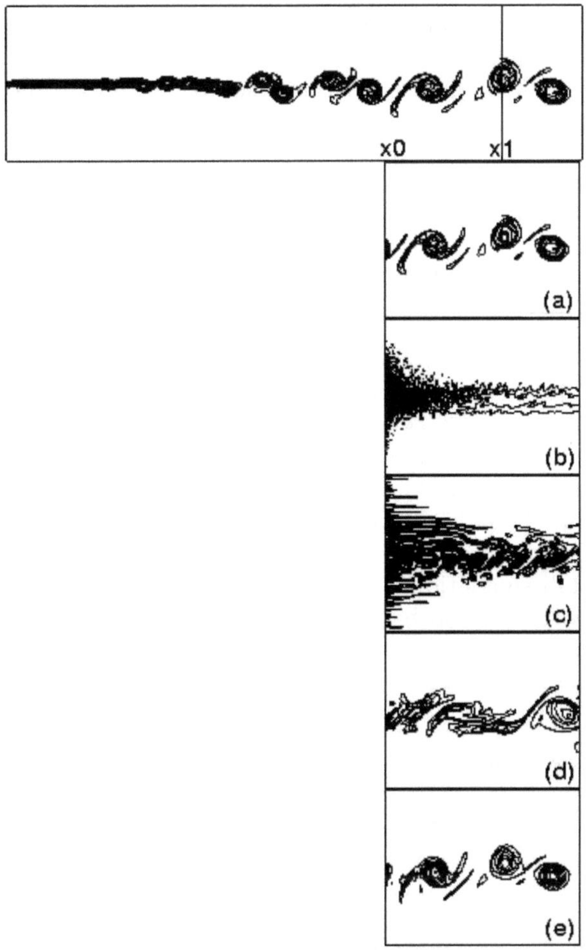

Fig. 7.3. Relaxation of a 2D spatially developing mixing layer with different inflow reconstructions having the same Reynolds stress distribution. *Top*: baseline 2D direct numerical simulation. *Below*: Truncated simulations using as inflow conditions: (**a**) exact instantaneous velocity field stored at the x0 section; (**b**) random velocity fluctuations, spatially and temporally uncorrelated but having the same Reynolds stress tensor components as the exact inlet velocity fluctuations; (**c**) instantaneous velocity field preserving temporal two-point correlation tensor of exact fluctuations; (**d**) instantaneous velocity field preserving spatial two-point correlation tensor of exact fluctuations; (**e**) reconstructed velocity field with Linear Stochastic Estimation having realistic two-point correlations. Courtesy of Ph. Druault

complex since correlations corresponding to each Kovasznay mode along with their inter-correlations must be prescribed. A very difficult point here is that acoustic/entropy and acoustic/vorticity correlations are expected to be very small, since acoustic modes are the only ones which travel at the speed of sound. In subsonic cases, most of the kinetic energy is borne by the vorticity mode, and the emphasis can be put on the sole vorticity mode if the acoustic field and the entropy field are not important output of the LES. Let us note that, to the knowledge of the authors, no fully general method which includes explicit reconstruction of fluctuations associated to each of the three Kovasznay modes is available in the literature.

7.3.2 Precursor Simulation: Advantages and Drawbacks

The precursor technique consists of performing an auxiliary simulation whose results are stored and used to generate the inlet conditions for the main simulation (see Fig. 7.4). Data are usually extracted from the auxiliary simulation on a plane, and are sometimes rescaled (see next subsection). This approach is used when the flow upstream the main simulation computational domain can be reasonably approximated as a simple generic flow, such as a zero-pressure gradient flat-plate boundary layer, a plane mixing layer, the flow in a infinite duct etc. The main advantage of the precursor technique with respect to the single-domain full simulation is to allow for a splitting of the computational domain. A theoretical drawback is that it does not allow to account for a feedback of the solution of the main simulation onto the precursor solution (e.g. by upstream traveling acoustic waves in the subsonic case). Therefore,

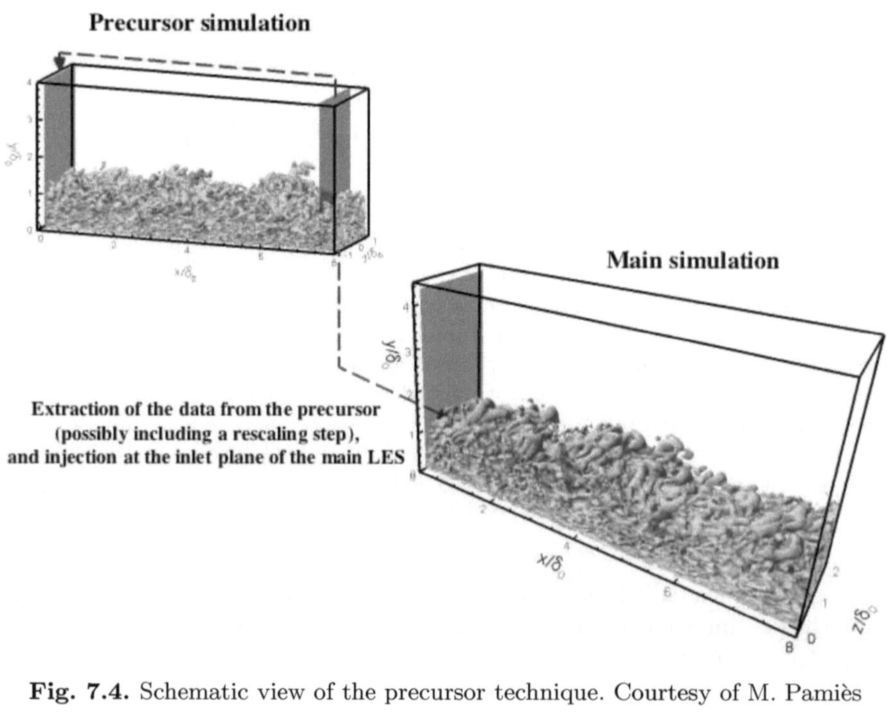

Fig. 7.4. Schematic view of the precursor technique. Courtesy of M. Pamiès

from the theoretical point of view, this technique is valid for strictly hyperbolic/parabolic flows only. But practical experience show that it can yield reliable results in all cases, at least if the emphasis is put on the vorticity modes.

Two different uses of the precursor simulations can be found [242]: one can extract either the full field at the inlet from the precursor simulation, or the fluctuating part only, the mean flow at the inlet being prescribed separately (and be deduced from RANS simulations, experimental measurements, ...). Let us note that, in the second case, the fluctuating field may be not fully adapted to the mean flow, leading possibly to the existence of a relaxation process downstream the inlet plane.

Two types of precursor simulations can be found in the literature: spatially developing simulations or time-developing ones. In the later case, the precursor simulation cost is very low since the size of the computational box is minimized thanks to the use of periodic boundary conditions. The inlet data of the main simulation are obtained slicing the precursor data in a series of plane, the distance between two consecutive planes being taken equal to $U_p \Delta t$, where U_p and Δt are a typical advection velocity and the time step of the main simulation, respectively. An additional problem arises here, since a single advection velocity for the solution in the precursor simulation is used, while there should be at last three: one for the upstream traveling acoustic modes, one for the downstream traveling acoustic modes and one for the vorticity/entropy modes. The problem is rendered even more complex if one takes into account that, in some shear flows, the advection velocity is scale-dependent. As an example, let us recall that in a mixing layer, the very large vortical structures whose diameter scales with the shear layer thickness travel at a speed which is roughly equal to the average speed across the shear layer, while very small scales are advected at the local mean speed. Not accounting for the existence of these different propagation speeds induces a disequilibrium at the inlet of the main simulation. But practical experience show that the use of a single advection speed for all vorticity modes, a problem which is already present in the incompressible case, does not preclude the recovery of reliable results, if the emphasis is put on aerodynamic variables. The induced errors on the entropy and the acoustic fields have not been investigated up to now.

7.3.3 Extraction-Rescaling Techniques

For incompressible flows, the turbulent inflow conditions can be generated by a rescaling technique proposed by Lund et al. [183]. The main idea of the rescaling-recycling method is to use rescaled data taken at a downstream station as inflow data (see Fig. 7.5). Therefore, the simulation also works as its own precursor. The main problem is to rescale the data extracted at the sampling plane. To be able to do that, one must know how the flow evolves between the inlet and extraction plane, in order to apply the relevant rescaling, which must be designed to obtain new perturbations that will effectively correspond to the targeted flow at the inlet location.

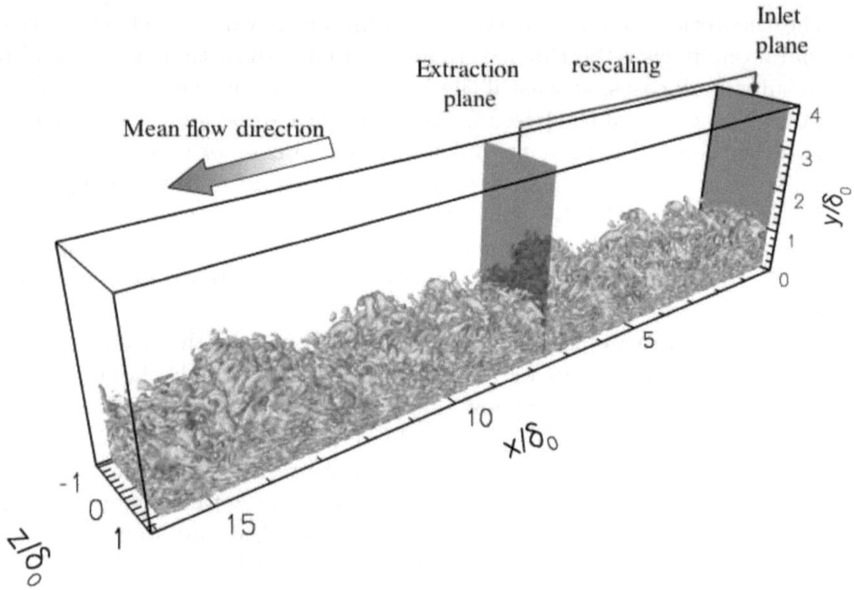

Fig. 7.5. Schematic view of Lund's extraction-rescaling technique. Courtesy of M. Pamiès

Therefore, the anticipation of the evolution law of the flow is a mandatory element to use this technique. In practice, only flows which exhibit a self-similar evolution have been treated. Let us first note that most of the time self-similar evolution laws are known for the mean velocity field only, and that instantaneous fluctuations do not exhibit such a behavior. The problem is a priori much more complex in compressible flows, since the acoustic and entropy fields may obey behaviors which may differ significantly from the one of the vorticity mode.

We now illustrate this point considering the case of the turbulent boundary layer, which is certainly the case that has been the most often addressed up to now. The idea is to decompose each flowfield component into a mean and a fluctuating part and then apply the appropriate scaling law to each one separately. A variable $u_i(x,y,z,t)$ is decomposed as the sum of an average in the spanwise direction and in time $U_i(x,z)$ and an instantaneous fluctuating part $u'_i(x,y,z,t)$ according to

$$u'_i(x,y,z,t) = u_i(x,y,z,t) - U_i(x,z). \tag{7.58}$$

The index i is equal to 1, 2 and 3 and we denote the streamwise, spanwise and wall-normal velocity components by u_1, u_2 and u_3, the corresponding coordinates being x, y and z. Then, in the rescaling method, the velocity at the downstream station to be rescaled (index *re*) can be linked with that at the inlet (index *in*) via the relations presented in Table 7.2. The mean

Table 7.2. Relation between inlet and rescaled velocities

Inner region
$U_{1,in}^{inner} = \beta U_{1,re}(z_{in}^+)$
$U_{2,in}^{inner} = 0$
$U_{3,in}^{inner} = U_{3,re}(z_{in}^+)$
$(u_i')_{in}^{inner} = \beta (u_i')_{re}(y, z_{in}^+, t)$
Outer region
$U_{1,in}^{outer} = \beta U_{1,re}(\eta_{in}) + (1-\beta)U_\infty$
$U_{2,in}^{outer} = 0$
$U_{3,in}^{outer} = U_{3,re}(\eta_{in})$
$(u_i')_{in}^{outer} = \beta (u_i')_{re}(y, \eta_{in}, t)$

and fluctuating velocities are decomposed into inner and outer variables. The rescaling factor $\beta = u_{\tau,in}/u_{\tau,re}$ is the ratio of friction velocities at the inlet station and at the recycled station. $z_{in}^+ = z u_{\tau,in}/\nu$ and $\eta_{in} = z/\delta_{in}$ are inner and outer coordinates at the inlet station.

Velocity profiles that hold across the entire inflow boundary layer are obtained thanks to a weighted average of inner and outer profiles

$$
\begin{aligned}
(u_i)_{in} = {} & [(U_i)_{in}^{inner} + (u_i')_{in}^{inner}][1 - W(\eta_{in})] \\
& + [(U_i)_{in}^{outer} + (u_i')_{in}^{outer}]\, W(\eta_{in}).
\end{aligned}
$$

The weighting function W is defined as

$$
W(\eta) = \frac{1}{2}\left(1 + \frac{\tanh[\frac{\alpha(\eta-b)}{(1-2b)\eta+b}]}{\tanh(\alpha)}\right), \tag{7.59}
$$

where $\alpha = 4$ and $b = 0.2$.

The rescaling operation requires the knowledge of the skin friction velocity u_τ and the boundary layer thickness δ, both at the rescaling station and at the inlet. These quantities can be determined easily from the mean velocity profile at the recycled station, but they must be specified at the inlet. Lund et al. propose to fix δ at the inlet, whereas u_τ is evaluated through the relation

$$
u_{\tau,in} = u_{\tau,re}\left(\frac{\theta_{re}}{\theta_{in}}\right)^{\frac{1}{[2(n-1)]}}, \tag{7.60}
$$

where θ is the incompressible momentum thickness and the exponent n is set equal to 5.

For compressible flows, extensions of the technique by Lund et al. have been presented in Refs. [295, 280, 252, 242]. The problem is more difficult since a rescaling law for two thermodynamic variables must be provided. The rescaling methods for compressible flow published in the literature are summarized in Table 7.3. According to Urbin et al. and Schröder et al., the pressure is assumed to be constant and a rescaling for ρ is not to be specified. S.1

Table 7.3. Summary of rescaling methods for compressible flow

	U	T	ρ	β
Stolz et al. [280]	S.1	S.3	S.4	$(\frac{\delta_{re}}{\delta_{in}})^{\frac{1}{8}}$
Urbin et al. [295]	S.2	S.3	no	$(\frac{\delta_{re}}{\delta_{in}})^{\frac{1}{10}}$
Schröder et al. [252]	S.1	S.5	no	$\sqrt{\frac{\rho_{(wall)re}}{\rho_{(wall)in}}}(\frac{\delta_{1re}}{\delta_{1in}})^{\frac{1}{8}}$

denotes the same rescaling strategy as Lund et al. for the mean streamwise velocity whereas S.2 uses the van Driest transformation of the velocity U_{VD} for the rescaling of this quantity

$$U_{VD,in}^{inner} = \beta U_{VD,re}(z_{in}^{+})$$

and

$$U_{VD,in}^{outer} = \beta U_{VD,re}(\eta_{in}) + (1-\beta)U_{VD}^{\infty},$$

where

$$U_{VD} = \frac{U_\infty}{A}\sin^{-1}\left(A\frac{U}{U_\infty}\right),$$

with

$$A = \sqrt{\frac{[(\gamma-1)/2]M_\infty^2 Pr_{tm}}{1+[(\gamma-1)/2]M_\infty^2 Pr_{tm}}}.$$

$Pr_{tm} = 0.89$ is the mean turbulent Prandtl number. The rescaling of the wall-normal velocity and fluctuating velocities is the same as the one of Lund et al. for all authors. Moreover, to obtain β, Urbin et al. impose the ratio δ_{re}/δ_{in} according the following classical law:

$$\frac{\delta_{re}}{\delta_{in}} = \left[1+\left(\frac{x_{re}-x_{in}}{\delta_{in}}\right)0.27^{\frac{6}{5}}Re_{\delta_{in}}^{-\frac{1}{5}}\right]^{\frac{5}{6}},$$

where $Re_{\delta_{in}}$ is the Reynolds number based on the inflow boundary-layer thickness.

Rescaling strategies S.3, S.4, S.5 are detailed in Table 7.4 where $C_1 = (1+A)(1-\beta^2)$, $C_2 = 2A\beta(1-\beta)$ and $C_3 = (1-\beta)[1+\beta+2A\beta]$. A similar technique, based on the decomposition of the turbulent boundary layer into four regions (instead of 2 in the methods discussed above) has also been proposed [316].

The extraction-rescaling method has several drawbacks. First, it holds only for flows which exhibit a self-similar evolution. Second, the self-similarity law should also be valid for instantaneous motion, which is not true. Third, self-similar scalings should be available for the three Kovasznay modes, and one must be able to extract and rescale each mode separately. At last, even in the incompressible case, spurious couplings between the inlet and the extraction planes have been reported, leading to unphysical periodic forcing of the flow

Table 7.4. Rescaling strategies for density and temperature

	Inner region	Outer region
S.3:	$T_{in}^{inner} = T_{re}(z_{in}^+)$	$T_{in}^{outer} = T_{re}(\eta_{in})$
	$T_{in}^{\prime inner} = T'_{re}(y, z_{in}^+, t)$	$T_{in}^{\prime outer} = T'_{re}(y, \eta_{in}, t)$
S.4:	$\rho_{in}^{inner} = \rho_{re}(z_{in}^+)$	$\rho_{in}^{outer} = \rho_{re}(\eta_{in})$
	$\rho_{in}^{\prime inner} = \rho'_{re}(y, z_{in}^+, t)$	$\rho_{in}^{\prime outer} = \rho'_{re}(y, \eta_{in}, t)$
S.5:	$T_{in}^{inner} = \beta^2 T_{re}(z_{in}^+) + C_1 T_e$	$T_{in}^{outer} = \beta^2 T_{re}(\eta_{in}) - C_2 \frac{U_{re}(\eta_{in})}{U_e} T_e + C_3 T_e$
	$T_{in}^{\prime inner} = \beta^2 T'_{re}(y, z_{in}^+, t)$	$T_{in}^{\prime outer} = \beta^2 T'_{re}(y, \eta_{in}, t) - C_2 \frac{u'_{re}(y,\eta_{in},t)}{U_e} T_e$

[209], even for very large separation distances between the inlet plane and the extraction location (up to 70 pipe radius for a turbulent pipe flow). The situation may be expected to be even worse in the compressible case, because of the acoustic modes.

In the methods described above, both mean and fluctuating fields are extracted and rescaled. A mean flow profile drift has been reported by several authors, which originates in the fact that the mean flow profile is not fixed, but extracted and rescaled. Therefore, it is subjected to error accumulation. To alleviate this problem, it is proposed to prescribe the mean flow component at the inlet plane, and to evaluate the fluctuating component by the extraction-rescaling method.

7.3.4 Synthetic-Turbulence-Based Models

The synthetic turbulence approach consists in generating a random field with prescribed statistical features (e.g. arbitrary fixed Reynolds stress distribution) to generate the fluctuating part of the flow at the inlet plane, the mean flow component being arbitrarily prescribed. Up to now, no work devoted to a fully compressible extension (i.e. including the acoustic and the entropy modes in a consistent way) of these methods has been published. Existing methods can be grouped into two families.

The first one encompasses all methods based on an initially white noise, which is post-processed to obtain prescribed statistical properties. Many procedures have been proposed during the last two decades. Many of them can be interpreted as variant of Kraichnan's original synthetic turbulence model [149]. In the first step, an isotropic velocity field is generated as the sum of N monochromatic fields according to the following formula

$$u(x) = 2 \sum_{n=1}^{N} \hat{u}_n \cos(k_n \cdot x + \psi_n) \sigma_n, \tag{7.61}$$

where $\hat{u}_n$, k_n, ψ_n and σ_n are the amplitude, wave vector, phase and direction in physical space of mode n. The three last parameters are random variables with uniform distribution. The amplitude parameter $\hat{u}_n = \sqrt{E(k_n)\Delta k_n}$ is computed by to the prescribed kinetic energy spectrum $E(k)$. Δ_k is the thickness of the associated shell in the Fourier space.

The resulting field is statistically isotropic and time-independent. Several methods have been proposed to obtain a time-dependent synthetic turbulent field. A first method consists in adding convective effects in the definition of the synthetic field [12]

$$u(x,t) = 2\sum_{n=1}^{N} \hat{u}_n \cos(k_n \cdot (x - u_c t) + \psi_n + \omega_n t)\sigma_n, \tag{7.62}$$

where u_c is a local convection velocity vector and ω_n a random frequency, respectively. The N random fields are generated once for all at the initial time, and are then advanced in time by changing t in the formula.

A second method consists in generating a new set of N modes every time step and then to filter it in time to enforce the targeted time-correlation [17, 22]

$$u(x, n\Delta t) = au(x, (n-1)\Delta t) + b(v_n(x) + v_{n-1}(x)), \tag{7.63}$$

in which $v_n(x)$ denotes the new random field generated at time step n according to (7.61). Denoting t_c the time separation for which the autocorrelation is reduced to $\exp(-1)$, one has $a = \exp(-\Delta t/t_c)$. To preserve a constant kinetic energy after the filtering step, one must take $b = \sqrt{(1-a)/2}$.

Let us now address the issue of generating an anisotropic field with Reynolds stress tensor R. Let the superscript $*$ denotes the principal coordinate system related to R. One has

$$R^* = Q^T R Q, \tag{7.64}$$

in which R^* is a diagonal matrix and Q is a matrix whose columns are the three eigenvectors of R. In a similar way, starting from the dimensionless stress tensor

$$A = -\frac{3}{2}\frac{1}{\overline{\rho}\bar{k}} R, \tag{7.65}$$

where $\bar{k}$ is the mean turbulence kinetic energy, one can compute $A^* = Q^T A Q$. The anisotropic velocity field $u(x)$ with associated Reynolds stress R is then generated according to the following formula

$$u(x) = Q u^{a*}(x^*), \tag{7.66}$$

with

$$u^{a*}(x^*) = 2\sum_{n=1}^{N} \hat{u}_n \cos(A^{*-1/2} k_n^* \cdot x^* + \psi_n) A^{*1/2} \sigma_n^*. \tag{7.67}$$

The second family gathers all methods which rely on a random combination of deterministic coherent events. An illustrative example is the approach developed by Sandham and coworkers to generate vorticity modes at the inlet plane in a turbulent compressible boundary layer [248]. To the knowledge of the authors, other methods belonging to this family have not yet been used for compressible flow simulation. This method is based on the observation that fluctuations in the inner and outer parts of the turbulent boundary layer have different characteristic scales. Therefore, ad hoc disturbances are introduced in each part of the boundary layer. The inner-part fluctuations, $\mathbf{u}'^{\text{inner}}$ are designed to represent lifted streaks with an energy maximum at altitude $z^+_{p,j}$:

$$u_i'^{\text{inner}}(y,z,t) = c_{ij} \exp(-z^+/z^+_{p,j}) \sin(\omega_j t) \cos(k_{y,j} y + \phi_j). \tag{7.68}$$

The outer-part fluctuation is assumed to be of the following form (with a peak at $z_{p,j}$)

$$u_i'^{\text{outer}}(y,z,t) = c_{ij} \frac{z}{z_{p,j}} \exp(-z/z_{p,j}) \sin(\omega_j t) \cos(k_{y,j} y + \phi_j), \tag{7.69}$$

where subscripts $i = 1, 2$ and j are related to the velocity component and to the mode indices, respectively, and c_{ij} are constants. The + superscript refers to inner coordinates (wall units). The ϕ_j, ω_j and $k_{y,j}$ are phase shifts, forcing frequencies, and spanwise wave numbers, respectively. These parameters are tuned using information on the boundary-layer dynamics. The spanwise velocity component is deduced from the continuity constraint.

In the inner region of the boundary layer, it is assumed that the disturbances travel downstream for a distance of 1000 wall units at a convective velocity $U_c \approx 10\, u_\tau$, where u_τ is the average friction velocity. The wave numbers $k_{y,j}$ are chosen such that there will be four streaks with a typical characteristic length of 100 wall units. In the outer region, the downstream traveling distance is taken equal to 16 and the convection velocity is $U_c \approx 0.75\, U_\infty$, where U_∞ is the external velocity. The spanwise wave number is chosen to be of the order of the spanwise extent of the computational domain.

This procedure was applied to a turbulent boundary layer, taking one mode in the inner region and three in the outer region. Corresponding parameters are given in Table 7.5. Sandham and coworkers also add a random noise with a maximum amplitude of 4% of the external velocity to preclude possible spurious symmetries.

Table 7.5. Coefficients of the four-mode Sandham model of fluctuations for boundary-layers

	j	c_{1j}	c_{2j}	ω_j	$k_{y,j}$	ϕ_j	$z^+_{p,j}$	$z_{p,j}$
inner region	0	0.1	−0.0016	0.1	π	0	12	–
outer region	1	0.3	−0.06	0.25	0.75π	0	–	1
outer region	2	0.3	−0.06	0.125	0.5π	0.1	–	1.5
outer region	3	0.3	−0.06	0.0625	0.25π	0.15	–	2.0

This approach has been further improved by the Synthetic Eddy Models. The concept of Synthetic Eddy Model was first proposed by Sergent and

Bertoglio [255] and then further developed by Jarrin et al. [132]. Following this approach, fluctuations are reconstructed generating a random array of coherent vortices, whose shape is given analytically. These vortices are passively advected through the inlet plane at an arbitrarily prescribed velocity, and both their size and intensity are tuned so that the targeted Reynolds stress profiles are recovered. This technique was further improved in the case of subsonic turbulent boundary layer by Pamiès and coworkers [211]. The modification is twofold: the boundary layer is split into four region (while only one region was considered in previous variants), and the coherent vortices definition is changed in each region to mimic coherent structures which are observed in direct numerical simulations and laboratory experiments.

The general formulation for the inlet velocity fluctuation is

$$u_i'(y,z,t) = \sum_j A_{ij}\tilde{u}_j(y,z,t), \tag{7.70}$$

where the matrix A is related to the Cholesky decomposition of the Reynolds stress tensor: $R = A^T A$. The synthetic field $\tilde{u}$ is defined as follows:

$$\tilde{u}_j(t,y,z) = \sum_{p=1}^{P} \frac{1}{\sqrt{N(p)}} \sum_{k=1}^{N(p)} \varepsilon_k \Xi_{jp}\left(\frac{t - t_k - \ell_p^t}{\ell_p^t}\right) \Phi_{jp}\left(\frac{y - y_k}{\ell_p^y}\right) \Psi_{jp}\left(\frac{z - z_k}{\ell_p^z}\right), \tag{7.71}$$

where:

- ε_k: random sign.
- P: number of modes ($P = 4$ in the turbulent boundary layer case, as in Sandham's model).
- $N(k)$: number of structures generated which correspond to mode $k = 1, P$. In practice, one uses $N(k) = S_k/S_S$, where S_k is the area of the region in which mode k is defined and S_S is the area of the cross-section of the vortex associated with mode k.
- $z_p^{\text{low}} - z_p^{\text{up}}$: Wall-normal extent of mode p.
- c_p: Convection velocity of mode p.
- ℓ_p^x (or ℓ_p^t): Streamwise (or temporal) scale for mode p.
- ℓ_p^z: Wall-normal scale for mode p.
- ℓ_p^y: Transverse scale for mode p.
- Ξ_{jp}, Φ_{jp}, Ψ_{jp}: Shape functions of mode p in time, direction y and direction z, respectively.
- σ_p: Scale parameter of the Gaussian shape function for the temporal support of mode p.
- $(t_k; y_k; z_k)$: randomly chosen coordinates within the domain $[t; t + \ell_p^t] \times [z_p^{\text{low}}; z_p^{\text{up}}] \times [-L_y/2; L_y/2]$. The region in which mode p exists is $[z_p^{\text{low}}; z_p^{\text{up}}] \times [-L_y/2; L_y/2]$. t_k is a random time of appearance assigned to each vortex.

All these parameters must be tuned to mimic physical coherent vortices. Following Pamiès, the four modes are:

Table 7.6. Locus of the center, sizes and convection velocity of turbulent structures associated with modes 1, 2 (expressed in wall units), 3 and 4 (in external units). δ and U_∞ are the boundary layer thickness and the external velocity, respectively

Mode	$(z_p^{\rm low})^+$	$(z_p^{\rm up})^+$	$(\ell_p^x)^+$	$(\ell_p^y)^+$	$(\ell_p^z)^+$	c_p^+
$p=1$	0	60	100	30	20	15
$p=2$ (legs)	60	$0.5\delta^+$	120	60	60	15
$p=2$ (head)	60	$0.5\delta^+$	60	120	60	15
Mode	$z_p^{\rm low}$	$z_p^{\rm up}$	ℓ_p^x	ℓ_p^y	ℓ_p^z	c_p
$p=3$	0.5δ	0.8δ			0.1δ	$0.8U_\infty$
$p=4$	0.8δ	1.5δ			0.15δ	$0.8U_\infty$

- $p=1$: Streamwise elongated vortices observed in the viscous sublayer and the buffer layer. Associated geometric and kinematic parameters are given in Table 7.6. The shapes functions are given in (7.72):

$$\Xi_{1p}(\tilde{t}) = A(\sigma_p)e^{-\frac{\tilde{t}^2}{2\sigma_p^2}}, \qquad \Phi_{1p}(\tilde{y}) = A(\sigma_0)e^{-\frac{\tilde{y}^2}{2\sigma_0^2}}, \qquad \Psi_{1p}(\tilde{z}) = \frac{cos(2\pi\tilde{z})-1}{2\pi\tilde{z}\sqrt{C}}, \tag{7.72}$$

$$\Xi_{2p}(\tilde{t}) = A(\sigma_p)e^{-\frac{\tilde{t}^2}{2\sigma_p^2}}, \qquad \Phi_{2p}(\tilde{y}) = A(\sigma_0)e^{-\frac{\tilde{y}^2}{2\sigma_0^2}}, \qquad \Psi_{2p}(\tilde{z}) = -\frac{cos(2\pi\tilde{z})-1}{2\pi\tilde{z}\sqrt{C}} \tag{7.73}$$

$$\Xi_{3p}(\tilde{t}) = A(\sigma_p)e^{-\frac{\tilde{t}^2}{2\sigma_p^2}}, \qquad \Phi_{3p}(\tilde{y}) = \frac{cos(2\pi\tilde{y})-1}{2\pi\tilde{y}\sqrt{C}}, \qquad \Psi_{3p}(\tilde{z}) = A(\sigma_0)e^{-\frac{\tilde{z}^2}{2\sigma_0^2}} \tag{7.74}$$

where $C \approx 0.214$ is a normalization factor set and $\sigma_0 = \frac{1}{3}$ is the reference scale parameter.
- $p=2$: hairpin vortices observed in the logarithmic layer and the wake region. Geometric and kinematic parameters are displayed in Table 7.6. Hairpin vortices are split into leg component (in the region $60 < z^+ < 0.4\delta^+$) and head component (in region $0.4\delta^+ < z^+ < 0.5\delta^+$). Related shapes functions are displayed in (7.72).
- $p=3$ and $p=4$: nearly isotropic structures in the wake region. Associated parameters are displayed in Table 7.6, while shape functions are defined as follows:

$$\Xi_{1p}(\tilde{t}) = A(\sigma_p)e^{-\frac{\tilde{t}^2}{2\sigma_p^2}}, \qquad \Phi_{1p}(\tilde{y}) = -\frac{cos(2\pi\tilde{y})-1}{2\pi\tilde{y}\sqrt{C}}, \qquad \Psi_{1p}(\tilde{z}) = A(\sigma_0)e^{-\frac{\tilde{z}^2}{2\sigma_0^2}}, \tag{7.75}$$

$$
\begin{aligned}
\Xi_{2p}(\tilde{t}) &= -\tilde{t}A(\sigma_p)e^{-\frac{\tilde{t}^2}{2\sigma_p^2}}, \qquad \Phi_{2p}(\tilde{y}) = A(\sigma_0)e^{-\frac{\tilde{y}^2}{2\sigma_0^2}}, \\
\Psi_{2p}(\tilde{z}) &= A(\sigma_0)e^{-\frac{\tilde{z}^2}{2\sigma_0^2}},
\end{aligned}
\tag{7.76}
$$

$$
\begin{aligned}
\Xi_{3p}(\tilde{t}) &= A(\sigma_p)e^{-\frac{\tilde{t}^2}{2\sigma_p^2}}, \qquad \Phi_{3p}(\tilde{y}) = A(\sigma_0)e^{-\frac{\tilde{y}^2}{2\sigma_0^2}}, \\
\Psi_{3p}(\tilde{z}) &= \tilde{z}A(\sigma_0)e^{-\frac{\tilde{z}^2}{2\sigma_0^2}},
\end{aligned}
\tag{7.77}
$$

8

Subsonic Applications with Compressibility Effects

This chapter is dedicated to the presentation of subsonic applications including compressibility effects. Canonical flows such as homogeneous turbulence, channel flow and mixing layer flows which are essentially studied for the validation of new developments are first presented. Boundary layers, jets and cavity flows are discussed subsequently. It should be noticed that this presentation is not intended to be complete. The intention is to establish a link with the techniques presented in the former chapters and to summarize the required knowledge before a simulation can be carried out successfully. In most of the case, some general details about numerical methods and SGS models used in referenced papers are provided. For the sake of conciseness, these information are gathered in tables which make use of the nomenclature proposed in Table 8.1.

8.1 Homogeneous Turbulence

8.1.1 Context

Isotropic homogeneous turbulence is the simplest turbulent flow for which subgrid models can be validated. The physical description of this flow is precisely the one on which the vast majority of these models are based on. Moreover, the statistical homogeneity of the flow makes it possible to use periodicity boundary conditions for the computation, and highly-accurate numerical methods. Here, we restrict the discussion to freely decaying homogeneous turbulence since the derivation of forcing terms for compressible isotropic turbulence is still an open issue. The main difficulty here is that vorticity, entropy and acoustic waves can be forced independently, yielding an infinite number of possible flow regimes. At the initial time of these simulations, the turbulent kinetic energy is contained in the largest scales and then, as the energy cascade sets in, is directed toward the small scales and finally dissipated at the cutoff by the subgrid model.

E. Garnier et al., *Large Eddy Simulation for Compressible Flows*,
Scientific Computation,

Table 8.1. Abbreviation used this chapter

Numerical schemes for the convective fluxes	
Cn	nth-order accurate centered scheme
Con	nth-order accurate compact centered scheme
UBn	nth-order accurate upwind biased scheme
Jam	Jameson scheme
FCT	Flux corrected Transport scheme
ENO	Essentially Non Oscillatory scheme
TVD	Total Variation Diminishing scheme
CULD	Compact Upwind with Low Dissipation scheme
DRPn-$m$pt	nth-order accurate Dispersion Relation Preserving scheme using m points
MC(n, m)	m is the spatial order for Mc-Cormack scheme
Spec	Spectral method
Filters (generally added to convective fluxes)	
+eFn	nth order-accurate explicit filter
+iFn	nth order-accurate implicit filter
Time integration schemes	
RKn	n steps Runge-Kutta scheme
BWn	nth-order accurate backward scheme
MC(n, m)	n is the temporal order for Mc-Cormack scheme
SGS models	
SM	Smagorinsky Model
DSM	Dynamic Smagorinsky Model
Sim	Similarity model
TDM	Tensor Diffusivity Model (gradient model or Clark model)
DCM	Dynamic Clark Model
MSM	Mixed Scale model
SMS	Selective Mixed Scale model
HSS	Hybrid Similarity Smagorinsky (mixed model)
DHSS	Dynamic Hybrid Similarity Smagorinsky (dynamic mixed model)
HSMS	Hybrid Similarity Mixed Scale model
HSSMS	Hybrid Similarity Selective Mixed Scale model
FSF	Filtered Structure Function model
HPFS	High Pass Filtered Smagorinsky model

The physics of compressible homogeneous turbulence has been investigated using DNS and LES has been mostly dedicated to the analysis of essential ingredients of such simulations i.e. the SGS model and the scheme used for the computation of the convective fluxes. The dissipative contribution of the latter can be considered as a model within the MILES framework (see Chap. 6).

8.1.2 A Few Realizations

Compressibility effects are characterized to first order by the turbulent Mach number defined as $M_t = q/\langle c \rangle$ where q is twice the turbulent kinetic energy

$\langle u'_i u'_i \rangle$, and $\langle c \rangle$ is the average speed of sound. The rms Mach number can also be used [323, 91]. The difference between these two Mach numbers is not significant in practice and their definitions coincide at the initial time if the temperature fluctuations vanish. We employ these two quantities interchangeably in Table 8.2. Two other parameters can drastically influence the temporal evolution of such simulations: the ratio of the compressible turbulent kinetic energy over the turbulent kinetic energy $\chi = \mathcal{K}_d/(\mathcal{K}_s + \mathcal{K}_d)$ and the initial thermodynamical state. The effect of the latter parameter is discussed extensively in Ref. [247]. The choice of lower bounds for M_t and χ from which compressibility effects become appreciable is subject to interpretation. Here, we restrict the presentation to the *low Mach number quasi isentropic regime* and to the *nonlinear subsonic regime* both defined in Sects. 3.5.1 and 3.5.2 respectively. This means that shocklets may appear but their contribution to the dissipation remains reasonably weak. We have then reported simulations satisfying one of the two following conditions: $0.2 \leq M_t \leq 0.6$ or $\chi \geq 0.1$. The *supersonic regime* will be addressed in Chap. 9.

Table 8.2. Characteristics of LES of homogeneous turbulence

Ref.	M_t	Re	χ	Grid
[201]	0.4	?	0.2	32^3
[75]	0.1	26.3 (λ)	0.2	32^3
[323]	0.1	27 (λ)	0.2–0.8	32^3
[305]	0.6	50 (λ)	0.1	48^3
[274]	0.2; 0.4; 0.6	735; 2157; 2742; 6170 (L)	0; 0.1; 0.2	32^3
[276]	0.4; 0.6	2157; 2742; 6170 (L)	0; 0.2	32^3
[91]	0.2–0.5	∞	0.–0.05	64^3, 128^3
[303]	0.4	2157 (L)	0.2	32^3, 64^3, 128^3
[147]	0.4–0.488	153–175	≈ 0	32^3, 48^3, 64^3
[121]	0.3	50 to ∞	0	32^3, 64^3

The length scale used in the Reynolds number definition of Table 8.2 can be either the integral scale L or the Taylor microscale (λ). The resolution of the computations was generally low (typically 32^3 or 64^3) to permit comparisons with DNS simulations performed at reasonable computational cost.

8.1.3 Influence of the Numerical Method

Table 8.3 summarizes the numerical methods employed in each of the aforementioned paper.

The effect of the temporal integration scheme was only studied by Visbal and Rizzetta [303]. It was observed comparing an explicit second-order accurate Runge-Kutta scheme and first and second-order accurate implicit schemes that a first-order discretization significantly affects the dissipation rate of turbulent kinetic energy. The same authors have also investigated the spatial dis-

Table 8.3. Numerical methods used for LES of homogeneous turbulence

Ref.	Spatial scheme	Temporal scheme	SGS model
[201]	Co6	RK3	DSM
[75]	Spec	RK3	HSS
[323]	Spec	RK3	HSS
[305]	C2[1] and C4	RK4	SM
[274]	Spec	RK3	DSM
[276]	Spec	RK3	ADM
[91]	C4; ENOs; TVD; Jameson;	RK3	SM; DSM
[303]	Co6+iF10; Co4+eF8; C2; C4+eF8; Roe	RK4; BW1; BW2	SM; DSM
[147]	C2; Co4; Co6; Co10	?	Nonlinear model
[121]	Co6	RK3	DSM

[1] 4 different implementations of the convective terms were assessed

cretization effect. Some filtered compact, centered and upwind biased schemes of different order of accuracy were compared. Figures 8.1 and 8.2 show respectively the discretization effect on the turbulent kinetic energy decay and on turbulence spectra. It is found that filtered compact schemes of fourth and sixth order fit well with the reference DNS data obtained with a spectral method. As expected, third-order accurate upwind Roe scheme dissipates too much energy but centered schemes with explicit filters (C2+Jameson dissipation and C4+eF8 denoted respectively E2 and E4F8 in Figs. 8.1 and 8.2) are also over-dissipative. As shown in Fig. 8.2, these schemes significantly affect the smallest scales. One should note that, despite the demanding initial conditions ($M_t = 0.4$, $\chi = 0.2$), the very low Reynolds number chosen in this study guarantees the stability of the solution. Furthermore, the influence of filters of different order of accuracy associated with a sixth-order accurate compact scheme was also investigated. An over-dissipation was registered for filter orders lower than 6, second order filters being to preclude.

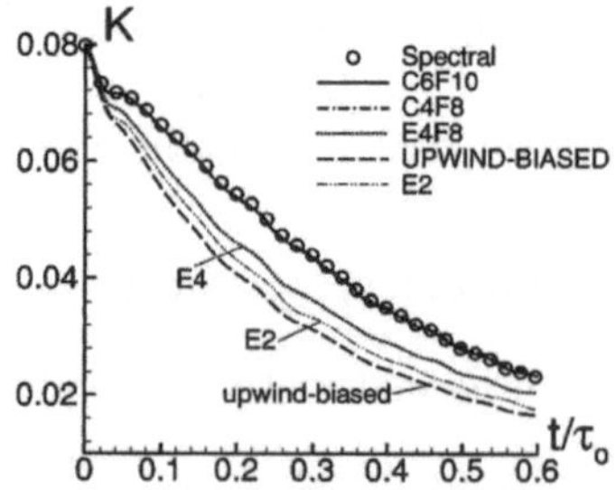

Fig. 8.1. Effect of spatial discretization on the time history of turbulent kinetic energy. 32^3 grid (from [303] with permission)

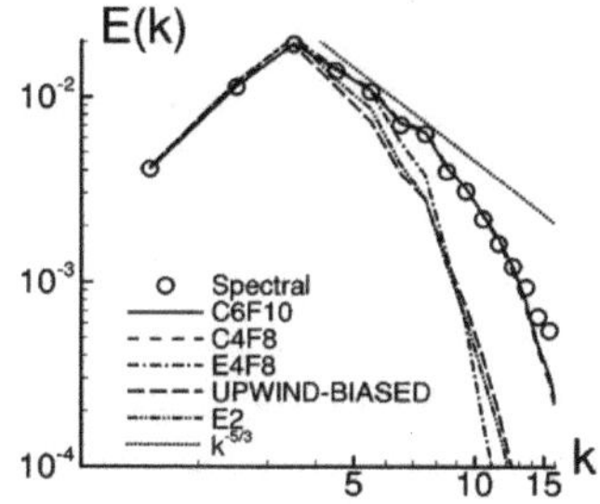

Fig. 8.2. Effect of spatial discretization on the instantaneous three dimensional energy spectra at $t/\tau_0 = 0.2985$. 32^3 grid (from Ref. [303] with permission)

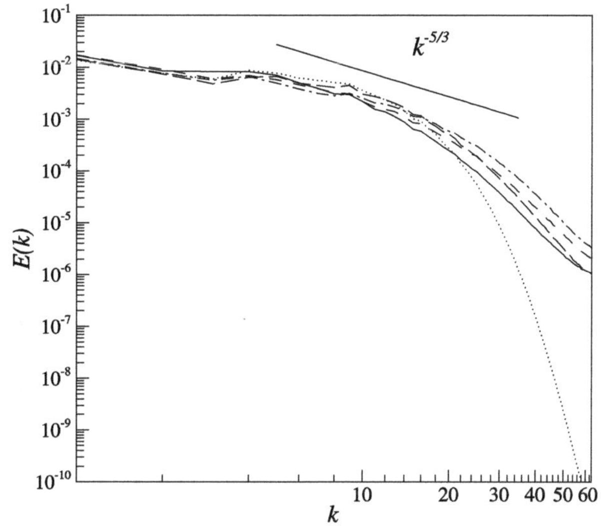

Fig. 8.3. Effect of some classical shock capturing scheme on the instantaneous three-dimensional energy spectra from Ref. [91]. *Continuous line*: ENO, *Dashed line*: WENO, *Dashed-dotted line*: Modified ENO, *Dotted line*: Jameson, *Long dashed line*: Roe-MUSCL compression factor = 4, *Dashed dotted dotted line*: Roe-MUSCL compression factor = 1, 128^3 grid

The capacity of some common shock-capturing schemes for reproducing the proper dynamic of homogeneous turbulence was studied in reference [91]. Jameson, TVD MUSCL, third-order ENO, fourth-order weighted ENO (WENO), and fifth-order modified ENO (MENO) schemes were employed to compute such flow without any physical viscosity. One can then expect to find an inertial range with the theoretical $-5/3$ slope. Nevertheless, it is observed in Fig. 8.3 that despite the presence of a limited $-5/3$ subrange at medium scale, the small scales are strongly affected by the intrinsic dissipation of these schemes. This is particularly true for the Jameson scheme which does not leave any energy at the cut-off. We will consider the consequence of this observation on SGS modeling in the next subsection.

Two papers discuss the numerical formulation of the convective term. Dealing with second-order accurate centered schemes, Vreman et al. [305] recommend the use of a Simpson cell-vertex discretization for both convective and viscous terms in order to reproduce accurately the $-5/3$ slope of the Kolmogorov law. Honein and Moin [121] have assessed different implementations of the skew-symmetric form of the convective terms. A technique that conserves intrinsically the entropy flux remains stable for arbitrary large Reynolds numbers whereas the other implementations leads to a divergence of simulations for finite values of Re_λ (typically 300).

8.1.4 SGS Modeling

Zang et al. [323] and Erlebacher et al. [75] have assessed the SEZH model [273] (which can be understood as a mixed model). In particular, Zang et al. have studied the effect of large variations of C_I (see discussion Sect. 4.3) on a case where $M_t = 0.1$ and $\chi = 0.2$. They did not find any influence of C_I in the range $[0.0066 - 0.066]$. The recommended value of the square of the Smagorinsky constant is $0.012/\sqrt{2} \approx 0.0085$.[1] The SGS Prandtl number is fixed to 0.7. Multiplying it by 1.5 or taking half of its value has little effect on the solenoidal energy spectrum. Nevertheless, both the compressible spectrum and the temperature fluctuations are affected by a change in SGS Prandtl number value. Unfortunately, an improvement on one of these quantities is paid by a loss on the other. Nearly simultaneously, Moin et al. [201] have proposed an extension of the dynamic model for compressible flows. This model gives results which agree well with their DNS. In a $M_t = 0.4$ case, their values of the dynamic constants converge toward $C_d = 0.0125$, $C_I = 0.0175$ and $Pr_{sgs} = 0.4$. These values depend both on the resolution and on the implementation of the test (second level) filter.

Spyropoulos and Blaisdell [274] have investigated the effect of some important parameters related to the implementation of the dynamic model. First, they have demonstrated that the Lilly's contraction for the computation of C_d (see (4.41)) is superior to the original one (see (4.40)) which is slightly too dissipative for the smallest scales. As mentioned in Chap. 4, they have also investigated the influence of the ratio $\hat{\Delta}/\Delta$ in the range $[16/10; 16/6]$. A change of this parameter only affects the density fluctuations by 3%. The optimal value is found equal to 2. Furthermore, these authors have also studied the influence of the implementation of the second level filter F. They have found that low-order explicit filters such as (2.49) and (2.50) introduce an additional dissipation on the smallest scales by increasing the dynamic model 'constant'.

[1] One can notice that Erlebacher et al. [75] have chosen the same value of the Smagorinsky constant but they have neglected C_I and taken a different value of Pr_{sgs}: 0.5 instead of 0.7 in [323].

Implicit and high-order (7 points) explicit filters do not affect significantly the energy spectra with respect to a sharp cut-off filter.

More recent approaches like the ADM was proven by Stolz and Adams [276] to be efficient in reproducing both density fluctuations and spectra for large turbulent Mach number (0.4 and 0.6). Furthermore, advanced models like the nonlinear and stretched vortex models proposed by Kosovic et al. [147] which agree well with the filtered DNS but do exhibit an energy pile-up at the cut-off which may be related to the unfiltered spatial compact scheme.

In some important cases, it is not relevant to add a SGS model to the simulation. For example, it is found in [91] that SGS models should not be added to shock-capturing schemes which are already naturally over-dissipative. Besides, in the particular case of a very low Reynolds number LES for which their compact scheme is regularized by high order filters, Visbal and Rizetta [303] came to the same conclusion.

8.2 Channel Flow

8.2.1 Context

The time-evolving plane channel flow is a flow between two infinite parallel flat plates having the same velocity. The time character is due to the fact that we consider the velocity field as being periodic in both directions parallel to the plates. Since the pressure is not periodic, a forcing term corresponding to the mean pressure gradient is added in the form of a source term in the momentum and energy equations. This academic configuration is used for investigating the properties of a turbulent flow in the presence of solid walls, and is a popular test case. Turbulence is generated near each wall. This production mechanism must be accurately recovered to obtain reliable results. To do so, the grid has to be refined near the surfaces, which raises additional numerical problems with respect to the homogeneous turbulence. Moreover, the subgrid models must be able to preserve these driving mechanisms. The flow topology is illustrated by an iso-value of the Q criterion colored by the streamwise vorticity in Fig. 8.4. The near wall region is populated by quasi-streamwise vorticies whose dynamics has to be accurately computed.

8.2.2 A Few Realizations

Two papers devoted to LES of channel flow were published by Lenormand and coworkers [168, 169]. They have treated both a subsonic and a supersonic case. We will focus here on the second reference which is more comprehensive on the subsonic Mach number case. The supersonic case will be discussed in Chap. 9. The study was later complemented by Mossi and Sagaut [206] using the same flow conditions with the aim of investigating the properties of some numerical schemes used in industrial codes. Furthermore, Terracol

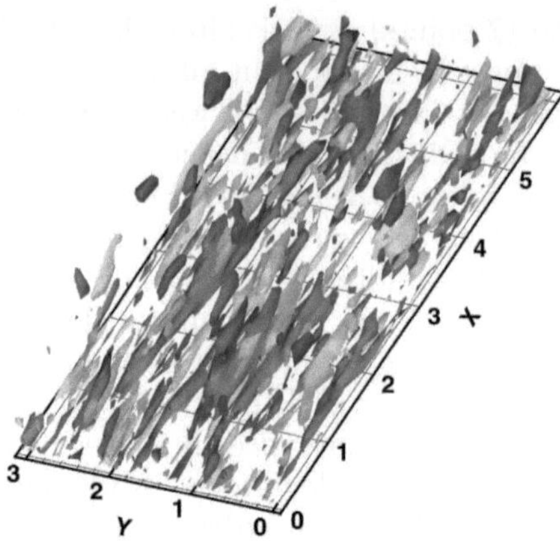

Fig. 8.4. Visualization of the Quasi Streamwise Vortices in a plane channel flow ($Re_\tau = 590$). *Dark* and *light gray* denote opposite sign of streamwise vorticity (Courtesy of M. Terracol)

et al. [287] have validated their multilevel approach on this test case at two different Reynolds numbers.

Table 8.4 summarizes the flow conditions of the aforementioned studies. In this table, the Mach number is based on the bulk velocity and the Reynolds number is based on the channel height. The low-Reynolds number case corresponds to a friction Reynolds number equal to 180 which permits a comparison with the incompressible results of Kim et al. [142], the high Reynolds number case can be compared with the DNS of Moser et al. [205].

Table 8.4. Characteristics of LES of subsonic channel flows

Ref.	Ma	Re_h	Wall boundary condition
[169]	0.5	3000	Isothermal
[287]	0.5	2800	Isothermal
[287]	0.5	11000	Isothermal
[206]	0.5	3000	Isothermal

Each presented study makes use of isothermal wall boundary condition. The interested reader can refer to [312] for more information about LES of channel flow with wall injection which is connected to application issues concerning rocket internal flows.

8.2.3 Influence of the Numerical Method

Some general elements about numerical methods used in channel-flow LES are summarized in Table 8.5.

Table 8.5. Numerical methods used for LES of subsonic channel flows

Ref.	Spatial scheme	Temporal scheme	SGS model
[168]	C4	RK3	SM, MSM, HSS, HSMS, SMS, HSSMS
[287]	C2	RK3	DSM, Multilevel
[206]	C2, Jam, Roe	RK4	MILES, DSM

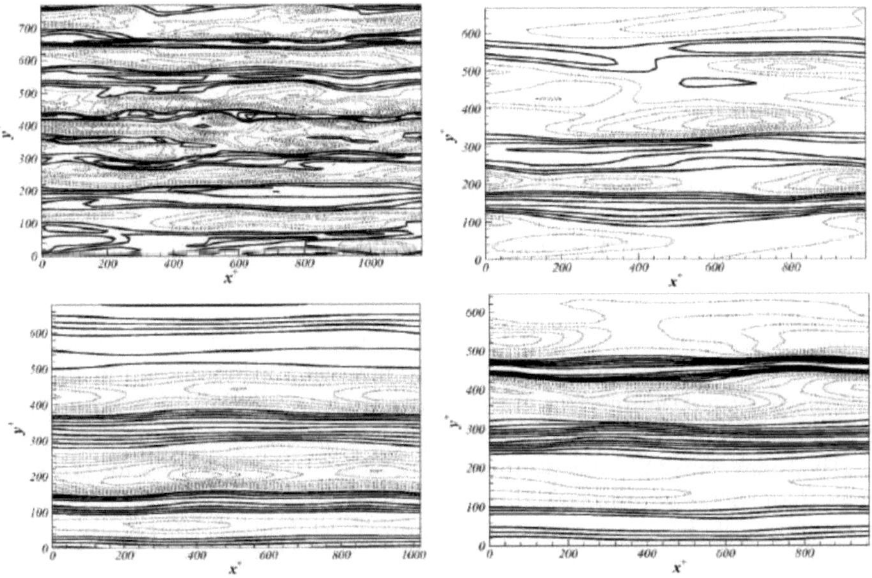

Fig. 8.5. Iso-contours of the fluctuating longitudinal velocity in the $y^+ = 10$ plane. *Top left*: C2+DSM, *top right*: Jameson, *bottom left*: Jameson +DSM, *bottom right*: Roe. Reproduced from [206] with permission

In their study, Mossi and Sagaut put the emphasis on the characterization of the dissipative behavior of classically employed shock-capturing schemes. A dynamic model is added to a centered scheme, a Jameson scheme with matrix dissipation, a Jameson scheme in its standard form and a Roe scheme. Roe TVD and classical Jameson scheme plus dynamic model suffer from a pathological behavior with very coherent streaky structures which stretch longitudinally over the whole computational domain (see Fig. 8.5). Additionally, the authors show that the addition of a SGS model to the Jameson scheme (in standard or matrix dissipation forms) does not affect turbulent kinetic energy distribution in the turbulence production zone. This shows that the level of dissipation introduced in these simulations is almost independent of the presence of a SGS model. This effect is also evident when using an extension to inhomogeneous flows of the generalized Smagorinsky constant concept [91]. The authors then recommend not to add a SGS model to an intrinsically dissipative numerical method. Such straightforward implicit LES with shock-capturing schemes generally are found to be inferior to "classical" explicit LES, but on this particular flow case, the authors show that even better results are achieved suppressing the model.[2]

In the low-Reynolds-number case, it is shown that a second-order accurate centered scheme [168] can give better results than a fourth-order accurate one [206]. This counter-intuitive finding is not explained. In the supersonic case discussed in Sect. 9.2 fourth order accurate and second-order accurate schemes perform equally well. It is furthermore noteworthy that with the fine grid resolution employed ($\Delta x^+ \approx 27$ and $\Delta y^+ \approx 12$) the performance of the different SGS models cannot be differentiated.

8.2.4 Influence of the SGS Model

Figure 8.6 presents the fluctuating longitudinal velocity profiles obtained on the coarser grid of Ref. [169] ($\Delta x^+ \approx 57$ and $\Delta y^+ \approx 20$). The Smagorinky, Mixed scale and Selective Mixed scale models are compared to their Hybridized counterparts (the hybridization is performed with a classical scale similarity model). On the longitudinal fluctuations, the hybridization improves the results of all the models. This conclusion can be generally extended to other fluctuating quantities but noticeable exceptions were found. For example, the Smagorinsky and SMS models are closer to DNS results for the shear stress than their hybridized counterpart. Furthermore, it is generally observed that the use of the selection function improves the results of the mixed scale

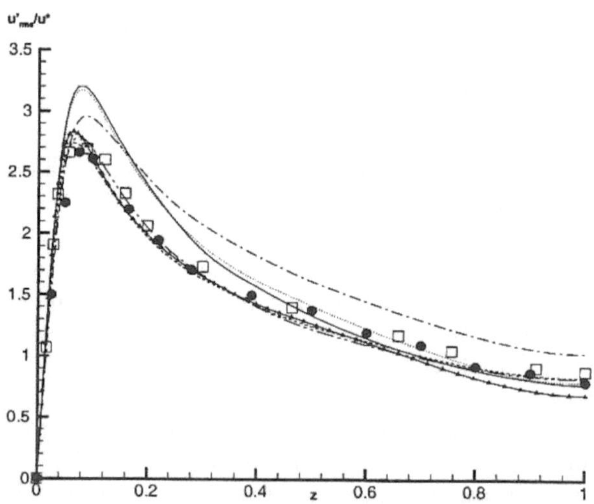

Fig. 8.6. Profile of longitudinal velocity fluctuation for a Mach = 0.5 channel flow. •: DNS; □: Exp.; *dotted line*: SM; *dashed line*: HSS; *dash-dotted line*: MSM; *dash-dot-dot line*: HSMS; *full line*: SMS; *triangle*: HSSMS (Courtesy of E. Lenormand)

[2] It is then an under-resolved DNS.

model. Once complemented with this function, the SMS model is found to be more accurate than the Smagorinsky model without any need for a damping function accurate which is generally added to this model to limit its over-dissipative behavior near walls.

Furthermore, Terracol et al. have shown that their multilevel approach gives results of equivalent quality as the one obtained with a dynamic Smagorinsky model for a computational cost 2 to 4 times smaller. An adaptive dynamic algorithm proposed in [288] which guarantees the validity of the quasi-static approximation used to justify the time advancement on the coarse grid trough a multigrid V-cycles allows up to a fivefold CPU-time reduction while controlling the accuracy of the method.

Depending on the resolution, the implementation of the convective terms, the effective precision of the method, conclusions on the relative performance of SGS models can be difficult to draw unambiguously. Furthermore, care should be taken to the normalization of the results which can also bias the analysis.

8.3 Mixing Layer

8.3.1 Context

The mixing layer is formed by two parallel flows with different velocities. No wall is present and this type of flow is the simplest free shear flow. The study of such a mixing zone is of application interest since it concerns every separated flows (base flow, backward facing step, cavity to quote a few) or free flows (jets and any type of wakes). The interface between the two streams contains an inflection point in the velocity profile, and it is therefore inviscidly unstable. Eddies are formed typical of a Kelvin-Helmholtz-type instability. In subsonic flows, the eddies grow rather quickly which leads to large spreading rate of the mixing layer. In the compressible regime, the spreading rate can be significantly reduced for values of convective Mach numbers larger than typically 0.6. This quantity is defined as: $M_c = (U_1 - U_2)/(a_1 + a_2)$ where the U_i and a_i are respectively the mean longitudinal velocities and the speeds of sound in each side of the interface. This effect remained unexplained for years but DNS performed in the nineties [212] have resulted in an explanation which accounts for the observation of reduced spreading rates at large convective Mach number.

8.3.2 A Few Realizations

Large Eddy Simulations of mixing layers in the compressible regime are quite rare since most of the work aiming at clarifying the role of compressibility have been performed using DNS. Moreover, the study of Vreman et al. [309] is very comprehensive and has only left a narrow place for further developments at least in terms of SGS model assessment. The study of Ragab et al. [226] worth to be mentioned since it gives results at higher convective Mach number. The work by Terracol et al. [288] is motivated by the assessment of the multilevel approach introduced by the authors. Table 8.6 summarizes the flow conditions of the aforementioned studies.

Table 8.6. Characteristics of LES of turbulent mixing layers

Ref.	Development	Ma	s	$Re(\delta_0)$
[226]	Temporal	0.4	1	∞
[309]	Temporal	0.2	1	100
[309]	Temporal	0.2	1	500
[58]	Spatial	0.64	0.57	$\approx$12000
[288]	Temporal	0.2	1	100
[288]	Temporal	0.2	1	∞

The temporal framework has been employed by most of the simulations to date. In this case periodic boundary conditions are used in streamwise direction, and the mean flow evolves in time. The Reynolds number is based on the initial vorticity thickness (δ_0). In all cases but [58], the density ratio s between the two sides of the mixing layer is equal to one.

8.3.3 Influence of the Numerical Method

Table 8.7 gathers some general elements about numerical methods used in the aforementioned papers.

Table 8.7. Numerical methods used for LES of plane mixing layers

Ref.	Spatial scheme	Temporal scheme	SGS model
[226]	MC(2, 4), Roe, C2	MC(2, 4), RK3	HSS
[309]	C4	RK4	SM, Sim, TDM, DSM, DHSS, DCM
[58]	TVD, WENO	RK3	MSM
[288]	C2	RK3	Multilevel, DSM

In the aforementioned temporal studies, the flow is initialized with a linear combination of 2D and 3D modes inferred from a stability analysis. The most unstable mode wavelength determines the numbers of rollers that will appear in the box in the early phase of the simulation.

An important difference between the studies of Vreman et al. and Terracol et al. comes from the fact that in the latter reference the authors have chosen a computational domain twice longer than the one of Vreman *et al.* It allows to initialize the flow such that 3 successive vortex pairings can be observed. This is an important point since a developed turbulent state is reached only after the second pairing with the establishment of a full turbulent spectrum. In their work, Vreman et al. had to stop their computations after the second pairing but before the fully turbulent state. This might have affected their conclusions on the relative merit of the different tested SGS models.

The paper by Ragab et al. [226] is mainly devoted to the comparison of numerical schemes (second-order accurate central difference, third order Roe scheme and fourth order Mac-Cormack scheme). This paper distinctly shows the influence of the numerical dissipation on the small scales. Nevertheless, in the particular case of the fourth-order accurate Mac-Cormack scheme, the use of a SGS model has been found mandatory to prevent the divergence of the solution. This scheme was found clearly less dissipative than the Roe scheme. A demonstration that the choice of the constant of the Smagorinsky model affects the slope of the energy spectrum can also be found in this paper. An other illustration of the spatial scheme dissipation can be found in [58] where a TVD scheme is observed to affect significantly the shear layer spreading rate whereas a less dissipative WENO scheme gives a slope in agreement with an existing experiment.

8.3.4 Influence of the SGS Model

In their paper, Vreman et al. have compared 6 popular SGS models with filtered DNS data. These models are the Smagorinsky, the similarity, the tensor diffusivity (TDM), the dynamic Smagorinsky, the dynamic mixed and the dynamic Clark (TDM+Smagorinsky) models. In the gradient model, a limiter insures that the model is always dissipative. The authors have concluded that dynamic models in general are superior to non-dynamic ones. Within dynamic models, dynamic mixed and dynamic Clark models are observed to be slightly more accurate than the dynamic Smagorinsky model in the $Re(\delta_\omega) = 100$ case. This conclusion is not general since the flow obtained with the latter model is found closer to a self similar state than the ones resulting from other dynamic models in the case of simulations carried out at much higher Reynolds number. The Smagorinsky model is observed to be too dissipative in the transitional regime. Conversely, the tensor diffusivity and similarity models are found not sufficiently dissipative for the smallest scales when the flow is fully developed.

These observations were later explained by Shao et al. [260] who have introduced a decomposition of the SGS stresses into a rapid component which depends explicitly on the mean velocity gradient and a slow component that does not. The scale similarity model reproduces the anisotropic backscatter features of the energy transfer mechanism associated with the rapid part, while the transfer related to the slow part, usually purely dissipative forward energy transfer, is adequately captured by the Smagorinsky model. These findings give an explanatory framework to justify the success of mixed models which behave appropriately both in transitional and fully turbulent flows.

8.4 Boundary-Layer Flow

8.4.1 Context

Boundary layer flows are very important from an application point of view. Nevertheless, both for historical and practical reasons, studies concerning channel flows are much more numerous than studies concerning boundary layers in the incompressible regime. In the subsonic regime, the change in the physics of the flow with respect to the incompressible case is not significant enough to justify computations dedicated to the study of compressibility effects. Nevertheless, some studies are of practical interest. For example, in refs. [291–293] Tromeur et al. have simulated a boundary layer flow at a Mach number of 0.9 in order to document aero-optic effects around airborne systems for aerial reconnaissance or target designation.

8.4.2 A Few Realizations

The first LES in the subsonic compressible regime was published by Ducros et al. in 1996 [62]. They have focused on the physical analysis of the transition process but in the last part of the simulation the turbulence is fully developed. Since the inflow is a perturbed laminar boundary layer, they did not deal with the problem of feeding the simulation with realistic turbulent fluctuations.

Later, after a first paper where temporal simulations[3] were employed [291], Tromeur et al. have implemented a compressible version of the Lund et al. [183] rescaling procedure to simulate spatially evolving turbulent boundary layers in spatial development [292, 293]. Their data are compared with available experiments.[4] Furthermore, Pascarelli et al. [213] have computed boundary layers at Mach numbers equal to 0.3 and 0.7. Nevertheless, their analysis is essentially focused on the evaluation of multi-block strategies capable of reducing the cost of a LES. It is not further detailed here.

Table 8.8 summarizes briefly the flow conditions for Refs. [62] and [292]. Some general elements about numerical methods used in the aforementioned papers are summarized in Table 8.9. As mentioned by Ducros et al. the resolution of their computation is "minimal". 32 points in the wall normal direction and only 20 points in the transverse one (which represent only one displacement thickness at the end of the domain) are only justified by the limited computational power available at that time. As a result, the agreement with canonical boundary layers is only qualitative (the friction coefficient is underestimated by 25%). Nevertheless, from a physical point of view, the transition scenario is clearly recovered. In particular, the role of the oblique subharmonic mode of the secondary instability is demonstrated. The first generation

[3] Simulations with periodic conditions in the streamwise direction in which the boundary layer grows in time.

[4] The last paper being mainly devoted to the analysis of a model for the phase fluctuations of electromagnetic waves in turbulent flows, we focus here on Ref. [292].

Table 8.8. Characteristics of LES of subsonic turbulent boundary layers

Ref.	Inflow fluctuation	Ma	$Re(\theta)$
[62]	Forced transition	0.5	380–1350
[292]	Recycling	0.9	2917

Table 8.9. Numerical methods used for LES of turbulent boundary layers

Ref.	Spatial scheme	Temporal scheme	SGS model	Grid size
[62]	MC(2, 4)	MC(2, 4)	FSF	4.2×10^5
[292]	C4	RK3	SMS	2.0×10^6

of SGS models (Smagorinsky, structure function model) is not able to tackle this kind of flow (they add to much dissipation and lead to a flow relaminarization). More advanced models, like the filtered structure function model are necessary to lead the transition process to its end.

The work by Tromeur et al. is more conservative in terms of resolution. 119 points are used in the wall normal direction and the spanwise extent of the domain is 1.3 times the boundary layer thickness. The grid sizes in each direction given in wall unit are $\Delta x^+ = 50$, $\Delta y^+ = 18$, $\Delta z^+_{min} = 1$. Independently from the application to aero-optics,[5] the study [292] illustrates the use of a rescaling method to introduce inflow turbulence (see Chap. 7). The Mach number is sufficiently high for large acoustic wave to be generated. The main issue in this case relies on the fact that the characteristics theory forbids the imposition of the 5 flow variables. Nevertheless, the passage in the characteristics space affects significantly the inflow field which drifts from the value computed by the rescaling procedure. This contradiction is difficult to solve. In the work by Tromeur et al., the velocity and the thermodynamic variables is forced to fit the values imposed by the rescaling procedure. This leads to small amplitude numerical oscillations which are damped by a tenth order explicit filter applied only in the longitudinal direction. The domain size has also been significantly lengthened with respect to supersonic simulations and a large sponge zone is added at the outflow to damp turbulent fluctuations and acoustic waves.

Figure 8.7 shows that density fluctuations are in reasonable agreement with experimental data obtained at much higher Reynolds number by the IUSTI team of J.P. Dussauge in the ONERA T2 wind tunnel. These data are difficult to measure and are rarely found in experimental papers in particular at transonic Mach numbers.[6]

[5] In terms of application, Ref. [293] has evidenced that the phase fluctuations of an electromagnetic wave which cross the boundary layer can be accurately estimated using LES.

[6] The incertitudes on these measurements were estimated to be about 10%.

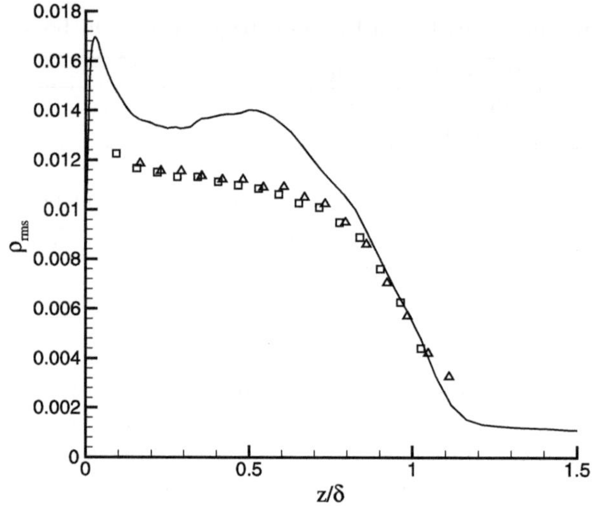

Fig. 8.7. Profile of density fluctuations in a $Ma = 0.9$ boundary layer. *Straight line*: LES; *square and triangle*: hot wire data (Courtesy of E. Tromeur)

8.5 Jets

8.5.1 Context

Jet noise reduction has become a critical issue for the aerospace industry. A large contributor to aircraft noise at takeoff is the noise produced by the jet exhausting from the turbofan engines. After significant gains obtained by means of a reduction of the jet velocity by an increase of the engine's bypass ratio, it appears clear that future strict noise regulation will require additional noise reduction strategies. Such strategies will act on the turbulent mixing characteristics of the jets which need to be computed accurately. Moreover, the noise by itself should also be computed as precisely as possible. RANS computations rely on acoustic Lighthill's analogy to estimate the radiated noise. The latter give reasonably good predictions for cold round jets but results are much less satisfactory for heated or non-axisymmetric jets. LES which accounts accurately for turbulent mixing and allows for a direct computation of the noise is intended to be the next step toward improved jet prediction.

The goal of most aeroacoustic studies is to evaluate the far field noise. Even if the acoustic disturbance propagation could in principle be computed solving the Navier-Stokes equations, it would result in grid sizes that are by far out of reach of current supercomputers capacities. In practice, the far field noise propagation is performed using the Kirchoff or Flowcs Williams-Hawkings techniques. The reader is referred to [310, 177] for more information on these approaches. In terms of application, the capacity of LES to capture the jet mixing can also be very useful for the evaluation and control of infra-red signature of military engines.

8.5.2 A Few Realizations

LES of compressible jets has been reviewed recently both by Debonis [50] and Bodony and Lele [24]. The discussion proposed here is partly inspired by these works and the reader is encouraged to refer to them for additional information.

Table 8.10 summarizes the characteristics of some existing LES simulations. Many other references can be found in [50, 24]. Axi-symmetric jets are considered by most of the authors.

Table 8.10. Characteristics of LES of subsonic jets

Ref.	Geometry	Ma	$Re(D)$	T_j/T_∞
[87]	Square Jet	0.6	3.2×10^3–2.2×10^5	1
[25]	Axi-symmetric Jet	0.9	6.5×10^4	?
[296]	Axi-symmetric Jet	0.9	4×10^5	1
[27]	Axi-symmetric Jet	0.9	4×10^5	0.96
[29]	Axi-symmetric Jet	0.9	4×10^5	0.96
[329]	Axi-symmetric Jet	0.9	3.6×10^3	0.86
[329]	Axi-symmetric Jet	0.4	5×10^3	0.97
[23]	Axi-symmetric Jet	0.5	27×10^3	1.8
[23]	Axi-symmetric Jet	0.5	79×10^3	0.95
[23]	Axi-symmetric Jet	0.9	13×10^3	2.7
[23]	Axi-symmetric Jet	0.9	84×10^3	0.86
[262]	Axi-symmetric Jet	0.36; 0.5; 0.57; 0.9	1×10^4	1
[262]	Axi-symmetric Jet	0.5; 0.9	1×10^4	3.4
[10]	Axi-symmetric Jet	0.75	5×10^4	1
[10]	Axi-symmetric Jet	0.75	5×10^4	2

The Reynolds number based on the jet diameter is defined as $Re(D) = \rho_j U_j D/\mu_j$ where the j subscript denotes an evaluation at the jet exit. The results of Bodony and Lele [23] are related to the experiments of Tanna [285]. Zhao et al. [329] have validated their data against the DNS of Freund [89]. For heated jets, the relevant quantity is the acoustic Mach number defined as U_j/a_∞. Such jets have been studied by Bodony and Lele [23], Shur et al. [262] and Andersson et al. [10]. Furthermore, co-flow jets and jets with "synthetic chevrons" presented in [262] are not discussed here.

8.5.3 Influence of the Numerical Method

Some general elements about numerical methods[7] used in the aforementioned papers are summarized in Table 8.11.

Table 8.11. Numerical methods used for LES of subsonic jets

Ref.	Spatial scheme	Temporal scheme	SGS model	Grid size (max.)
[87] [88]	FCT	Pred-Corr 2	MILES, SM	6.8×10^6
[25]	DRP4-7pt	RK4	SM	6×10^6
[296]	C6+iF6	RK4	MILES, DSM	16×10^6
[27]	DRP4-13pt	RK6	filtering (13 and 21pt)	12.5×10^6
[29]	DRP4-13pt	RK6	filtering (13 and 21pt)	16.6×10^6
[329]	(C6 (r, x); Spec(θ))+eF4	RK4	DSM, DHSS	3.6×10^6
[23]	(C6 (r, x); Spec(θ))+iF6	RK6	DSM	1×10^6
[262]	$(1-\sigma)$C4+σUB5	BE2	MILES	1.1×10^6
[10]	UB3	RK3	SM	3×10^6

The effect of the numerical scheme on the development of turbulent resolved structures is clearly highlighted by Shur et al. (Fig. 4 of Ref. [261]). It is demonstrated that the fifth-order upwind scheme dramatically delays the appearance of these structures compared with a hybrid scheme which combines a fourth-order centered scheme and fifth-order upwind scheme. Shur et al. have performed a grid sensitivity analysis. They have found that despite some delay of the transition process with the finest grid, the difference in the OASPL obtained on the two grids (5×10^6 and 1.2×10^6 points) is less than 1 dB for the whole range of observer angle. In their computations of round jets, Bogey and Bailly [25, 27, 29] have used a Cartesian code whereas Zhao et al. [329] and Bodony and Lele [23] have used cylindrical coordinates. High-order methods have been used to limit the influence of numerical errors both on the mechanisms of vorticity generation and the propagation of acoustic waves. A minimum amount of dissipation is introduced through filters which in the simulations of Bogey and Bailly play the role of the SGS model by extracting energy from the smallest scales. In their work, particular care has been taken to define a filter which affects only scales smaller than 4.4Δ [26].[8]

The inflow condition has a dramatic influence on the jet spatial development. Most of the authors uses a hyperbolic tangent function to specify

[7] Filters are either presented as an ingredient of the numerical scheme or as a model like in the work of Bogey and Bailly. Both classifications are possible since the ADM and an explicit filtering can be shown to be formally equivalent (see Sect. 5.3.3).

[8] A 13 point filter is used in the Cartesian directions while a 21 point filter is preferred in the diagonal directions.

the inflow mean velocity profile. The pressure is assumed to be constant and the density is deduced from the Crocco-Busemann relationship. Nevertheless, Shur et al. [261, 262] have imposed a boundary layer profile at the exit of a nozzle which is included in the simulations. Even if the turbulence is not resolved on the nozzle wall, it makes the simulations more representative of the experiments. The nozzle is also included in the computations by Andersson et al. [10]. An important parameter of the velocity profile is the ratio of the initial momentum thickness θ_0 to the jet radius r_0. Typical values used in the aforementioned references range from 0.03 to 0.1.[9]

Fluctuations are usually superimposed to the mean flow in order to promote a fast transition mimicking in some sense the behavior of a turbulent jet. They have to be non-radiating in order to permit a non biased analysis of the jet generated noise. Two techniques have been used. The first strategy consists in introducing longitudinal and azimuthal modes solutions of a linear stability analysis. In practice, different variants of this method have been used. Zhao et al. [329] imposed two longitudinal modes (the fundamental and one subharmonic) plus the most instable azimuthal mode. Bodony and Lele [23] have employed no less than 48 modes (a combination between 6 longitudinal and 8 azimuthal modes). The phase of these modes evolves randomly in time to prevent any phase-locking among them which could force periodic large scale structures. Alternatively, Bogey et al. [25] imposes solenoidal perturbations which are mathematically expressed as a sum of azimuthal modes localized in the jet shear layer. In Ref. [27], Bogey and Bailly performed a reference computation with 16 azimuthal modes. Suppressing the first 4 modes leads to noticeable changes in the flow. In particular, the length of the potential core passes from about 10 r_0 to 11.9 r_0. This effect can be seen in Fig. 8.8. Turbulence statistics and noise level are also significantly affected. This illustrates the great sensitivity of such flows to the initial perturbation spectrum. The amplitude of the perturbations has much less influence. Finally, it is worthwhile to note that Shur et al. discourage the use of imposed inflow perturbations arguing that it introduces many uncontrolled arbitrary parameters into the simulation. Moreover, the natural flow instabilities are found sufficiently strong to quickly three-dimensionalize the flow. Finally, one has to mention attempts to maximize the jet spreading rate using different types of forcing (flapping and varicose-flapping excitation) [188].

As for every aeroacoustic applications, boundary conditions require particular attention. In Refs. [27, 29, 329], a sponge layer with grid stretching and additional dissipation is combined with non reflective boundary conditions to damp most of vorticity and pressure fluctuations. In a more original way, Bodony and Lele [23] have used a volume force in a sponge layer to force the LES toward the solution of a RANS computation on the domain boundaries.

9 Whereas experimental values are generally about one order of magnitude smaller.

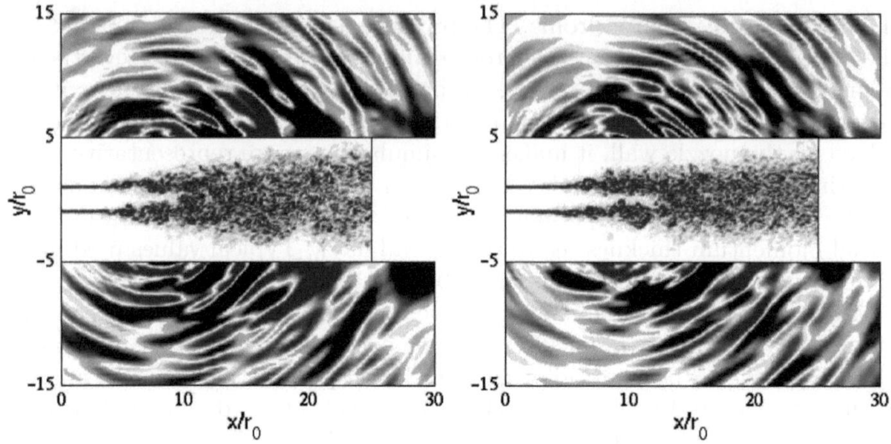

Fig. 8.8. Snapshots of vorticity and of fluctuating pressure. *Left*: initial perturbations with 16 modes and *right*: initial perturbations with 12 modes (from Ref. [27] with permission)

8.5.4 Influence of the SGS Model

The SGS model effect is discussed by Zhao et al. who have noticed that the dynamic mixed model (denoted DHSS in Table 8.1) produces larger turbulence intensity and noise than the classical dynamic Smagorinsky model. Nevertheless, the resulting flows and sound radiation patterns remain very similar. Fureby and Grinstein have investigated the MILES concept in reference [87]. They conclude that adding a Smagorinsky model to an intrinsically dissipative fourth-order FCT scheme has no significant effect on the kinetic energy spectra as well as on the shape of Cumulative Distribution Function (CDF) of vorticity magnitude in the developed part of the jet. As a consequence, they recommend the use of the FCT scheme without any model in this free shear flow. In a paper dedicated to the study of the SGS model effect, Bogey and Bailly [28] have performed computations with an explicit filter alone and with both the filter and a dynamic Smagorinsky model. The later simulation was found equivalent to a lower Reynolds computation performed only with the filtering. The Sound Pressure Level (SPL) was found to be damped with the dynamic model from moderate to high frequencies. The authors then suggest not to add any supplementary viscosity by means of a SGS model for such flows. Uzun et al. [297] have found similar results. They assume that the additional dissipation provided by the dynamic Smagorinsky model limits the energy transfer to the smallest scales with respect to a case without model.[10] As a result, the energy is magnified in the small wavenumber range with the SGS model. This mechanism has also been observed in

[10] A high order filter is employed to prevent energy pile-up at the cut-off.

[91] with shock capturing schemes providing different level of numerical dissipation. Nevertheless, its theoretical foundation remains to be brought to light.

8.5.5 Physical Analysis

LES and experiments differ both by their θ_0/r_0 ratio and by the physical nature of the inflow perturbations. The jet initial annular shear layers are thick relative to experimental jets and quasi-laminar velocity profiles with superimposed disturbances are enforced at the inflow. As a result, axial coordinates must be shifted to match experimental data. A possible consequence of the jet forcing by deterministic perturbations is the existence of privileged directions in the emitted sound diagram of directivity as reported by Zhao et al. [329]. Of course, DNS experiences the same issues, and it is worth noticing that attempts to recover DNS results with LES were generally successful even on coarse grids.[11]

Furthermore, Bodony and Lele [23] have found that the limited resolution of the computations impacts the radiated sound by yielding effectively low-pass filtered versions of the experimental data with a maximum frequency of $St \approx 1.2$. This upper bound in the frequency range is identified by Shur et al. [262] as the principal limitation of the Large Eddy Simulation.

The agreement of Bodony and Lele's data with Tanna's measurements [285] depends on the Mach number, the lowest velocity jets being the least accurate. The jet heating generally leads to additional discrepancies with respect to unheated jets.[12] The evolution of the jet mean field with respect to a change in jet velocity and temperature is in qualitative agreement with the experimental data: increasing the jet Mach number increases the potential core length while heating the jet at constant velocity decreases its length.

In order to illustrate the current level of accuracy reached by LES, Fig. 8.9 (left) shows that the discrepancies between better LES computations and experiments remain lower than 2 dB on the far-field Overall Acoustic Sound Pressure Levels (OASPL) for an unheated Mach 0.9 jet. Conversely, in the case of a heated jet at the same Mach number, the targeted accuracy of 2–3 dB retained by Shur et al. [262] as an objective from an application point of view is not reached (see Fig. 8.9 (right)).

---

[11] Bodony and Lele [23] reproduce the Freund's DNS data with a 25-fold reduction of the grid size.

[12] Even if Andersson et al. [10] obtained results of equivalent quality with heated and unheated jets.

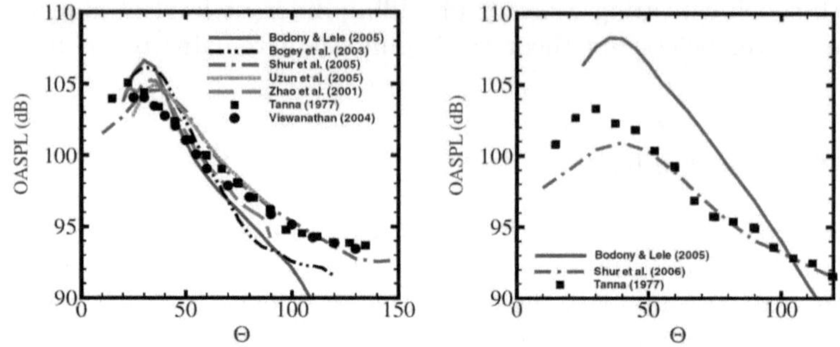

Fig. 8.9. Far field Overall Acoustic Sound Pressure Levels (OASPL) for $Ma = 0.9$ unheated jets (*left*) and heated jets (*right*) (Adapted from Bodony and Lele [24])

8.6 Flows over Cavities

8.6.1 Context

High speed flows past open cavities, such as weapon bays, wheel wells or measurement windows in aeronautical configurations, are mainly driven by a feedback mechanism between the shear layer instabilities and the acoustic waves. The large density and pressure fluctuations induced by these unsteady mechanisms may lead to aero-optical aberrations, high sound levels and strong vibrations with resulting structural fatigue. After some early attempts using URANS, LES is now considered as the best tool to compute cavity flows even for three-dimensional configurations at high Reynolds number. It is a configuration for which LES can satisfy application requirements at a reasonable computational cost.

8.6.2 A Few Realizations

The cavities treated using LES are of parallelepiped shape. They are generally classified by their length to depth and length to width ratios, denoted L/D and L/W respectively.

Table 8.12 summarizes the characteristics of some LES computations. An outstanding contribution to this field has been provided by Larchevêque and

Table 8.12. Characteristics of few LES of subsonic cavity flow

Ref.	L/D	L/W	Ma	$Re(L)$
[152]	0.42	0.42	0.8	8.6×10^5
[153]	5	5	0.85	7×10^6
[154]	2	0.42	0.8	8.6×10^5
[100]	3	0.03	0.8	9.5×10^5
[289]	0.42	0.42	0.8	8.6×10^5

coworkers [152–154]. The test case of Ref. [152] was chosen by Thornber and Drikakis [289] to assess the MILES formalism with a fifth-order accurate HLLC (Harten Lax van Leer Contact) scheme. Additionally, Gloerfelt et al. [100] have investigated the acoustic radiation emitted by such a flow. A previous paper by Dubief and Delcayre [59] mentions the realization of a LES of cavity flow in the compressible regime. Unfortunately, the description of this computation is not developed since this contribution is essentially dedicated to the analysis of turbulent structures visualization criteria on different geometries.

In all cases, the numerical results are related to existing experiments. This may explain the fact that the Reynolds number is typical for what is obtained in a wind tunnel at transonic Mach numbers for atmospheric stagnation conditions. A large variety of L/D ratios has been studied (from 0.42 to 5).

8.6.3 Influence of the Numerical Method

Table 8.13 gathers information about numerical methods used in these studies.

Table 8.13. Numerical methods used for computations of subsonic cavity flows

Ref.	Spatial scheme	Temporal scheme	SGS model	Inflow turbulence	Grid size
[152]	AUSM-like [193]	RK3	MILES, SMS	filtered random	1.6×10^6
[153]	AUSM-like [193]	BW2	SMS	filtered random	6×10^6
[154]	AUSM-like [193]	RK3	MILES, MS, SMS	filtered random	6×10^6
[100]	DRP4-11pt	RK6	SM	RFM [11]	4.9×10^6
[289]	HLLC	RK3	MILES	no fluctuations	0.8, 1.4, 3×10^6

Since in all cases the Reynolds number is large, classical boundary layer requirements in terms of grid resolution leads to prohibitive computational costs. In practice, wall-laws have been used in most of the presented papers with a first cell size in the wall-normal direction ranging from about 10 in Ref. [100] to about 70 in Ref. [153].

In terms of inflow conditions, Larchevêque and coworkers have added random fluctuations to the mean profile. These fluctuations are modulated by the RMS velocity profiles extracted from the experiment, space-filtered and time correlated in order to create a rough synthetic turbulence. Such a process is able to recover some kind of streamwise-oriented structures that can mimic streaks despite an over-estimated length ($\approx 2000x^+$) and height (up to $400y^+$) [242]. However, a case without fluctuation gives the same statistics in the cavity. This result is quite general when treating flows for which the separation point is fixed by the geometry since in such cases the turbulence is quickly regenerated by the Kelvin-Helmholtz instability and the three-dimensional structures of the recirculating flow. Gloerfelt et al. have preferred the Random Fourier Mode (RFM) procedure [11] which is able to reconstruct more realistic fluctuations. Additionally, they used a finer grid than Larchevêque et al. to convect the inflow velocity fluctuations.

Finally, it appears that a crucial parameter in term of grid construction is the number of points within the initial vorticity thickness which should be larger than 10.

Larchevêque et al. have also pointed out in Ref. [152] that the domain span is a key parameter of such simulations. In the case of a simulation performed with periodicity conditions in the transverse direction, they recommend to adopt a domain span equal to its length (L). For span smaller than L, they have noticed fewer harmonics in the pressure spectra and discrepancies in the mean velocity profile, almost independently of the number of cells used in the spanwise direction.

8.6.4 Influence of the SGS Model

In Ref. [154], Larchevêque et al. have studied the influence of the subgrid modeling. They have found that the use of the Mixed Scale model leads to spurious harmonics of the main peak on the pressure spectra. These peaks are due to the coalescence of incoming and reflected waves. The significant indirect effect of MS model over-dissipative behavior is a change in the aero-acoustic phase relation which has a strong influence on the whole flow. Conversely, Selective Mixed Scale model and MILES offer an accurate description of these pressure spectra which compare well with experimental data. In the same paper, the authors have demonstrated that the asymmetry of the whole flow field inside the cavity results from an inviscid confinement effect induced by the lateral walls. The branch of the bifurcation can be selected by slightly altering the incoming mean flow.

The work of Thornber and Drikakis [289] in which an implicit LES was performed with a fifth-order accurate shock-capturing scheme offers the opportunity of comparing this computational strategy with the one of Larcheveque et al. [152] who have employed a second-order accurate AUSM-like scheme [193] and a selective mixed scale model. With similar grid resolution, results are of comparable accuracy on first- and second-order statistics. Going into details, one can notice that a better prediction of the pressure spectrum second peak is found with conventional LES. No information which might have helped in evaluating the computational efficiency of these two strategies is given in [289].

8.6.5 Physical Analysis

In the papers presented here, the simulations have demonstrated their ability to reproduce accurately not only first and second order statistics but also phase averaged data. The more important findings concern the dynamic of turbulent structures which is well reproduced by the simulations. As an illustration, Fig. 8.10 extracted from [152] shows an isovalue of the Q criterion which visualizes the 3 main coherent structures present in the cavity. The trajectory of each of these structures correlates outstandingly well with the experimental data of Forestier et al. [84].

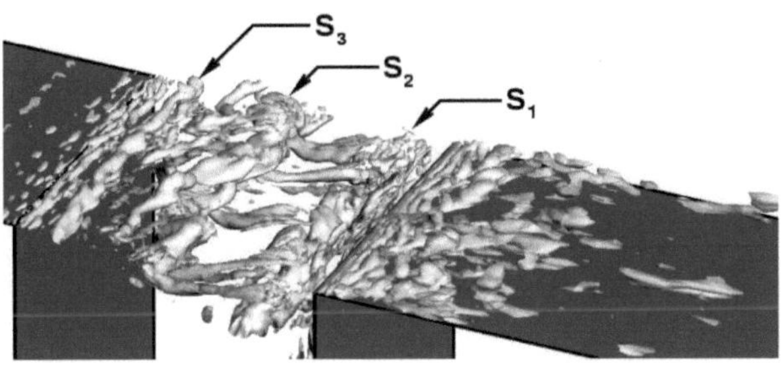

Fig. 8.10. Instantaneous view of coherent structures educed using the Q criterion (Courtesy of L. Larchevêque)

Additionally, an auxiliary simulation wherein the wind tunnel upper wall has been removed resulted in the finding that the upper wall presence enhances harmonics of pressure waves but does not change the nature of the a flow-acoustic resonance.

Besides, Larchevêque et al. [153] have identified a small recirculation bubble located at the origin of the shear layer as a possible cause of the Kelvin-Helmholtz instability forcing. Furthermore, analysis of the pressure fluctuations on the floor of the cavity has revealed that the Rossiter modes are independent of the spanwise location. These modes are subjected to streamwise modulations. These examples show that it is possible to perform physical analyses with such simulations which have now reached a high level of maturity.

9

Supersonic Applications

This chapter is dedicated to the presentation of supersonic applications for which the mean field does not contain flow discontinuities. Nonetheless, shocklets which are related to turbulent fluctuations can occur in particular in supersonic homogeneous turbulence computations. Shock/turbulence interactions will be treated in the next chapter. This chapter follows essentially the same outline than the previous one. No LES of supersonic shear layers was found in the literature. Moreover, on the subject of supersonic cavity flow only one paper was found ([235]), and no extensive discussion of this application case is included. The abbreviations defined in Table 8.1 are also used in this chapter.

9.1 Homogeneous Turbulence

Supersonic homogeneous turbulence has been computed by Porter et al. [221–224]. They have considered a turbulent Mach number equal to one, a flow condition which is unusual in "classical" aerodynamics but rather occurs under astrophysical flow conditions. The modeling is based on the MILES approach. The physical viscosity is set to zero and numerical dissipation is provided by the PPM (Piecewise Parabolic Method) algorithm. This scheme offers a good compromise between an acceptable numerical dissipation and good shock-capturing properties. High resolution computations were performed with as much as 1024^3 grid points. As reported is Sect. 3.5.3, these simulations have allowed to study of the supersonic regime where the initial flow field is populated by many small shocks which interact, generate vortex sheets which roll-up due to Kelvin–Helmholtz-type instabilities, eventually generating vortex tubes. The usual kinetic energy cascade develops from stretching of these vortex tubes. During the supersonic phase, the rms Mach number decreases from 1 to about 0.55, the ratio ρ_{max}/ρ_{min} reaches values as large as 20. At the end of this phase, both dilatational and solenoidal velocity spectra exhibit an inertial range with a -2 slope. The self-similar decay sets in at times

E. Garnier et al., *Large Eddy Simulation for Compressible Flows*,
Scientific Computation,

larger than 2.1 times the acoustic time scale. At this time, the enstrophy peak has already occurred. The remarkable achievement of these simulations lies in their large resolution which allows to establish clearly a $k^{-5/3}$ inertial range spectrum both for dilatational and solenoidal kinetic energy spectra [224]. Just before the dissipative range, a k^{-1} behavior is found.

Besides their interest in terms of physical description of such flows, Porter et al. give valuable information about the dissipative properties of their code. The numerical viscosity of the PPM scheme affects scales ranging from 2 to 12 times the width of a computational cell. From 12 to 32 cell widths, the flow is observed to be affected indirectly by the numerical dissipation. It gives qualitatively the equivalent filter width imposed in the simulation by the numerical method. Furthermore, this indicates the need of high resolution simulations for recovering unambiguously an inertial range (at least 512^3).

9.2 Channel Flow

9.2.1 Context

The main objective of plane-channel-flow simulations is that they allow for the computation of supersonic wall-bounded flows without taking care of the generation of realistic inflow turbulence since periodic boundary conditions are enforced both in the longitudinal and in the spanwise directions. Up to now, they have been used to evaluate SGS models and, to a lesser extent, to assess some hypothesis concerning the behavior of wall-bounded flows when increasing the Mach number.

9.2.2 A Few Realizations

The following Table 9.1 summarizes the flow conditions for some existing LES computations of supersonic plane channel flows. Pipe flows are not considered here. A reference study of axisymmetric converging/diverging pipe can be

Table 9.1. Characteristics of LES of supersonic channel flows

Ref.	Ma	Re_h	Wall boundary condition
[168]	1.5	3000	Isothermal
[133]	1.5	3000	Isothermal
[134]	1.5	3000	Isothermal
[206]	1.5	3000	Isothermal
[194]	1.5	3000	Isothermal
[194]	3	6000	Isothermal
[195]	3	6000	Isothermal
[32]	1; 1.5; 2; 3	3000; 4880	Isothermal

found in [99]. It is worthwhile to note that each computation listed in Table 9.1 has been carried out with isothermal walls. The Mach number is based on the bulk velocity and the wall temperature.

For the $Ma = 1.5$ case, the first DNS which is often taken as a reference was performed by Coleman et al. [43], they have found a Reynolds number based on the friction velocity Re_τ equal to 224. With a better resolution, Mathew et al. [194] have found $Re_\tau = 220$ and, on a even finer grid, von Kaenel et al. [133] have established a value of Re_τ equal to 216. For the high Mach number cases, LES are compared to the DNS of Foysi et al. [85] at $Re_\tau = 565$.

9.2.3 Influence of the Numerical Method

Some general facts about the numerical method used in the aforementioned papers are summarized in Table 9.2. Reference [32] is not reported here since this paper gives very few information about the numerical method. It is however worthwhile to note that it proposes a new formulation of the driving terms which force the flow entrainment in the channel. This important aspect of channel flow computations will not be developed here but the reader interested in such simulations is encouraged to consider this reference.

Table 9.2. Numerical methods used for LES of supersonic channel flows

Ref.	Spatial scheme	Temporal scheme	SGS model
[168]	C4	RK3	SM, MSM, HSS, HSMS, SMS, HSSMS
[133]	C2	RK4	ADM
[134]	Jam	RK4	ADM
[206]	C2, Jam, Roe	RK4	MILES, DSM
[194]	CULD, Co6	RK3	explicit filter (ADM like)
[195]	Co6	RK3	explicit filter (ADM like)

Jameson and Roe schemes are widely used in industrial codes and the assessment of their suitability in the LES framework is of importance. With the standard version of these schemes, Mossi and Sagaut [206] observe that the flow remains turbulent but the friction coefficient is significantly underestimated. However, this underestimation is much less marked than in the subsonic case (see Sect. 8.2) despite a slightly finer resolution. Using the generalized Smagorinsky constant concept, they were able to demonstrate that the dissipation brougth by the Jameson scheme in its matrix dissipation version is close to the one provided by the dynamic Smagorinsky model. Nevertheless, this equivalence of the dissipation magnitude is not directly observed from the statistics of fluctuating quantities. Additionally, the authors have confirmed that adding a functional SGS model to a dissipative scheme is counter-productive in terms of results quality. Nevertheless, as shown by von Kaenel et al. [134], good results can be obtained when coupling a Jameson scheme with a (structural) ADM model. In this case, the Jameson scheme replaces the ADM regularization procedure.

A comparison of the studies of Lenormand et al. [168] and Mossi and Sagaut [206] demonstrates a quantitative agreement between the fourth-order accurate scheme used by the first authors and the second-order accurate scheme employed in the second paper. Furthermore, due to the fine grid resolution ($\Delta x^+ \approx 35$ and $\Delta y^+ \approx 14$), this test case is believed to be not discriminant enough for SGS models since the second-order accurate scheme without any model gives the best results. Besides, von Kaenel et al. [133] point out that, in their simulations, good results of the no model case can be attributed to the use of the skew-symmetric form of the convective fluxes. For such a simple cases, it provides the adequate level of flow regularization.[1]

In Ref. [194], Mathew et al. have compared a high-order accurate compact upwind scheme and a compact centered scheme. The results are virtually unaffected by changing the underlying scheme. It shows that high-order accurate upwind schemes can be used for wall-bounded LES.

9.2.4 Influence of the Grid Resolution

Table 9.3 summarizes facts concerning the grid resolution in each direction expressed in wall units for some of the aforementioned papers. Notation has been unified so that z stands for the wall-normal direction and y is oriented in the spanwise direction. This table is provided here to give the reader an overview of the grids employed in the literature. It is nevertheless difficult from the results analysis to establish recommendations in terms of grid resolution since the employed numerical methods are very different in each study. Mathew et al. [194] have performed a computation with a doubled spanwise spacing. The resulting grid is relatively coarse and leads to an underestimation of the friction velocity of 2%. Nonetheless, the peak of longitudinal velocity fluctuations is clearly overestimated.

Table 9.3. Example of grid resolutions used for supersonic channel flow computations

Ref.	Grid size	Δ_x^+	Δ_y^+	Δ_z^+
[168]	$41 \times 65 \times 119$	35	14	1
[133, 134]	$72 \times 40 \times 60$	39	23	1
[206]	$41 \times 65 \times 119$	≈ 32	≈ 13	1
[194] LESFxxx cases	$32 \times 64 \times 129$	≈ 80	≈ 14	≈ 1
[195]	$128 \times 64 \times 111$	55	37	1.2

[1] The divergence form is observed to be unstable.

9.2.5 Influence of the SGS Model

The conclusion drawn by Lenormand [168] for the subsonic case that has been presented in Sect. 8.2.4 can be extended to the $Ma = 1.5$ case. Some improvement is obtained with selective and/or hybridized models. Nevertheless, the superiority of these modifications cannot be demonstrated unambiguously for all quantities. E.g., the mean centerline temperature and the skin friction are overestimated by hybrid (whether selective or not) models. Reference [206] offers the opportunity to compare the HSSMS model with both a no model case and the dynamic Smagorinsky model. It is found that the no model case and the HSSMS model give results of similar quality, both being better than those of the dynamic model.

Using the ADM method on the same case, von Kaenel et al. [133] show that results of comparable quality as that of the best models used by Lenormand et al. can be obtained with a second-order accurate scheme. This finding is remarkable since the number of points is divided by 2 in the wall normal direction, the spanwise resolution being multiplied by about 1.6.

In Ref. [194], the contribution of the unresolved scales is accounted for by an explicit filter which is defined as the combination of the LES filter G and of its approximate inverse Q (see Sect. 5.3.3). With the appropriate filter strength, the results are of equivalent quality as that ones of the best models employed by Lenormand et al. but the used resolution is much coarser in the streamwise direction, the spanwise resolution being the same for both studies. The formalism was further improved in Ref. [195] and the results were found superior to the ones of the original method for the $Ma = 3$ case.

9.3 Boundary Layers

9.3.1 Context

LES allows to investigate physical aspects of supersonic boundary-layer flows at significantly larger Reynolds number than DNS. In this category, we can find for example reference [280] in which the Strong Reynolds Analogy (SRA) [207] was assessed. A second category concerns validation studies where the ability of inflow turbulence generation techniques are investigated [295, 280, 242], the underlying objective being generally a computation of a more complicated configuration such as shock/boundary layer interactions. The last class of paper concerns studies which give information about numerical methods in a wider sense (grid resolution, numerical scheme, SGS model) [275, 319]. The discussion is here restricted to the flat plate boundary layer, the boundary layer developing on a cylinder such as in Ref. [70] is not discussed.

9.3.2 A Few Realizations

Table 9.4 lists some LES computations of supersonic boundary layers. Computations which have been used to feed simulations of shock/boundary layer interaction have not been included in this table. The Reynolds number based on the compressible momentum thickness θ has been chosen to compare the simulations.

Table 9.4. Characteristics of LES of supersonic boundary layers

Ref.	Inflow turbulence	Ma	Re_θ	Wall BC	Grid
[275]	Forced transition	2.25	1000–4000	Isothermal	$416 \times 257 \times 55 \approx 5.8 \times 10^6$
[295]	Recycling	3	1400	Adiabatic	0.16×10^6–3.3×10^6
[319]	Recycling	2.88	1450	Adiab., isother.	1.4×10^6
[319]	Recycling	4	1100	Adiab., isother.	1.4×10^6
[280]	Recycling	2.5	3150	Isothermal	$251 \times 51 \times 101 \approx 1.3 \times 10^6$
[280]	Recycling	2.5	7130	Isothermal	$361 \times 73 \times 145 \approx 3.8 \times 10^6$
[292, 242]	Recycling	2.3	5300	Adiabatic	$159 \times 63 \times 115 \approx 1.2 \times 10^6$
[282]	Temporal simulation	4.5	≈ 1100	Isothermal	$32 \times 32 \times 101 \approx 0.1 \times 10^6$

These simulations can be differentiated into three categories depending on the way the turbulence is introduced at the inflow. In the work by Spyropoulos et al. which was inspired by the precursor DNS of Rai et al. [228], turbulence is triggered by a blowing/suction slot which perturbs the flow. The perturbation is three-dimensional and time-dependent. The second technique which is now the most popular is based on recycling methods [319, 280, 295, 242]. Such methods have been presented in details in Chap. 7. For specific studies, mostly focused on the transition process [282], it is also possible to use the temporal approach. In this case, deterministic fluctuation solutions of a linear stability analysis are superimposed to a laminar boundary-layer profile. The flow is periodic in the streamwise direction so that the boundary layer thickens continuously. The computation must be stopped when the upper part of the boundary layer overshoots the finely resolved zone of the computational grid. The listed simulations have been performed either with an adiabatic or an isothermal boundary condition on the wall. The wall temperature effect is discussed in the paper by Yan et al. [319] who have fixed a wall temperature value equal to 1.1 times the theoretical adiabatic wall temperature for isothermal wall computations.

9.3.3 Influence of the Numerical Method

Table 9.5 lists some general facts concerning numerical methods used for simulations of supersonic boundary layers.

The only paper which treats the effect of the numerical scheme is the one by Spyropoulos et al. [275]. The convective terms are computed both with fifth-order and third-order accurate upwind schemes. The latter scheme

Table 9.5. Numerical methods used for LES of supersonic boundary layers

Ref.	Spatial scheme	Temporal scheme	SGS model
[275]	UB5, UB3	BW2	DSM, MILES
[295]	UB2	RK2	MILES, SM
[319]	UB2	RK2	MILES
[280]	Co6	RK3	ADM
[292, 242]	FD4	RK3	SMS
[282]	Co6+Spec	RK3	ADM
[282]	FD4+ES	RK3	HPFS

results in a decrease of the friction coefficient by 20%. Moreover, the quality of results is similar to that employing a grid of 1.1×10^6 points with the fifth-order upwind scheme than using a grid of 2.9×10^6 points with a third-order scheme. The authors attribute most of the discrepancies to the dissipative errors intrinsic to upwing schemes which mask the effect of the SGS model. This observation is consistent with that has been found in incompressible flows in Ref. [150]. It suggests that low order dissipative (upwind) finite difference methods should be used with care for computing such flows. In the absence of shocks stable simulations [280] are obtained with centered scheme even at higher Mach number than that of Spyropoulos et al. [275]. The computations carried out with a second-order accurate upwind scheme in Refs. [295, 319] with good results may contradict the previous statement on upwind schemes, but one should note that a DNS-like resolution has been used.

9.3.4 Influence of the Grid Resolution

Table 9.6 summarizes facts concerning the grid resolution in each direction expressed in wall units for some of the aforementioned papers. Notations have been unified so that z stands for the wall-normal direction and y is oriented in the spanwise direction.

Figure 9.1 presents the friction coefficient evolution with respect to the Reynolds number based on the momentum thickness for cases A1, B1 and C1 of Ref. [275]. The friction coefficient extracted from this figure for $R(\theta) \approx 3800$ shows that increasing the grid size in the spanwise direction from 11 to 22 wall units leads to a friction coefficient underestimated by 14%. An additional doubling of the grid size in the spanwise resolution and an increase of the streamwise resolution from 59 to 88 wall units leads to an extra error of 22%. These errors must be added to the one of the reference LES (15%) to recover the DNS friction coefficient value. Additionally, increasing the grid size leads to an overestimation of the longitudinal fluctuations and to an underestimation of transverse and wall-normal fluctuations. The latter may lead to a reduced turbulent transport of momentum and consequently to a reduced skin friction.

Table 9.6. Example of grid resolutions used for supersonic boundary layers computations

Case	Grid size	Δ_x^+	Δ_y^+	Δ_z^+
[275] A1	$416 \times 257 \times 55 \approx 5.8 \times 10^6$	59	11.4	0.87
[275] B1	$416 \times 129 \times 55 \approx 2.9 \times 10^6$	59	22.7	0.83
[275] C1	$311 \times 65 \times 55 \approx 1.1 \times 10^6$	88	42	0.77
[295] Baseline	1.68×10^6 tetras	11	3.1	1.1
[295] Coarse	0.32×10^6 tetras	28	13	0.9
[295] YveryCoarse	0.16×10^6 tetras	28	26	0.9
[319] A2.88	1.4×10^6 tetras	20	7	1.8
[280] Case A	$251 \times 51 \times 101 \approx 1.3 \times 10^6$	41	21	2.7
[280] Case A (coarse)	$151 \times 31 \times 61 \approx 0.3 \times 10^6$	68	34	5.7
[280] Case B	$361 \times 73 \times 145 \approx 3.8 \times 10^6$	59	29	3.6
[292]	$159 \times 63 \times 115 \approx 1.2 \times 10^6$	50	18	1

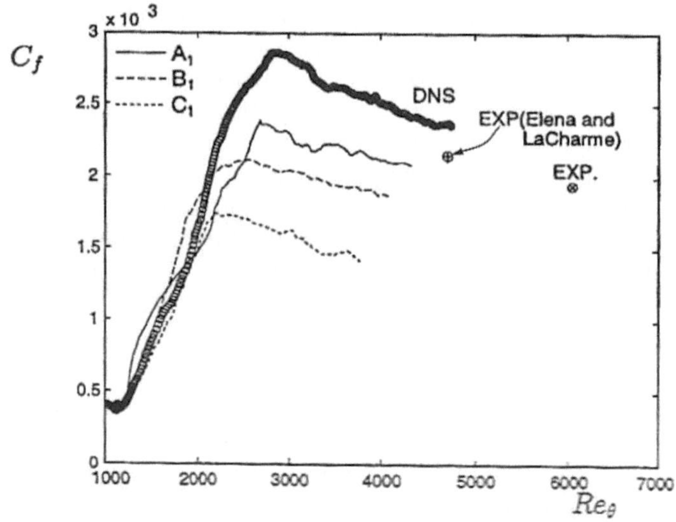

Fig. 9.1. Variation of the skin-friction coefficient with $R(\theta)$, Effect of grid resolution. See Table 9.6 for the definition of cases A_i (Reprinted from [275] with permission)

Before commenting on the results of Urbin et al. [295], it is worthwhile to note that the resolution of their LES is finer than the one used by Pirozzoli et al. [217] for their DNS ($\Delta x^+ = 14.5$, $\Delta y^+ = 6.56$ and $\Delta z^+_{min} = 1.05$). It is almost also true for the "coarse" LES which is nearly as fine as the DNS used as a reference in [275]. Nonetheless, going from the baseline grid to the coarse LES grid leads to an underestimation of the friction coefficient by 12%.[2] A further twofold coarsening in the spanwise direction leads to a

[2] The friction velocity values given in Ref. [295] have been converted in terms of friction coefficient to allow a direct comparison with other papers.

32% friction coefficient decrease. It then appears that second-order accurate upwind methods must be employed together with very high resolution grids.

Stolz and Adams [280] have employed a grid spacing larger than one wall unit in the wall-normal direction as a consequence of the large Reynolds numbers considered. For their 'coarse' grid spacing (Case A in Table 9.6) the friction coefficient is in good agreement with a regression curve based on experimental data. A further coarsening in each direction (Case A coarse) leads to a 15% increase of the friction coefficient. Increasing the resolution in the wall-normal direction from 101 to 151 grid points ($\Delta z^+ \approx 1.6$) for case A gives almost the same results for the skin friction and the shape factor, indicating that the employed model (ADM) may allow to relax slightly the wall-normal resolution requirements. We also note that in Ref. [292] the friction coefficient is underestimated by 15% with respect to an experiment performed in at same flow conditions.

This brief review shows that it is not possible to give guidelines for grid resolution without discussing of the choice of the numerical method. It appears that with high-order accurate numerical methods $\Delta x^+ \approx 40$, $\Delta y^+ \approx 20$ and $\Delta z^+_{min} \approx 2$ may be sufficient. With a second-order accurate upwind schemes the resolution constraint is very stringent and the lower computational cost with respect to high-order schemes is annihilated by the extra resolution requirement.

The choice of the domain extent in the spanwise direction L_y is an additional issue. As an example, values of 1.3δ and 1.6δ can be found in Refs. [292] and [280] respectively. The influence of this parameter has been studied in Ref. [295]. The authors have demonstrated that the friction coefficient is nearly insensitive to the choice of L_y/δ within the range [1.1–4.4].

9.3.5 SGS Modeling

Spyropoulos and Blaisdell [275] have investigated some important aspects of SGS modeling. It should be kept in mind that this study was performed with upwind schemes. On their fine grid (A_1) no significant differences on the friction coefficient were found between computations with and without model. Furthermore, they have pointed out that the use of dissipative schemes prevents a good estimation of the energy transfer at the cut-off. It affects the functioning of the dynamic Smagorinsky model. In order to avoid this problem, they recommend to prefilter the results.

Additionally, the SGS tensor isotropic part ($1/3\tau_{kk}$) is modeled dynamically and it is found that this quantity represents less than 8% of the dynamic pressure even on a relatively coarse grid (case B1) and can then be neglected. Finally, they have observed that the test filter can be applied in the 3 directions or in the wall-parallel plane without significant effects on the results.

In Ref. [282], Stoltz et al. mentioned that the simulation stability can be affected by a change of SGS model. When replacing their ADM model by the HPFS one, they needed to modify the convective scheme from a spectral+compact method to a fourth-order accurate centered one employing an entropy-splitting formulation. With the latter scheme, the results are less accurate than with the HPFS model than with ADM. Eventually, the authors conclude that in their application (a transitional flow) the influence of the SGS model predominates over the effect of the numerical discretization.

It maybe worthwhile to note that the ADM model [280] is anti-dissipative from the wall to $z^+ = 10$. Even the relaxation term contributes a slight amount of backscatter at the first grid-point off the wall, which may be related to fact that non-symmetric filter kernels have to be used near the wall.

9.4 Jets

9.4.1 Context

For realistic conditions in aircraft propulsion a jet coflow with a velocity which reaches about 60% of the jet velocity should be present. This situation has considered by Shur et al. [262]. In this section, we limit the discussion to perfectly adapted jets for which jet exit static pressure is equal to the external pressure. From a technical point of view, supersonic adapted jets are not more difficult to compute than subsonic jets. Shock-containing jets (over or under-expanded jets) require some form of numerical stabilization which may affect the development of turbulent structures. Moreover, for such flows, new physical phenomena like the screech tone must be taken into account.

9.4.2 A Few Realizations

Table 9.7 lists some LES computations of supersonic jets. As in the Sect. 8.5, the Mach number reported in this table is the acoustic Mach number defined as $Ma = U_j/a_\infty$. The Reynolds number $Re(D)$ is based on the jet diameter (D). Morris et al. [204] is particular in the sense that it is one of the rare studies which relies on the Non Linear Disturbance Equation (NLDE) formalism. Moreover, they assume that, for this free shear flow, viscous terms can be neglected.

Table 9.7. Characteristics of LES of supersonic jets

Ref.	Geometry	Ma	$Re(D)$	T_j/T_∞	P_j/P_∞
[90]	Axi-symmetric Jet	2	3×10^4	2.71	1
[49]	Axi-symmetric Jet	1.39	1.2×10^6	0.72	1
[204]	Axi-symmetric Jet	1.47	∞	0.56	1
[23]	Axi-symmetric Jet	1.47	8.4×10^4	2.3	1
[23]	Axi-symmetric Jet	1.47	3.36×10^5	0.56	1
[262]	Axi-symmetric Jet	1.37	5×10^5	1	1

9.4.3 Influence of the Numerical Method

Table 9.8 summarizes some general facts concerning numerical methods used for simulations of supersonic jets. Adapted jets are treated with the same kind of numerical method than in subsonic cases. Debonis and Scott [49] find that it is necessary to include the nozzle lip within the simulation for improving significantly the agreement with experimental data. In terms of boundary conditions, Debonis and Scott [49] and Gamet et al. [90] impose a small value of the freestream Mach number (typically 0.05).

Table 9.8. Numerical methods used for LES of supersonic jets

Ref.	Spatial scheme	Temporal scheme	SGS model	Grid size (max.)
[90]	MC(2,4)	MC(2,4)	FSF	2.88×10^5
[49]	C4+F6	RK5	SM	2.3×10^6
[204]	DRP4-7pt+F6	RK4	SM, DSM	4.5×10^6
[23]	$(\text{Co6}(r,x); \text{Spec}(\theta)) + \text{iF6}$	RK6	DSM	1×10^6
[262]	$(1-\sigma)\text{C4} + \sigma\text{UB5}$	BW2	MILES	1.6×10^6

Morris et al. [204] have assessed an original way to impose inflow perturbations. An azimuthal spectrum defined in the Fourier space is prescribed in a frequency band ranging from $St = 0.1$ to 1. The energy introduced in the perturbation is a fraction of the RANS turbulent kinetic energy equally distributed on the three velocity components. The use of this technique affects the sound pressure level (SPL) by less than 1 dB with respect to a classical method in which the axisymmetric and the ± 1 helical modes are forced.

9.4.4 Influence of the SGS Model

Morris et al. [204] have reported a significant improvement in the azimuthal distribution of SPL when using the dynamic Smagorinsky model instead of its constant coefficient counterpart. A 3 dB reduction of the SPL is given with the dynamic model improving significantly the agreement with the experimental data.

9.4.5 Physical Analysis

The study of Debonis and Scott [49] has focused on two-point correlations which constitute an input for the noise evaluation in the Lighthill theory. These correlations allow a quantification of the integral length scales which are found equal to about half of the jet diameter. The convection velocity of the turbulent structures are estimated to be within 57 to 71% of the jet velocity.

Both for heated and unheated jets, the results of Bodony and Lele [23] fit within 2 dB with the experimental OASPL for observer angles lower than 90

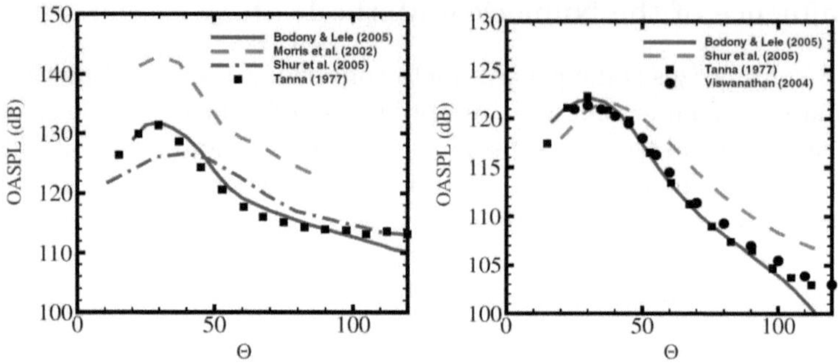

Fig. 9.2. Far field Overall Acoustic Sound Pressure Levels (OASPL) for supersonic unheated jets (*left*) and heated jets (*right*) (Adapted from Bodony and Lele [24])

degrees. Beyond this angle, the simulations underestimate the radiated sound amplitude. These observations are illustrated in Fig. 9.2 which demonstrates the predictive capacities of LES for supersonic heated jets. The quality of results is significantly better than that obtained for subsonic jets (see Fig. 8.9 for a comparison).

Shur et al. [262] compare the OASPL obtained for two jets at the same acoustic Mach number, one is underexpanded while the other one is perfectly expanded. The presence of shock cells in the case of the (sonic) underexpanded jet leads to a strong amplification of the noise in the lateral direction with respect to the ideal (supersonic) situation as shown in Fig. 9.3. As already mentioned in Sect. 9.4.1, the presence of shock leads to additional numerical constraints.

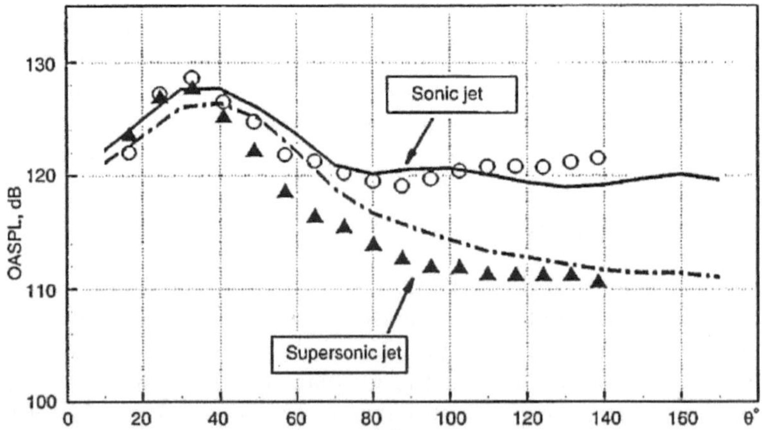

Fig. 9.3. Noise directivity of perfectly expanded supersonic and under-expanded sonic jets (from Ref. [262] with permission)

10

Supersonic Applications with Shock-Turbulence Interaction

In most practically relevant supersonic flows, shocks and flow turbulence are prevalent entities. A few examples for such flow configurations are transonic airfoils, supersonic air intakes of jet engines, propelling nozzles of rockets at non-adapted exit pressure, and deflected control surfaces of vehicles at transonic or supersonic speeds. Research on turbulence follows the strategy of a stepwise increase of geometrical and physical complexity as illustrated for the case of shock-turbulence interaction for unbounded flows in Fig. 10.1.

The simplest case is that of a shock interacting with isotropic homogeneous turbulence, e.g. grid-generated turbulence. Isotropy refers to rotational and reflectional invariance of the probability-density functions (PDF) of turbulent fluctuations and therefore also of resulting statistical averages, and homogeneity refers to translational invariance of the PDF [118]. Isotropy is lost once turbulence passes through a shock. By the one-dimensional compression of a plane shock incoming isotropic turbulence becomes axisymmetric. Physical complexity increases further if one allows for (constant) mean shear and shock obliqueness with respect to the mean flow. Another step in complexity is reached when the mean shear is inhomogeneous in one or more spatial directions, such as in jets, mixing layers or wakes, and in boundary layer flows. For the latter again several cases commonly are distinguished, as shown in Fig. 10.2:

- Compression ramp: the wall along which the boundary layer develops is deflected by a positive angle at a corner.
- Impinging shock: an oblique shock is generated outside of the boundary layer and impinges at an angle onto the boundary layer.
- Normal shock: a normal shock appears with one end at the boundary-layer edge or within the boundary layer, and the other end is either fixed in the exterior (to the boundary layer) stream or at an opposite wall.

The normal shock is a generic case for a shock pattern that is typical, e.g., for the flow in an overexpanded supersonic nozzle, see Fig. 10.3.

E. Garnier et al., *Large Eddy Simulation for Compressible Flows*, Scientific Computation,

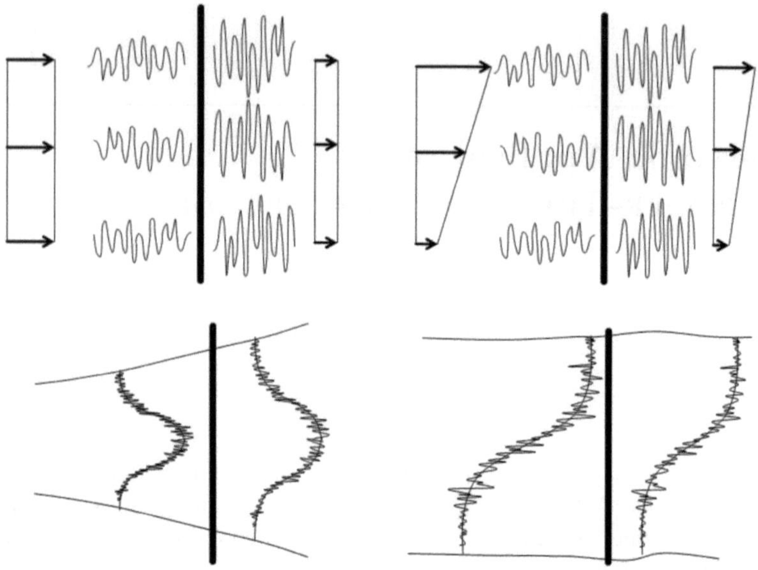

Fig. 10.1. Canonical configurations for the investigation of unbounded flows with shock-turbulence interaction, *from top left to bottom right*: isotropic homogeneous turbulence, homogeneous turbulence with constant shear, turbulent jet, turbulent mixing layer

10.1 Shock-Interaction with Homogeneous Turbulence

10.1.1 Phenomenology of Shock-Interaction with Homogeneous Turbulence

Qualitative predictions of shock-interaction with homogeneous turbulence can be obtained by linearized theories such as (RDT) and (LIA), see e.g. Ref. [159] and Chap. 3. The predictions have been confirmed by direct numerical simulations (DNS) which show that Reynolds normal stresses are amplified across the shock, whereas Reynolds shear stresses vanish before the shock and after the shock passage. A different amplification of velocity components normal and tangential to the shock wave, however, indicates that isotropy is lost by the interaction. After the shock passage isotropic turbulence becomes axisymmetric. This reflects also in unaffected shock-normal vorticity fluctuations, whereas tangential components are amplified. It was found that the energy spectrum is stronger amplified at large wavenumbers, implying an overall decrease of turbulence length scales, which is consistent with LIA predictions. This decrease of length scales can be appreciated from instantaneous vorticity contours for a Mach number of 2 at two subsequent time instants, taken from Ref. [160], see Fig. 10.4.

The shock-front distortion depends on the integral length scale and the turbulence Mach number. The Taylor microscales consistently decrease across the

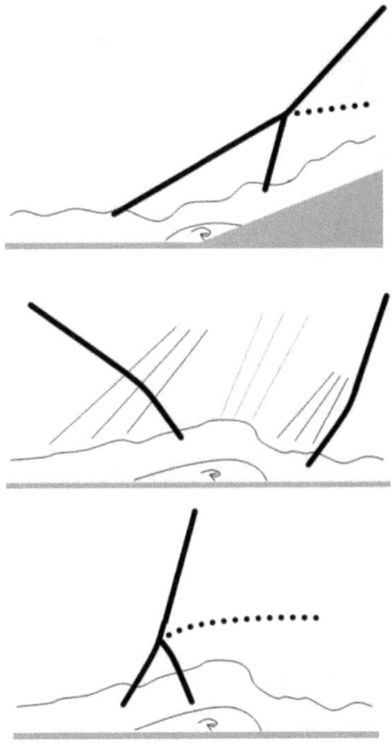

Fig. 10.2. Canonical configurations of shock-boundary-layer interaction *from top to bottom*: compression ramp, impinging shock, and normal shock

shock, whereas the dissipation length scale increases for Mach numbers less than 2 but decreases for higher Mach numbers [160]. The degree of compressibility of the incoming turbulence determines the turbulence amplification across the shock, the larger the part of compressible velocity fluctuations the smaller the turbulence amplification [108]. The effect of normal-shock strength on isotropic turbulence has been assessed by Mahesh et al. [185] and Lee et al. [160]. They show that DNS and LIA agree well within reasonable margins concerning the prediction of fluctuation amplification. A decrease of all turbulence length scales was observed across the interaction also for strong shocks. For higher Mach numbers (around 2 to 3) the thermodynamic fluctuations become non-isentropic, unlike at lower Mach numbers. Entropy fluctuations interacting with a normal shock have been investigated by Mahesh et al. [187] who conclude that the temperature-velocity correlation in the oncoming turbulence strongly affects turbulence evolution across the shock. This effect has been explained by the action of bulk compression and baroclinic torque. It was also found that shock-front oscillations invalidate Morkovin's hypothesis [266] across the shock.

Homogeneous turbulence with constant mean shear interacting with a normal or oblique shock has been studied by Mahesh et al. [186] with the fol-

Fig. 10.3. Shock-induced separation in a 2D nozzle. Reproduced from Ref. [229] with permission

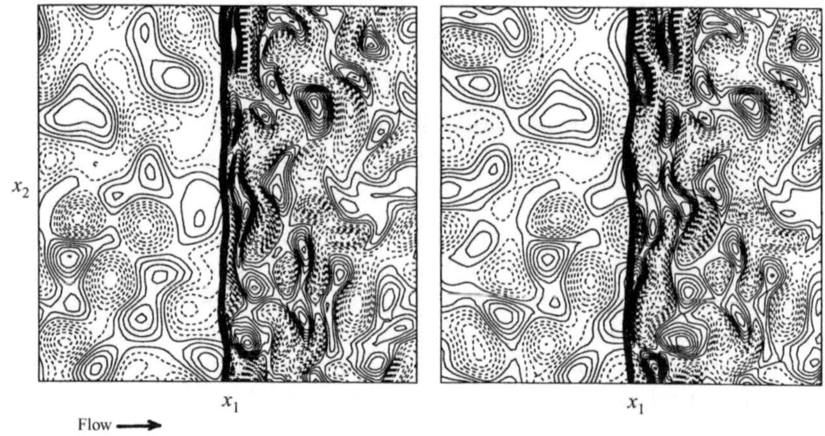

Fig. 10.4. Contours of one transversal vorticity component for isotropic turbulence interacting with a Mach 2 normal shock at two subsequent time instants. *Dashed lines* indicate negative vorticity, *solid lines* positive vorticity. Reproduced from Ref. [160] with permission

lowing main findings. The concept of the strong Reynolds analogy (SRA), see Ref. [266], is not valid across the shock. Reynolds stresses, normalized by the turbulent kinetic energy, decrease for normal shocks. The amplification of turbulent kinetic energy decreases with increasing shock obliqueness, whereas the normalized Reynolds shear-stress is less damped with increasing shock obliqueness and eventually is amplified for large obliqueness angles. It has also been demonstrated by Mahesh et al. [186] that traditional turbulence models, even second order closures, are unable to predict the correct turbulence structure.

A particular phenomenon arising in homogeneous turbulence with sufficiently large turbulence Mach number

$$M_t = \frac{1}{\langle a \rangle} \sqrt{\frac{\langle \rho u_i'' u_i'' \rangle}{\langle \rho \rangle}} \tag{10.1}$$

and with increasing Reynolds number, is the occurrence of shocklets [167]. In the definition of the turbulence Mach number $\langle a \rangle$ is the mean speed of sound, $\langle \rho \rangle$ the mean density and u_i'' the fluctuating i-component of the velocity versus the Favre average. Shocklets occur spontaneously within a flow field, and they are strong compression waves satisfying the Rankine-Hugoniot conditions with respect to the ambient flow and typically have normal Mach numbers only slightly larger than unity. Detailed investigations of the physical properties of shocklets due to Lee et al. [157] followed the initial discovery of shocklet existence by Passot & Pouquet [214]. Shocklets are characterized by a high correlation of pressure and dilatation and exhibit an entropy peak, indicating the conversion of kinetic energy into internal energy. Their appearance probability increases with increasing turbulence Mach number.

The effect of the existence of shocklets on turbulence statistics has been investigated by Samtaney et al. [247]. Figure 10.5, taken from Ref. [247], shows a volume-rendered snapshot of the velocity divergence for one of their simulations, where red and yellow indicate strong negative velocity divergence

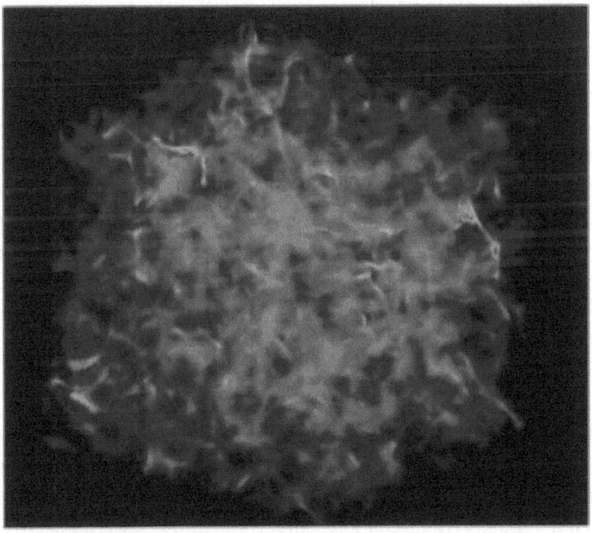

Fig. 10.5. Compressible isotropic turbulence with initial $M_t \approx 0.5$. Volume rendering of velocity divergence. Reproduced from Ref. [247] with permission

in regions where shocklets occur. Two main results of this investigations are that the shocklet strength scales proportional to $M_t/\sqrt{Re_\lambda}$, where Re_λ is the micro-scale Reynolds number, and that the most probable shocklet thickness is on the order of the Kolmogorov length scale. Based on the latter result, shocklets have to be considered as subgrid scales for LES.

10.1.2 LES of Shock-Interaction with Homogeneous Turbulence

Probably the first attempt of LES of shock-turbulence interaction was due to Lee [158] where numerical problems related to the formulation of the underlying equations were reported. The shock-capturing scheme used in combination with the compressible version of the dynamic Smagorinsky model caused excess dissipation away from the shock. The main problem which arises is that turbulence and shocks constitute SGS of different type. This can be illustrated by comparing the Fourier transform of a step function, which has a spectrum $\sim 1/\xi$, ξ being the wavenumber magnitude, and the Fourier transform of inertial-range turbulence which follows $\sim \xi^{-5/3}$. The former is consistent with a vanishing-viscosity concept for modeling the energy transfer [231] whereas the latter is consistent with a spectral eddy viscosity [173].

For SGS modeling in the computations of Ducros et al. [63] the filtered structure-function model is used, along with common simplifications concerning the viscous terms. The considered flow parameters are matched to the case considered by Lee et al. [159], albeit inflow data are taken from an isotropic-turbulence LES and convected through the inflow based on Taylor's assumption. Several cases with different resolution and filter combinations employing the filtered-structure-function model have been simulated. Generally, the LES recover the trends of the DNS of Lee et al. Finer grid resolution in the shock region leads to improved results, the use of a special shock sensor allows for decreasing spurious dissipation away from the shock.

In a subsequent study the performance of different SGS models was assessed by Garnier et al. [93]. A skew-symmetric form of the convective fluxes improves numerical stability by reducing aliasing errors. The shock was treated by hybridizing a 4th-order central finite-volume scheme with an essentially non-oscillatory scheme near the shock. Hybridization was performed on the numerical fluxes between centered scheme (index c) and ENO scheme (index eno) at the cell faces by

$$F_{i+1/2} = \beta F^{c}_{i+1/2} + (1-\beta) F^{eno}_{i+1/2},$$

where β is set to unity in the flow region away from the shock and to zero in a region near the shock. Two different shock Mach numbers were considered, $Ma = 1.2$ and $Ma = 2$, while the Taylor-microscale Reynolds number was very low in both cases, $Re_\lambda = 11.9$ and $Re_\lambda = 19$, respectively. Tested were the Smagorinsky model, the mixed-scale model, the dynamic Smagorinsky model, and the dynamic mixed model. For the low-Mach number case best results in comparison with DNS were obtained with the dynamic Smagorinsky model and the mixed-scale model, whereas for the higher-Mach-number case the dynamic mixed model was somewhat better. A mesh-refinement study showed that a sufficient resolution of the shock-front corrugation is crucial to obtain good LES results. A typical result is shown in Fig. 10.6 for the streamwise velocity-fluctuation variance. Best agreement with the filtered DNS is observed for the dynamic Smagorinsky model and the mixed-scale model, whereas worst agreement is observed for coarse DNS.

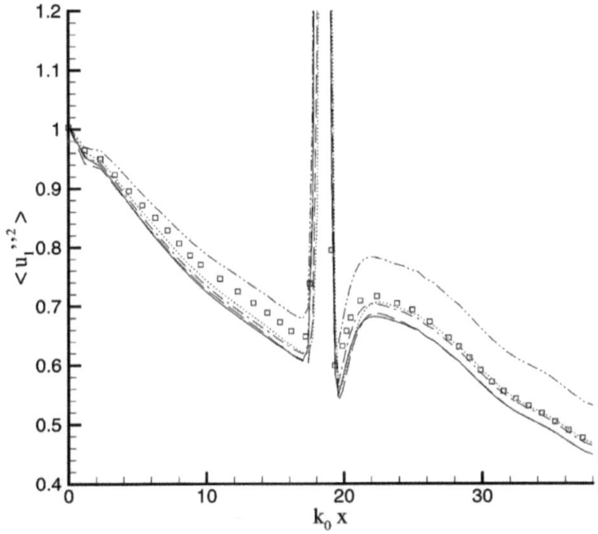

Fig. 10.6. Streamwise velocity-fluctuation variance. *Squares*: filtered DNS, *lines from bottom to top*: Smagorinsky model, dynamic mixed model, mixed-scale model, dynamic Smagorinsky model, coarse DNS. Reproduced from [93] with permission

A different approach was taken by Dubois et al. [61] who employed the subgrid-scale estimation model. The simulations are carried out with a 6th-order compact finite-difference scheme, hybridized with an ENO scheme near the shock. They consider a shock Mach number of $Ma = 1.3$ and slightly larger micro-scale Reynolds numbers of $Re_\lambda = 18$ and $Re_\lambda = 33$ than those of Garnier et al. [93]. The results of the estimation model were compared with that of the dynamic Smagorinsky model and with DNS. Generally both models give good results. The prediction of the estimation model shows an improvement compared to the dynamic Smagorinsky model which exhibits a small overprediction the resolved kinetic energy. A typical result is shown in Fig. 10.7. The shock-normal direction is denoted by x, the LES resolution is $231 \times 41 \times 41$ grid points, the DNS resolution is $231 \times 81 \times 81$. Table 10.1

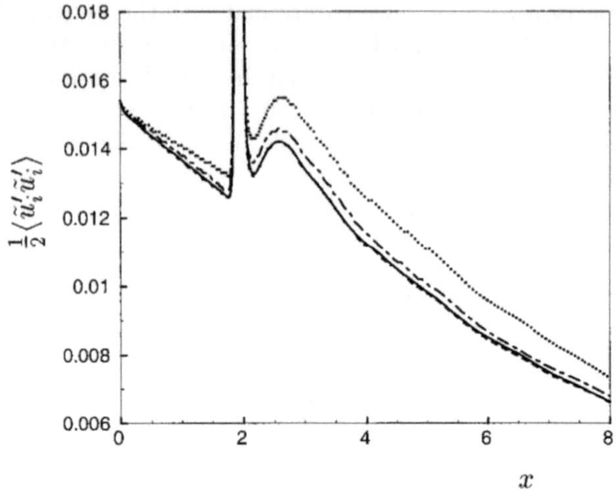

Fig. 10.7. Resolved kinetic energy averaged over planes parallel to the shock front. *Solid line*: full resolution DNS. *Dashed line*: estimation model. *Dot-dashed line*: dynamic Smagorinsky model. *Dotted line*: no model DNS. Reproduced from [61] with permission

Table 10.1. Summary of LES of shock-homogeneous-turbulence interaction discussed in this section

Ref.	Shock Mach number	Re_λ (before interaction)	M_t before interaction
[61]	1.3	17.29, 33.2	0.137, 0.152
[93]	1.2, 2	11.9, 19	0.136, 0.108
[63]	1.2	not specified	0.075
[158]	1.2, 2	12	0.095, 0.07

summarizes the LES of shock-homogeneous-turbulence interactions discussed in this section.

10.2 Shock-Turbulence Interaction in Jets

10.2.1 Phenomenology of Shock-Turbulence Interaction in Jets

Within an over expanded jet a shock-cell pattern develops. Interaction between turbulent fluctuations and the shocks can give rise to upstream-propagating acoustic waves, which can generate downstream-propagating instabilities at the nozzle lip [284]. This feedback cycle is responsible for the so-called screech noise in jets, which can cause sonic structural fatigue. For an efficient numerical simulation of such phenomena it is important that all relevant unsteady flow scales are accurately represented. Due to the coupling between shock, flow instabilities and sound waves a so-called direct noise computation is necessary. Whereas DNS is the appropriate tool, it cannot reach practically relevant Reynolds numbers or dimensions, leaving LES as the method of choice.

10.2.2 LES of Shock-Turbulence Interaction in Jets

Motivation of the LES of jets by Berland et al. [18] and Singh and Chatterjee [259] is the investigation of screech noise. A planar configuration, as visible from Fig. 10.8, has been studied by [18]. The jet Mach number is $M_j = 1.55$ and the jet Reynolds number based on the nozzle height $h = 3$ mm is $Re_h = 60000$. For spatial and temporal discretization low-dissipation and low-dispersion finite-difference and Runge-Kutta schemes, respectively are used. Instead of an explicit SGS model, filtering on the high-wavenumber scale content is performed, which is formally of second order. By the use of the explicit filter the LES scheme can be considered as implicit model, with some control on the wave-number range where numerical dissipation acts, and on the amount of numerical dissipation by how often the filtering is applied in terms of time steps. Periodic boundary conditions are used in the out-of-plane direction, non-reflecting conditions at the upper and lower truncation plane and at the outflow. The non-reflecting condition at the outflow is sup-

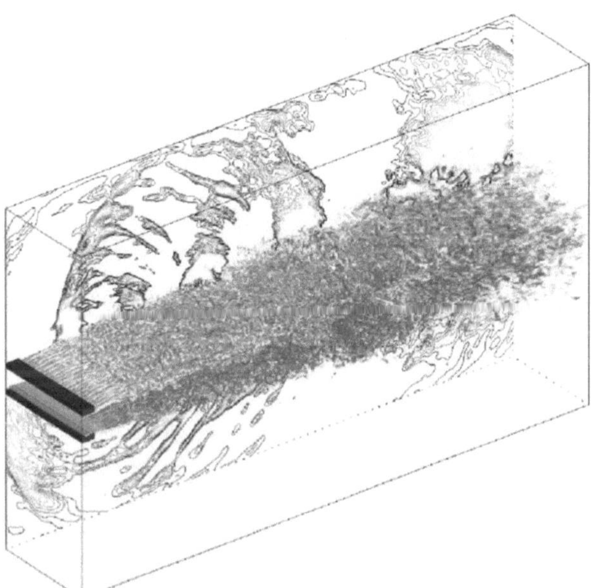

Fig. 10.8. LES of a planar jet. Isosurface of spanwise vorticity and pressure contours. Nozzle lips are indicated in *grey*. Reproduced from Ref. [18] with permission

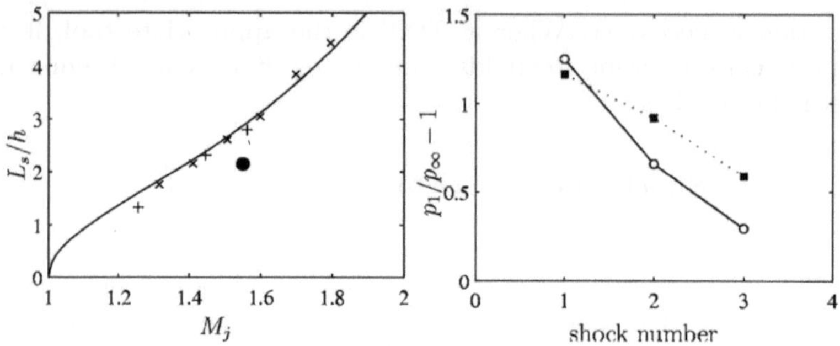

Fig. 10.9. (**a**) Normalized shock-cell spacing as a function of the fully expanded jet Mach number M_j (*bullet*) in comparison with reference experiments (*small symbols*) and a theoretical solution (*line*). (**b**) Normalized shock strength for the first three shock-cells (*full symbols*) in comparison with a reference experiment (*hollow symbols*). Reproduced from Ref. [18] with permission

plemented by a sponge-layer. The inflow is inside of the nozzle, part of which is included in the computational domain, the steady inflow data are imposed by a characteristic formulation. The shock-cell system is predicted in quite good agreement with a reference experiment for a rectangular jet at slightly larger Mach number, where the differences can be attributed to different inflow data and the side-wall effects in the experiment.

A circular jet geometry has been considered by [259]. The conservative form of the flow equations is solved and a constant-coefficient Smagorinsky model is used. As numerical discretization a WENO scheme is employed. A comparison with experimental results shows that essentially the shock-cell structure is recovered, the simulation is, however, more dissipative than desirable.

Table 10.2 summarizes the LES of shock-turbulence interaction in free-shear layers discussed in this section.

Table 10.2. Summary of LES of shock-turbulence interaction in free-shear layers discussed in this section

Ref.	Bulk-flow Mach number	Reference Re	Comment
[18]	1.55	60000	overexpanded planar jet, reference length: nozzle height
[259]	1.19, 1.49	not specified	under- and overexpanded round jet, reference length: nozzle height

10.3 Shock-Turbulent-Boundary-Layer Interaction

10.3.1 Phenomenology of Shock-Turbulent-Boundary-Layer Interaction

Research on shock-turbulent-boundary-layer interaction (STBLI) has been pioneered by Ackeret et al. [2] and Liepmann [176], who did the first systematic experimental studies on laminar and turbulent boundary layers interacting with normal or impinging oblique shocks. Figure 10.10 shows a historical Schlieren visualization of normal-shock turbulent-boundary-layer interaction taken from Ref. [2], the flow is from right to left.

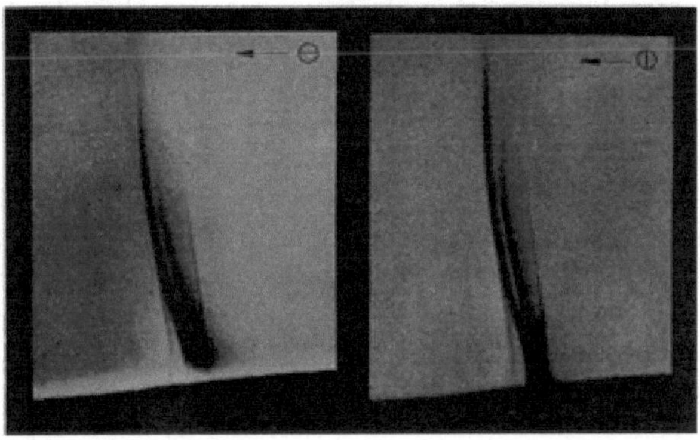

Fig. 10.10. Schlieren photograph of normal-shock interaction with a turbulent boundary layer. Reynolds number with respect to the boundary-layer momentum thickness is $Re_\theta = 2315$ and the Mach number of the free stream is $Ma = 1.3$. Reproduced from Ref. [2] with permission

Prompted by the need for accurate solutions of the RANS equations by computational fluid dynamics for design of supersonic aircraft, experimental data bases were established. Settles et al. [256] found that RANS computations with different turbulence models gave poor agreement with experimental wall-friction data in cases with large flow separation. Dolling & Murphy [53] detected a large-scale shock motion (LSSM) in their ramp data at a free-stream Mach number of $Ma_\infty = 3$ and a Reynolds number of $Re_{\delta_0} \simeq 10^6$. The observed shock motion spread over a region of 75% to 90% of the mean boundary layer thickness δ_0. Shock-motion frequencies were found between $f_{sh} = 0.035 f_c$ and $f_{sh} = 0.06 f_c$ where $f_c = U_\infty/\delta_0$ is the characteristic frequency (inverse time scale) of the oncoming boundary layer. These first quantitative results on LSSM have been corroborated later by other experimental investigations.

By detailed measurements of the fluctuating flow field Settles et al. [256] and Smits & Muck [267] found that flow fields were mainly two-dimensional for β up to 16°, where incipient separation was observed. Three-dimensional flow cells in the reattachment area were found for a larger deflection angle $\beta = 20°$. The fluctuation measurements showed large fluctuation-amplification factors of 5 to 10, along with a change in turbulence structure. The large-scale shock motion which was also observed by Smits & Muck prompted them to suggest a "pumping" mechanism to be responsible for turbulence-fluctuation amplification by a transfer of shock-motion energy to turbulent kinetic energy. In earlier investigations Plotkin [219] found a relation between shock motion and oncoming boundary-layer turbulence. This issue was taken up by Andreopoulos & Muck [8] who re-investigated the flow at the same parameters as [267]. They found that the shock-oscillation spectrum has a significant component at a frequency which corresponds to the incoming boundary layer bursting frequency. This lead them to the conclusion that shock-oscillation is indeed driven by the oncoming large-scale turbulent fluctuations. Erengil et al. [71, 72] established that the shock shows both large-scale and small-scale fluctuations. Both of them are random and seem not to be traceable clearly to turbulence events in the oncoming flow, although there is a correlation between shock motion and pressure fluctuations in the oncoming boundary layer. Also, the separation bubble was found to "breath", so that the overall flow structure transitions randomly between a state of small separation and a state of large separation, each of which results in a different shock position and pressure distribution.

Marshall & Dolling [191] revisit the experimental compression ramp data with the objective of addressing the question why even sophisticated second order closure turbulence models fail to predict the correct average wall pressure and wall friction. Concluding from the fact that turbulence models fail dramatically for significant separation only, but show a reasonable performance for attached flows, and from the fact that in most experiments a large-scale shock motion is going along with large-scale separation, Marshall & Dolling suspected this large-scale low-frequency unsteadiness to be a prime reason for the disagreement between RANS and experiments. Dolling [52] emphasizes again that the low-frequency expansion-contraction of the separated-flow area is critical in determining the correct time-averaged wall pressure, wall friction, separation location, and downstream velocity profiles.

Dupont et al. [65] investigate experimentally the shock-separation dynamics in turbulent boundary layers on the example of the impinging oblique shock, Fig. 10.11. They verified in their experimental setup that the incoming flow is free of disturbances in a frequency range in which LSSM can be expected. The moderate incoming-boundary-layer momentum-thickness Reynolds number of about 6900 is comparable to that of the compression-ramp experiments of [330]. Wall-pressure fluctuation spectra show a LSSM in the range of 100 Hz with an shock-excursion length of about 20% of the interaction-region extent. Downstream of the reflected shock the dynamics is dominated by large-scale unsteady structures characteristical for the detached shear-layer containing the separation region. The dynamics of the ini-

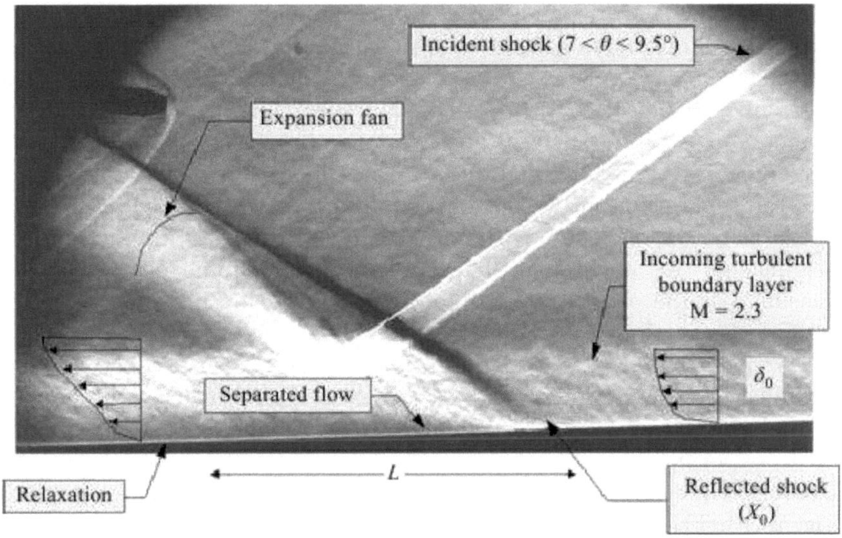

Fig. 10.11. Shock reflection with boundary-layer separation, impinging-shock angle of 8°. Reproduced from Ref. [65] with permission

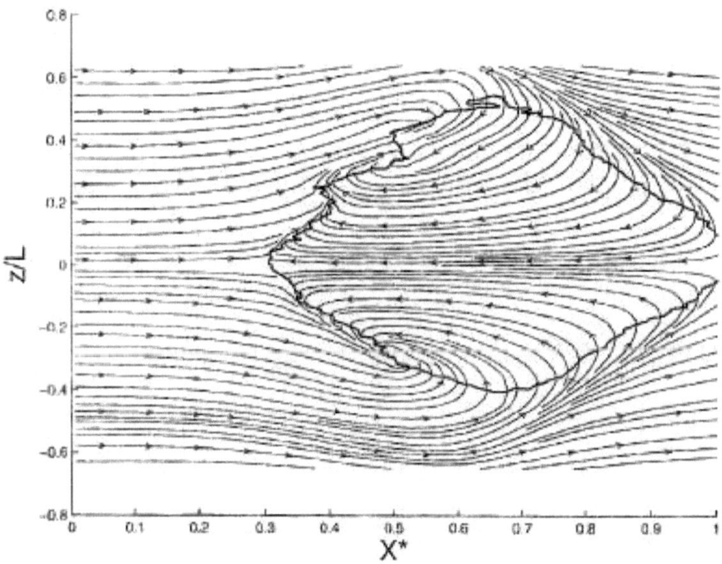

Fig. 10.12. Streamline pattern in a separation bubble caused by an impinging shock. Reproduced from Ref. [66] with permission

tial separation region depends on the impinging-shock strength. Dupont et al. report an amplitude of large-scale shock motion of $L_{ex}/L \approx 0.3$ where L is the mean separation length. For other flow configurations similar relations in the range of $0.3 \leq L_{ex}/L \leq 0.6$ have been found, see also [66]. Measurements of the power spectral densities of wall pressure indicate that there are significant residual fluctuations on the time-scales of the shock-motion within the separation-bubble flow. It is found that a simple outer scaling for the separation-region events with free-stream velocity and interaction length is insufficient. Dussauge et al. [66] emphasize the significance of the three-dimensional shape of separation bubbles which obviously would need to be recovered also in numerical simulations, see Fig. 10.12.

A particularly interesting configuration in Fig. 10.2 is the compression corner. Unlike the other configurations the shock originates from the boundary layer and the pressure increase is enforced by a deflection of the boundary layer. In experiments a compression corner is often followed by a decompression corner. For such a configuration the essential flow phenomena are sketched in Fig. 10.13. The undisturbed incoming turbulent boundary layer is deflected at the compression corner. The resulting compression shock penetrates into the boundary layer where the penetration depth depends on the local Reynolds number. For sufficiently large deflection angles the rapid compression within the boundary layer results in a region of mean-flow separation near the compression corner. The separation region is contained by a detached shear layer which reattaches at the deflected part of the compression ramp. A λ-shock sys-

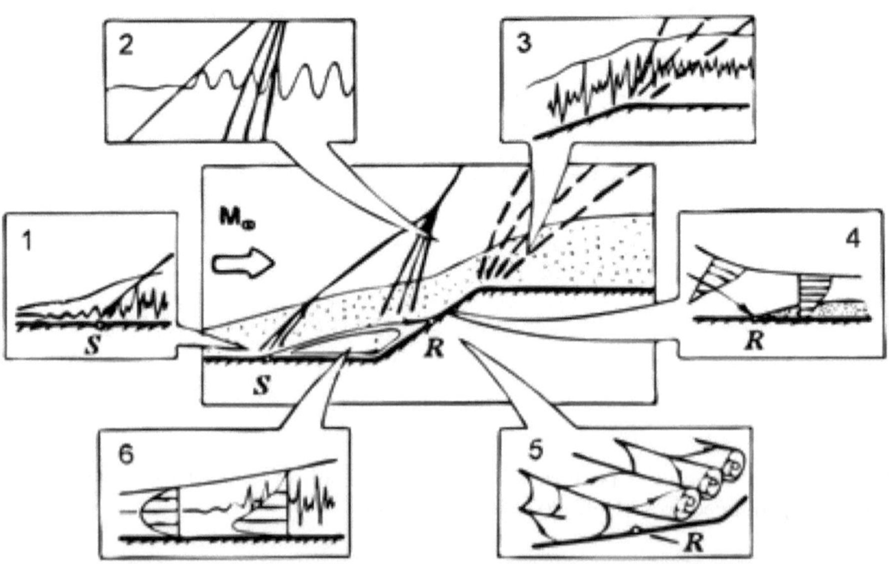

Fig. 10.13. Essential flow phenomena in compression ramp flows (see text). Reproduced from Ref. [181] with permission

tem is generated near the separation region. The forward foot of the λ-shock originates from the region of flow separation, the rearward foot from the region of flow reattachment. Further downstream, the reattached boundary layer reaches the decompression ramp and passes through the Prandtl-Meyer expansion. Even further downstream, the boundary layer relaxes again towards a developed zero-pressure-gradient boundary layer. Separation and reattachment lines are indicated by S and R, respectively. Turbulence is amplified (1) by interaction with a rapid compression within the boundary layer and (2) by direct interaction with the shock in the external flow. Note also that the shock foot spreads out towards the wall due to reduced local Mach number and due to turbulent diffusion. Turbulent fluctuations are damped by the interaction with the expansion wave (3) at the expansion corner. After reattachment at the deflected part of the compression ramp a turbulent boundary layer (4) is reestablished. Experimental results support the existence of pairs of large counter-rotating streamwise vortices (5) in the reattachment region as well as in the reverse flow of the separation zone. Within the area of flow separation the reverse mean flow has the character of a wall jet (6).

10.3.2 LES of Compression-Ramp Configurations

Probably the first attempt of an LES for compression ramp flow is due to Hunt & Nixon [130], who qualify their simulation as a very-large-eddy simulation. The flow parameters were chosen similar to the experiments of Dolling & Murphy [53]. The boundary-layer-thickness Reynolds number of about $Re_{\delta_0} = 10^6$, however, did not allow for a well-resolved LES. As subgrid-scale model a two-eddy-viscosity model, similar to that of Schumann [253], was used. The large Reynolds number required the use of a wall model, for that purpose a logarithmic wall function was imposed. The results show some qualitative agreement with the experiment of Dolling & Murphy [53], e.g. in terms of the shock-motion frequency.

For a weak interaction at a free-stream Mach number of $Ma_\infty = 3$ and a ramp deflection angle of $\beta = 8°$ no mean-flow separation was found in the LES of Urbin et al. [294], although instantaneous reverse flow regions may exist. In their LES a variable-density extension of the Smagorinsky model with a model constant of $C_S = 0.00423$ was used, and in the near-wall region a van-Driest damping was employed. A thin separation region was observed in the DNS of Adams [3] at $Ma_\infty = 3$, $\beta = 18°$. However, no LSSM was found. The observed small-scale shock-motion has a dominant frequency which is close to the inverse characteristic time scale of bursting events within the incoming boundary layer. An instantaneous Schlieren-type visualization exhibits compression waves shed by the main compression shock above the separated shear layer and downstream of the interaction. These DNS results were confirmed by the LES of Stolz et al. [278] and von Kaenel et al. [135] where the approximate deconvolution model (ADM) was employed for SGS modeling.

Rizetta et al. [233, 234] have performed LES for different compression ramp configurations. The LES equations are discretized in space by a 6th-order compact finite-difference scheme supplemented by a 10th-order non-dispersive filter. In the region near shocks the scheme is replaced by a MUSCL shock capturing scheme. For time integration an approximately factored Beam and Warming scheme is used. As subgrid-scale model the variable-density extension of the dynamic Smagorinsky model is employed. In a first set of LES parameters were matched to the DNS of Adams [3], inflow data were different from the DNS, however. For inflow-data generation Rizetta et al. [233, 234] use a separate flat-plate computation, for which again inflow-data are required. These inflow data are generated artificially from a laminar boundary-layer profile superimposed with disturbances taken from a half channel flow. For the DNS of Ref. [3] inflow-data were generated with a separate boundary-layer DNS, for which inflow data came from a temporally evolving DNS of the laminar-turbulent boundary-layer transition. Due to the narrow domain in the DNS no fully developed turbulent boundary-layer profile was obtained, but rather a profile with momentum deficit near the wall, which is typical for narrow spanwise domains. As result of the different inflow conditions, no quantitative agreement between the LES and the DNS is obtained. Also it was found by Rizetta et al. that with the chosen numerical method the effect of an explicit SGS model is small, since results with a dynamic Smagorinsky model or without model almost agree with each other. Rizetta et al. apply their approach to compression-ramp flows for experimental flow geometries and flow Mach numbers, albeit at much smaller momentum-thickness Reynolds numbers. As can be expected, essentially no quantitative agreement is found between simulations and experiments, see Fig. 10.14.

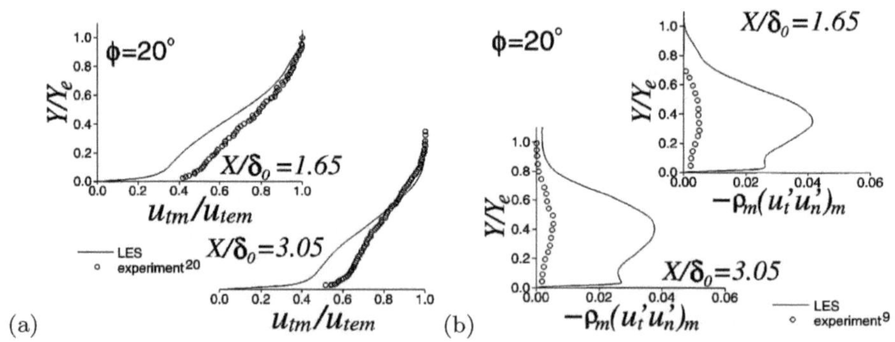

Fig. 10.14. Streamwise mean velocity profiles (**a**) and Reynolds-stress profiles (**b**) at two stations after reattachment for 20° compression ramp LES, symbols from experiments at higher Reynolds number. Reproduced from Ref. [234] with permission

An assessment of these LES is given in Fig. 10.15. Similarly as with RANS computations there is a rather large scatter in the predicted separation length.

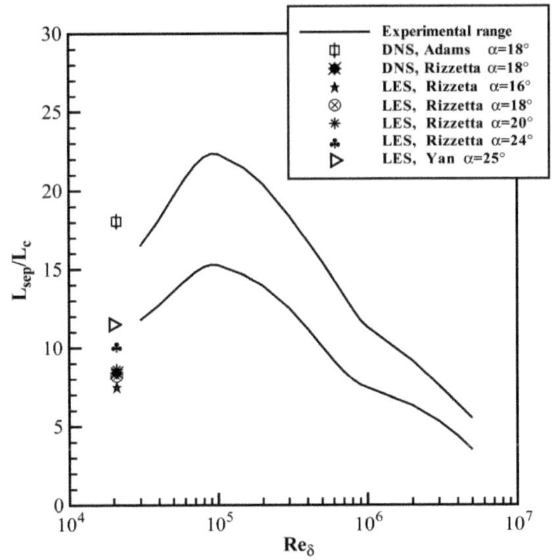

Fig. 10.15. Separation lengths for LES of compression ramp. Reproduced from Ref. [144] with permission

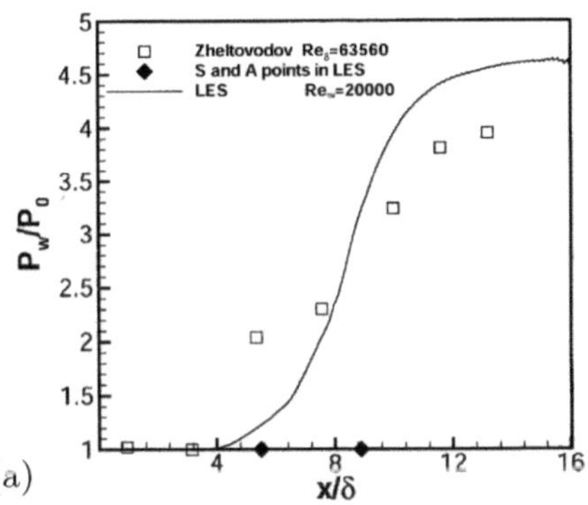

(a)

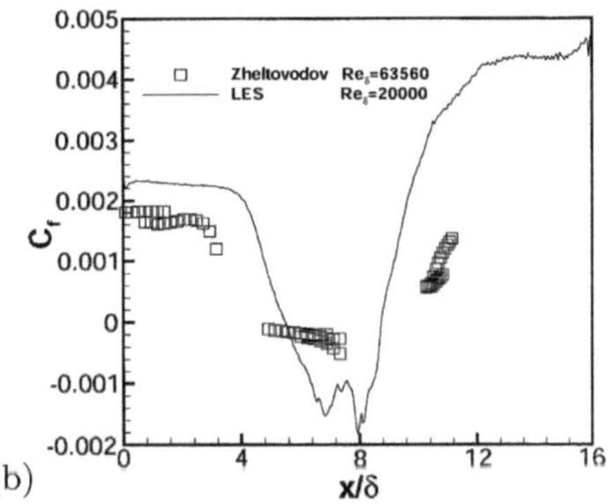

(b)

Fig. 10.16. Comparison of surface pressure (**a**) and wall friction (**b**) between experiment and LES. Reproduced from Ref. [144] with permission

Typical for these early LES is that the Reynolds number does not reach the experimental range and that the separation length is underpredicted, indicating an overly strong dissipation. A typical result for a comparison of wall pressure and wall friction between the experiment of Zheltovodov et al. [330] and the LES of Yan et al. [318] is given in Fig. 10.16. The underprediction of separation length and overprediction of attached-flow recovery is evident.

Another important question is the existence of large-scale streamwise structures in the reattaching flow and their origin. Experimental oil-flow patterns, see e.g. in Ref. [181], suggest the presence of pairwise counter-rotating vortices in the reverse flow of the separation zone and near reattachment. For a laminar interaction Comte & David [46] found Görtler-like vortices for an LES of the boundary-layer along a generic body-flap configuration at $Re_{\delta_0} \approx 840$, where δ_0 is the incoming boundary-layer thickness. It was shown that these vortices have a strong effect on local wall friction and heat transfer.

In the following we revisit the LES results for a compression-decompression-ramp configuration as shown in Fig. 10.17 in some more detail. The simulation parameter are matched to the experiment of Zheltovodov et al. [330]

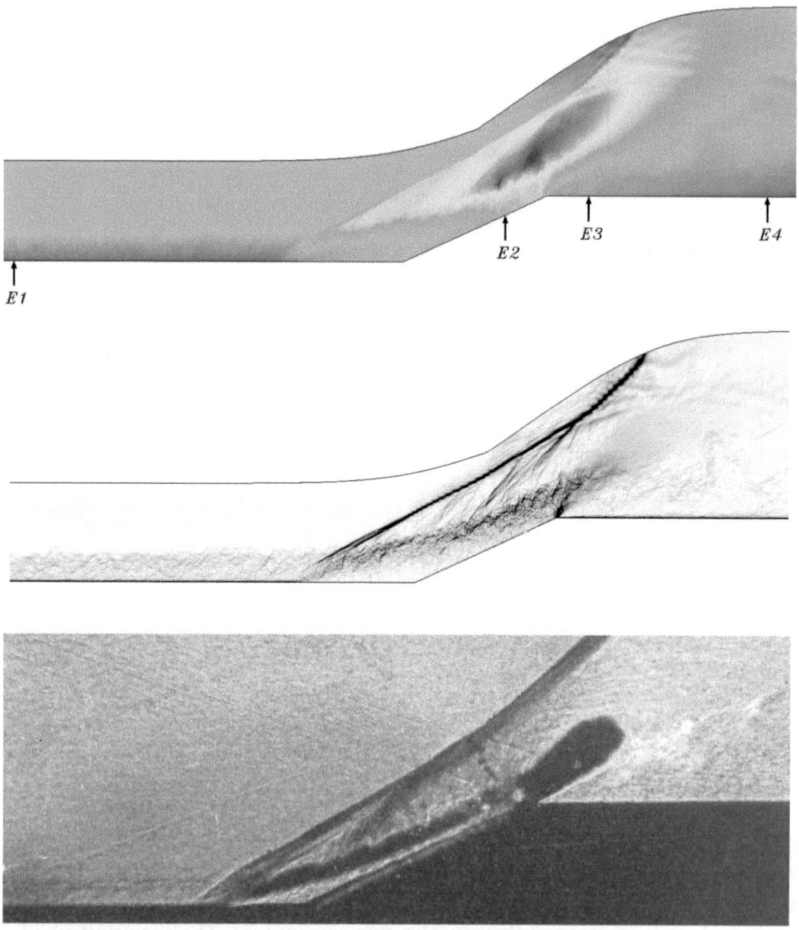

Fig. 10.17. Supersonic flow along a compression-decompression ramp with parameters as in Table 10.3: Instantaneous density contours (*top*), computational Schlieren imitation (*middle*), experimental Schlieren photograph (*bottom*). See also [180]

Table 10.3. Simulation parameters of the compression-decompression ramp flow of Ref. [330]

Parameter	Value	Comment
β	25°	
Ma_∞	2.95	
Re_{δ_0}	63560	
Re_{δ_1}	22120	
Re_θ	4705	using ν_∞
Re_{δ_2}	2045	using ν_w

as summarized in Table 10.3. Herein β is the deflection angle of the compression and the decompression ramp, respectively, Ma_∞ is the incoming-flow Mach number, Re_{δ_0} is the Reynolds number with respect to the boundary-layer thickness of the incoming boundary layer, Re_{δ_1} is the Reynolds number with respect to the displacement thickness of the incoming boundary layer, Re_{δ_2} is the Reynolds number with respect to the displacement thickness of the incoming boundary layer, Re_θ is the same as Re_{δ_2} except that the wall viscosity is used. The incoming boundary layer was generated in a separate flat-plate DNS using the recycling and rescaling technique for compressible flow [280]. Figure 10.17 shows that the computation recovers all relevant flow phenomena as sketched in Fig. 10.13. A more detailed qualitative comparison between computational and experimental data is made by Loginov et al. [181, 180].

Figure 10.18 shows a visualization of the average compression-ramp flow field. The computational-domain boundaries are indicated by thin black lines, crossflow-planes are colored according to values of the local mean temperature. A translucent isosurface at the mean-pressure level $\langle p \rangle = 0.1$ illustrates the mean compression shock originating from mean-flow separation. It is obvious that despite the fact that the flow geometry is nominally two-dimensional, STBLI breaks the spanwise translational symmetry. The temperature distribution in a cross-flow plane downstream of the interaction clearly shows a spanwise variation in the means, unlike that before the interaction. For an illustration of the mean-flow separation and the breaking of spanwise flow symmetry 10 colored mean streamlines are shown. Two pairs of counter-rotating streamwise vortices can be identified in the reattaching shear layer from isosurfaces of the contravariant streamwise vorticity (i.e. vorticity aligned with computational-grid lines, which corresponds roughly to a coordinate system aligned with the wall). A positively rotating vortex is indicated by cyan and a negatively rotating vortex by magenta.

Streamwise vortices reflect themselves in experimental oil-flow images in Fig. 10.19. It should be noted that this image was taken for a somewhat larger model than that of the experimental reference of the LES. This figure can be compared with Fig. 10.20 which shows the distribution of the computed mean wall-friction coefficient. The contour $C_f = 0$ is indicated by a thick solid line.

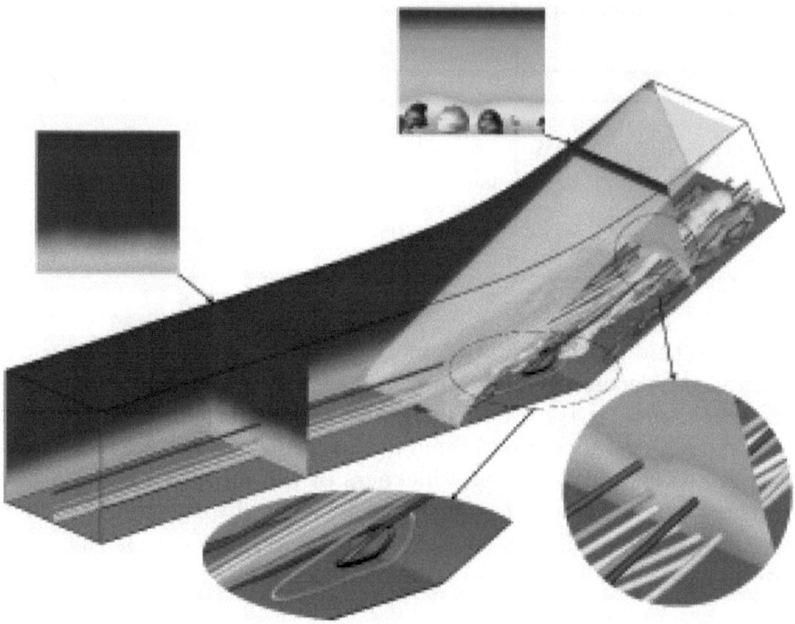

Fig. 10.18. Visualization of the average flow field along the compression ramp. Reproduced from Ref. [181] with permission

Also indicated by spanwise lines are mean separation and mean reattachment locations. Scaled by the respective incoming-flow boundary-layer thicknesses Fig. 10.19 and Fig. 10.20 agree with each other thus confirming the existence of two pairs of counter-rotating vortices separated by two incoming-flow boundary-layer thicknesses.

An impression of the instantaneous shock-wave structure can be obtained from a computed Schlieren-type visualization in Fig. 10.21. The flow snapshot is taken at a time instant of the simulation, a procedure which corresponds roughly to an experimental spark shadowgraph, although the shadowgraph exposure time is larger. Clearly visible are compression waves (5) above the separated shear layer (4) and the rearward stem of the λ-shock which originates from the reattachment region. Also the general shape of the forward shock appears to change slightly between the two time instants, a similar observation can be made for the experimental visualizations. This indicates an unsteady shock motion driven by the incoming turbulence. The rearward shock is highly unsteady and seems to disappear as an individual event at irregular time intervals. In the region of the rearward shock compression waves are shed which travel downstream. Some of these compression waves can be identified as shocklets and are the cause for a high level of fluctuations in the exterior flow downstream of reattachment. Although this particular computation covered only about $700\delta_0/U_\infty$, i.e. characteristic time units of the incom-

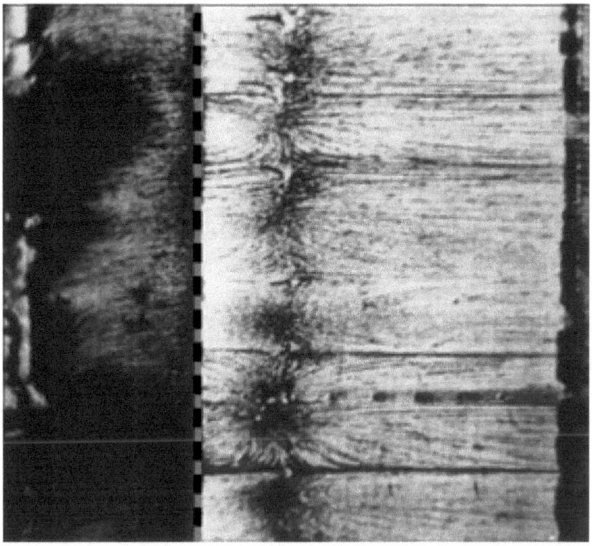

Fig. 10.19. Experimental oil-flow visualization, the *thick dashed vertical line* indicates the corner position. Reproduced from Ref. [181] with permission

ing boundary layer, indications for LSSM with one cycle spanning roughly over this time interval have been found. The spatial excursion of LSSM was estimated as about $1.3\delta_0$.

Table 10.4 summarizes the LES of shock-turbulence interaction at compression ramps discussed in this section.

Table 10.4. Summary of LES of shock-turbulence interaction at compression ramps discussed in this section

Ref.	Mach number	Reference *Re*	Deflection angle	Comment
[130]	$Ma_\infty = 1.55$	$Re_{\delta_0} = 10^6$	$\beta = 24°$	two-eddy-viscosity model, logarithmic wall model
[294]	$Ma_\infty = 3$	$Re_{\delta_0} = 20000$	$\beta = 8°$	Smagorinsky model
[278, 135]	$Ma_\infty = 3$	$Re_{\delta_0} = 21365$	$\beta = 18°$	ADM
[233, 234]	$Ma_\infty = 3$	$Re_{\delta_0} = 21000$	$\beta = 6°, 16°, 18°, 20°, 24°$	explicit filtering
[318]	$Ma_\infty = 2.88$	$Re_{\delta_0} = 20000$	$\beta = 25°$	Smagorinsky model
[181]	$Ma_\infty = 2.95$	$Re_{\delta_0} = 63560$	$\beta = 25°$	ADM
[46]	$Ma_\infty = 2.5$	$Re_{\delta_1} = 280$	$\beta = 20°$	transitional flow, selective structure-function model

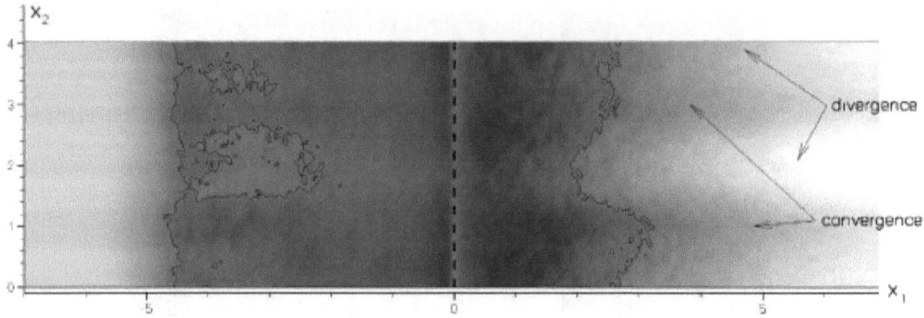

Fig. 10.20. Distribution of the mean wall-friction coefficient. Reproduced from Ref. [181] with permission

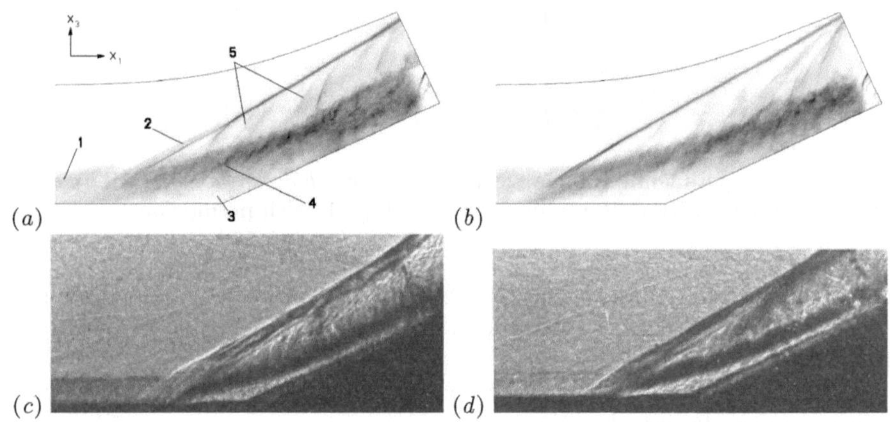

Fig. 10.21. Spark-shadowgraph imitation at two subsequent time instants of simulation (**a**) and (**b**); spark-shadowgraph of experiment (**c**) and (**d**); time instants of computation and experiment have each been chosen arbitrarily. Reproduced from Ref. [181] with permission

Normal Shock Configurations

The normal shock configuration at transonic Mach number was considered by Sandham et al. [248]. Figure 10.22 gives an overview of the flow configuration. The turbulent boundary layer along a flat plate at a displacement-thickness Reynolds number $Re_{\delta_1} = 2910$ encounters a bump where the flow is accelerated into a supersonic state. At the end of the bump, where the flow Mach number reaches $Ma = 1.16$ a compression shock appears. In the numerical setup the normal shock is established by prescribing an appropriate exit pressure and enforcing a slip boundary condition at the upper wall. A dynamic Smagorinsky model is used in connection with a high-order entropy-split finite-difference scheme. The acceleration of the subsonic flow along the bump causes initial fluctuations to decay by about a factor of 3, interaction with the nor-

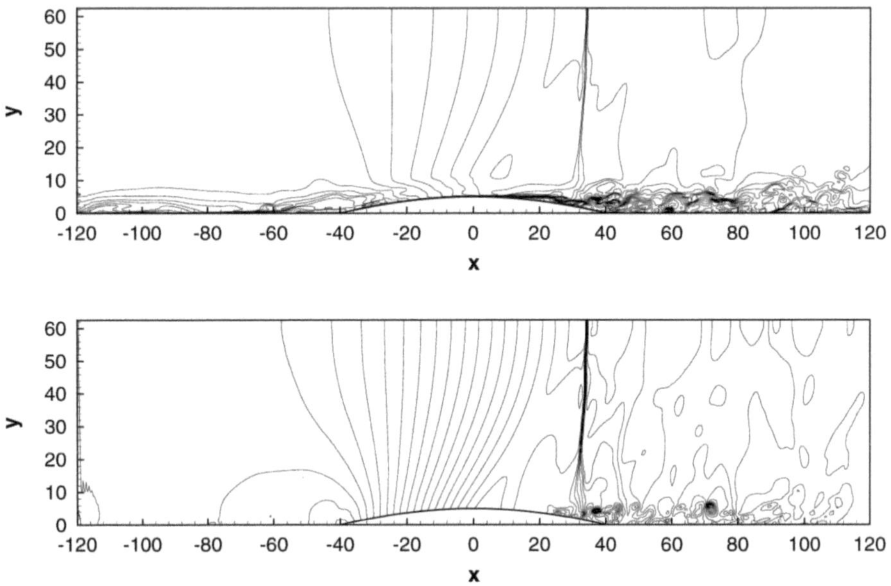

Fig. 10.22. Instantaneous Mach number (*top*) and pressure (*bottom*) contours for normal-shock-boundary-layer interaction. Reproduced from Ref. [248] with permission

mal shock amplifies fluctuations by about a factor of 8. In the simulation a steady shock was found. An illustration of turbulence amplification due to shock-turbulent-boundary-layer interaction is given in Fig. 10.23. Turbulent fluctuations are slightly damped due to flow acceleration before the interaction, which also causes a significant increase in turbulent fluctuations and in turbulence anisotropy.

Wollblad et al. [314] also consider the turbulent flow over a bump at transonic conditions. The Reynolds number with respect to the incoming boundary-layer thickness is $Re_{\delta_0} = 13400$. A Mach number of about $Ma = 1.3$ was reached at the shock location by adjusting the exit pressure, see Fig. 10.24. The inflow data are generated from a rescaled incompressible experimental mean-flow profile. The mean temperature profile is reconstructed based on the Crocco-Busemann relation. Whereas the geometry was adjusted to an available experiment the Reynolds number was reduced by about a factor of 10. A third-order upwind scheme with Jameson dissipation for shock capturing was used. SGS were modeled by compressible versions of the dynamic Smagorinsky model and the wall-adapting local eddy-viscosity model (WALE), respectively. The computations show a small separation on the downstream flank of the bump, see Fig. 10.25. The streamwise component of the resolved Reynolds stress show a significant amplification due to the interaction, see Fig. 10.26. A comparison of the mean streamlines and instan-

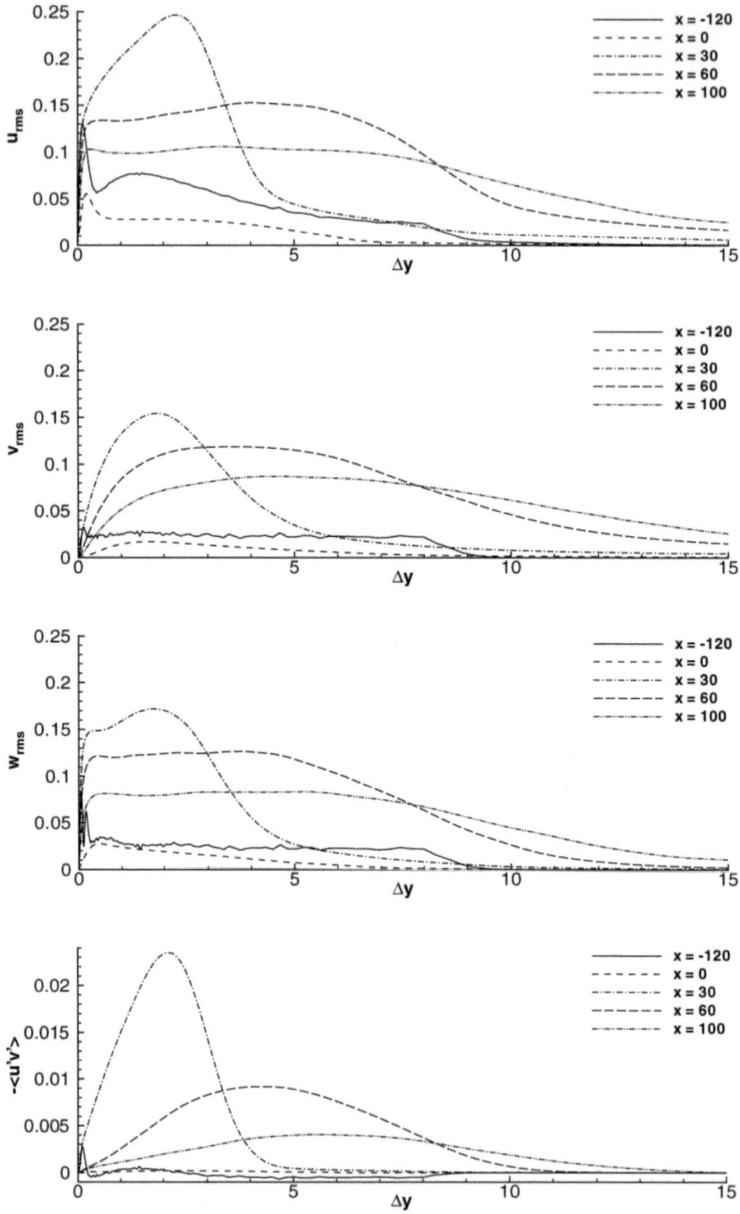

Fig. 10.23. RMS values for the streamwise, wall-normal and spanwise velocity fluctuations and for the streamwise-wall-normal Reynolds stress component (*top to bottom*) at several downstream stations. Reproduced from Ref. [248] with permission

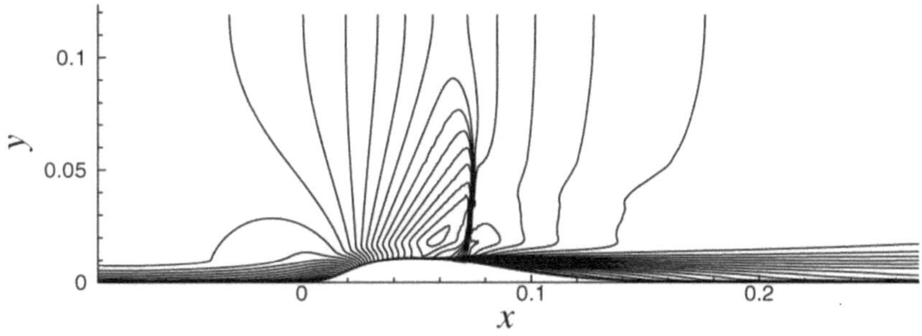

Fig. 10.24. Mean Mach number contours for the configuration investigated by Wollblad et al. Reproduced from Ref. [314] with permission

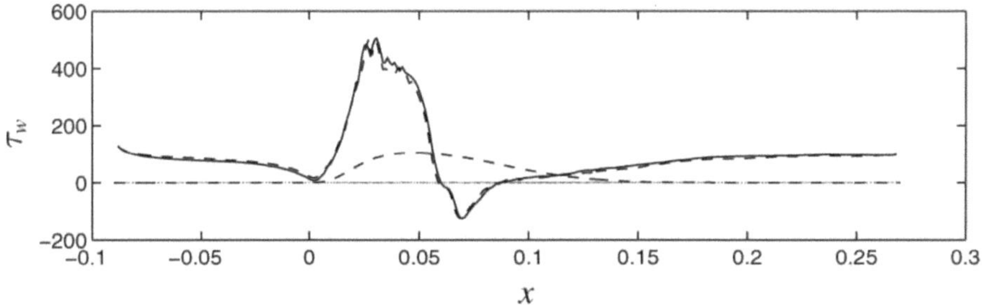

Fig. 10.25. Wall friction for the computations of Wollblad et al. at two different grid resolution. Reproduced from Ref. [314] with permission

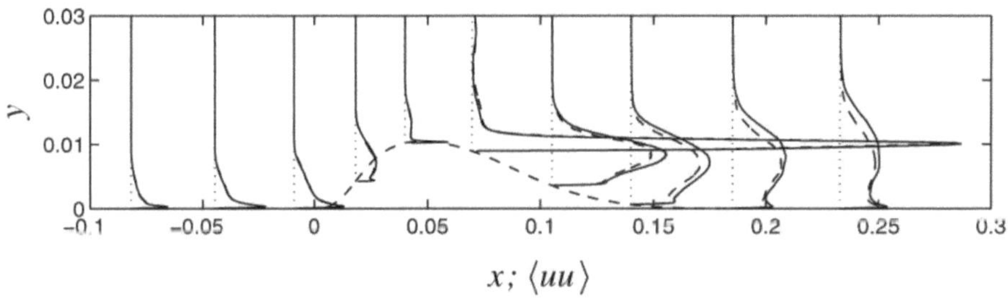

Fig. 10.26. Profiles of the streamwise Reynolds stress for the computations of Wollblad et al. at two different grid resolutions. Reproduced from Ref. [314] with permission

taneous streamlines for the separation region reveals several instantaneous recirculation regions, yielding a much more narrow separation region in wall-normal direction for the mean flow. Another important finding is that the computations did not show any indication of LSSM, unlike the reference experiment.

Table 10.5 summarizes the LES of shock-turbulent-boundary-layer interactions with normal shocks discussed in this section.

Table 10.5. Summary of LES of shock-turbulent-boundary-layer interactions with normal shocks discussed in this section

Ref.	Mach number	Reference Re	Comment
[248]	$Ma_{\max} = 1.16$	$Re_l = 233000$	Reynolds-number based on bump length, dynamic Smagorinsky model
[314]	$Ma_0 = 0.7$	$Re_{\delta_0} = 13400$	Mach number at inflow, Smagorinsky model + WALE

Impinging Shock Configurations

The case of an oblique shock impinging on a flat plate was studied by Garnier et al. [94]. For the case considered a reference experiment was available. The incoming-flow Mach number is 2.3, the Reynolds number measured by the incoming boundary-layer displacement thickness is about $Re_{\delta_1} = 19000$. The impinging shock is generated by an 8° deflection of the exterior flow. Turbulent inflow data have been generated by an inflow-rescaling and recycling method following that of [183]. As subgrid-scale model the mixed-scale model with a smoothed version of the selection function was employed. Compressibility-related terms in the energy equation were partially accounted for. As numerical discretization a WENO-filtered scheme, adapted from [92] was used. Figure 10.27 shows the configuration by the mean wall-pressure distribution in gray coding, also indicated is the mean sonic line within the boundary layer.

The results with explicit SGS models were compared with no-model computations where the nonlinear stabilization of the underlying numerical scheme was used to prevent nonlinear instability (i.e. instability due to underresolution). A satisfactory agreement with the experimental data was found in terms of mean-flow and velocity fluctuation profiles upstream and downstream of the interaction region. A mean-flow separation is present but quite weak. Accordingly no significant shock-separation dynamics was reported and the effect of the spanwise-domain length was found to be small. A resolution study exhibited a significant effect on the wall friction whereas for mean and fluctuation profiles the effect was rather small. From Fig. 10.28 it can be also noticed that the no-model computations (case D) appear to represent the incoming flow in better agreement with experimental hotwire-anemometry data (HWA) than the computations with SGS model (case A to C, with increasing resolution), these computations, however, fail to predict the typical two-peak structure of the separation (not resolved by the experimental data). Fluctuation data downstream of the interaction are in good agreement with LDA (laser-doppler anemometry) experimental data, with the highest resolution LES (case C) appears somewhat better than the other simulations in the near-wall region,

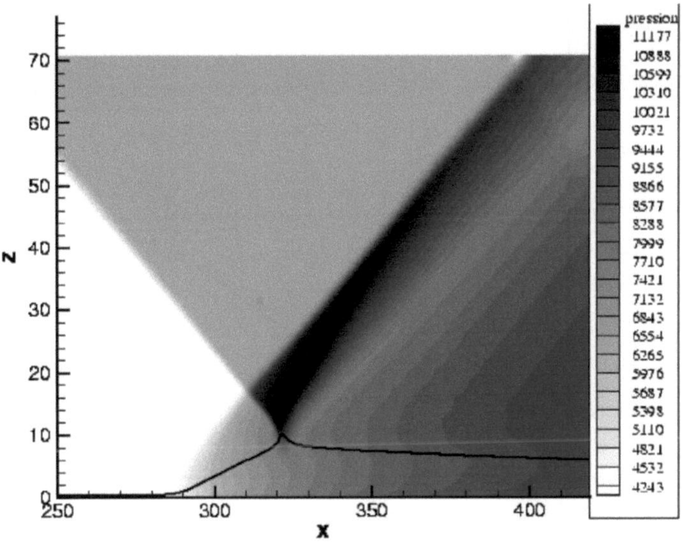

Fig. 10.27. Distribution of the mean wall-friction coefficient. Reproduced from Ref. [94]

and the no-model simulation (case D) appears to give the worst prediction in the outer part of the boundary layer, see Fig. 10.29 for a position about 30 incoming-boundary-layer displacements thicknesses downstream of the interaction.

Impinging-shock interaction with a transitional boundary layer is investigated by Teramoto [286]. The LES are aimed at reproducing an experiment at a Reynolds number of about 20000 based on the incoming laminar boundary-layer thickness. The incoming Mach number was 2 and the deflection angle of the imping shock about 35°. As numerical discretization a hybrid compact-Roe scheme was used, where the coupling spurious solutions were suppressed by a high-order compact filter. For SGS modeling the selective mixed-scale model was used. Boundary-layer instabilities are seeded by superimposing decaying isotropic turbulence to the free-stream of the incoming boundary layer at a specified intensity. The turbulence intensity providing the best match with the experiment was used for the simulation. An overview of the computation can be obtained from Mach-number contours in Fig. 10.30. A downstream sequence of mean-flow profiles is shown in Fig. 10.31, where the interaction takes place at about 60 mm downstream. The disagreement between experimental data and computation at the first station indicates that transition in the simulation takes place somewhat prematurely compared with the experiment. Several grid resolutions were tested, the finest (grid A) giving essentially the best agreement with the experiment, see Fig. 10.32 for the wall-friction coefficient. The results also show a rapid generation of near-wall streaks in

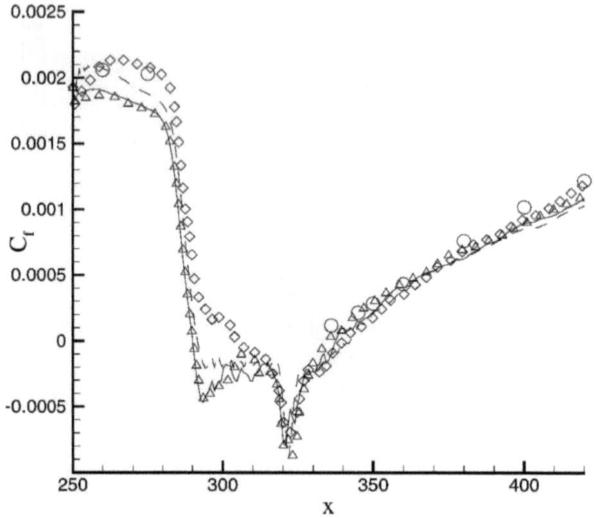

Fig. 10.28. Streamwise evolution of the wall-friction coefficient: *solid line*: case A, *triangle*: case B, *dashed line*: case C, *diamond*: case D, *circle*: HWA. Reproduced from Ref. [94]

the reattaching boundary layer. The computational domain having a spanwise extent of about 1.5 incoming-boundary-layer thicknesses is too narrow for the representation of Görtler-like vortices.

Table 10.6 summarizes the LES of shock-turbulent-boundary-layer interactions with impinging shocks discussed in this section.

Table 10.6. Summary of LES of shock-turbulent-boundary-layer interactions with impinging shocks discussed in this section

Ref.	Mach number	Reference Re	Shock generator	Comment
[94]	$Ma_0 = 2.3$	$Re_{\delta_1} = 19000$	$\beta = 8°$	mixed-scale model
[286]	$Ma_0 = 2.0$	$Re_l = 3.29 \cdot 10^5$	$\beta = 6°$	Reynolds number based on distance between leading edge and shock impingement, selective mixed-scale model

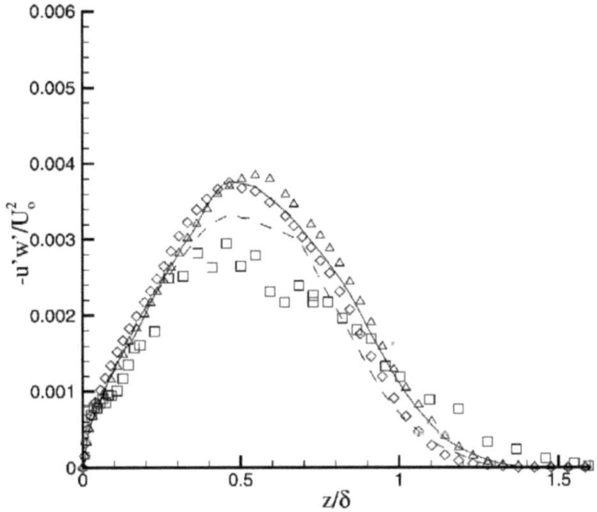

Fig. 10.29. RMS of the streamwise velocity fluctuation downstream of interaction: *solid line*: case A, *triangle*: case B, *dashed line*: case C, *diamond*: case D, *circle*: HWA, *square*: LDA. Reproduced from Ref. [94]

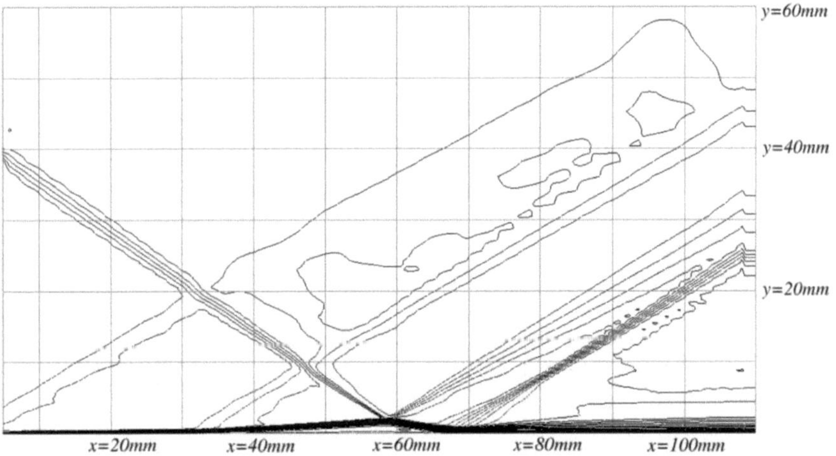

Fig. 10.30. Mach-number contours for transitional shock-boundary-layer interaction. Reproduced from Ref. [286]

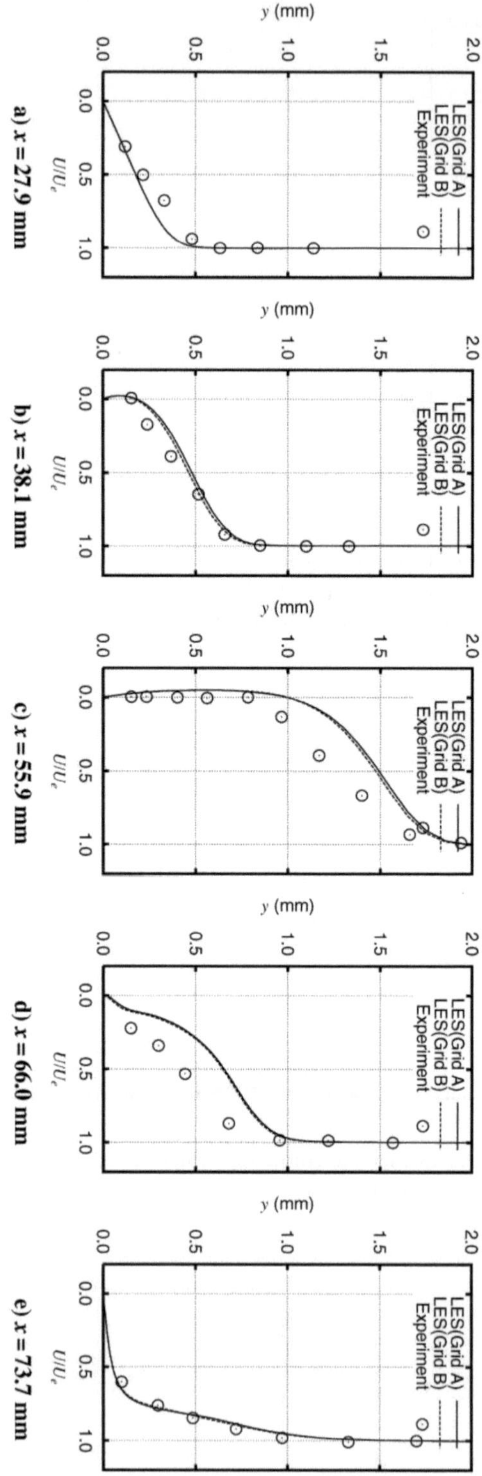

Fig. 10.31. Downstream sequence of mean-flow profiles for SWTI on the transitional boundary layer. Reproduced from Ref. [286]

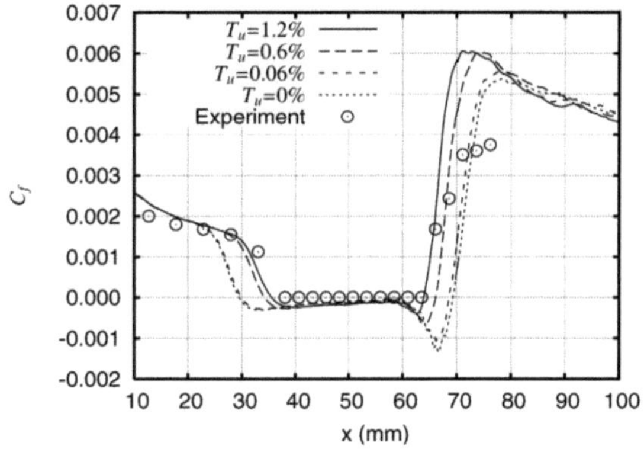

Fig. 10.32. Downstream distribution of the wall friction coefficient for different grid resolutions, experimental reference indicated by a *circle*. Reproduced from Ref. [286]

References

[1] Abbott, L. F., Deser, S. (1982): Stability of gravity with a cosmological constant, Nucl. Phys. B **195**, 76–96

[2] Ackeret, J., Feldmann, F., Rott, N. (1946): Untersuchungen an Verdichtungsstößen and Grenzschichten in schnell bewegten Gasen, ETH Zürich, Mitt. Inst. Aerodyn. **10**, Zürich, Switzerland

[3] Adams, N. A. (2000): Direct simulation of the turbulent boundary layer along a compression ramp at $M = 3$ and $Re_\theta = 1685$, J. Fluid Mech. **420**, 47–83

[4] Adams, N. A., Stolz, S. (2002): A subgrid-scale deconvolution approach for shock-capturing, J. Comput. Phys. **178**, 391–426

[5] Adams, N. A., Hickel, S., Franz, S. (2004): Implicit subgrid-scale modeling by adaptive deconvolution, J. Comput. Phys. **200**, 412–431

[6] Adamson, T. C., Messiter, A. F. (1980): Analysis of two-dimensional interactions between shock waves and boundary layers, Annu. Rev. Fluid Mech. **12**, 103–138

[7] Andreopoulos, Y., Agui, J. H., Briassulis, G. (2000): Shock wave-turbulence interactions, Annu. Rev. Fluid Mech. **32**, 309–345

[8] Andreopoulos, Y., Muck, K. C. (1987): Some new aspects of the shock-wave/boundary-layer interaction in compression-ramp flows, J. Fluid Mech. **180**, 405–428

[9] Ahmadi, G. (1988): Thermodynamically consistent k-Z models for compressible turbulent flows, Appl. Math. Model. **12**, 391–398

[10] Andersson, N., Eriksson, L.-E., Davidson, L. (2005): Large eddy simulation of subsonic turbulent jets and their radiated sound, AIAA J. **43**(9), 1899–1912

[11] Bailly, C., Lafon, P., Candel, S. (1995): A stochastic approach to compute noise generation and radiation of free turbulent flows, AIAA Paper 95-092

[12] Bailly, C., Juvé, D. (1999): A stochastic approach to compute subsonic noise using linearized Euler's equations, AIAA Paper 99-1872

[13] Bardina, J., Ferziger, J. H., Reynolds, W. C. (1983): Improved turbulence modeling based on large-eddy simulation of homogeneous, incompressible, turbulent flows, Technical Report TF-19, Thermosciences Division, Stanford University

[14] Bataille, F. (1994): Etude d'une turbulence faiblement compressible dans le cadre d'une modélisation quasi-normale avec amortissement tourbillonnaire, PhD Thesis, Ecole Centrale de Lyon (in French)

E. Garnier et al., *Large Eddy Simulation for Compressible Flows*,
Scientific Computation,

[15] Bataille, F., Zhou, Y. (1999): Nature of the energy transfer process in compressible turbulence, Phys. Rev. E **59**(5), 5417–5426

[16] Bayly, B. J., Levermore, C. D., Passot, T. (1992): Density variations in weakly compressible flows, Phys. Fluids **4**(5), 945–954

[17] Bechara, W., Bailly, C., Lafon, P. (1994): Stochastic approach to noise modelling for free turbulent flows, AIAA J. **32**(3), 455–463

[18] Berland, J., Bogey, C., Bailly, C. (2007): Numerical study of screech generation in a planar supersonic jet, Phys. Fluids **19**, 075105

[19] Berselli, L. C., Iliescu, T., Layton, W. J. (2006): Mathematics of Large Eddy Simulation of Turbulent Flows, Springer, Berlin

[20] Bertero, M., Boccacci, P. (1998): Introduction to Inverse Problems in Imaging, Institute of Physics Publishing, Bristol

[21] Bertoglio, J. P., Bataille, F., Marion, J. D. (2001): Two-point closures for weakly compressible turbulence, Phys. Fluids **13**(1), 290–310

[22] Billson, M. (2002): Computational techniques for jet noise predictions, PhD Thesis, Chalmers University of Technology, Göteborg, Sweden

[23] Bodony, D. J., Lele, S. K. (2005): On using large-eddy simulation for the prediction of noise from cold and heated turbulent jets, Phys. Fluids **17**, 085103

[24] Bodony, D. J., Lele, S. K. (2008): Current status of jet noise predictions using large eddy simulation, AIAA J. **46**(2), 364–380

[25] Bogey, C., Bailly, C., Juvé, D. (2003): Noise investigation of a high subsonic, moderate Reynolds number jet using a compressible large eddy simulation, Theor. Comput. Fluid Dyn. **16**, 273–297

[26] Bogey, C., Bailly, C. (2004): A family of low dispersive and low dissipative explicit schemes for flow and noise computations, J. Comput. Phys. **194**, 194–214

[27] Bogey, C., Bailly, C. (2005): Effects of inflow conditions and the forcing on subsonic jet flows and noise, AIAA J. **43**(5), 1000–1007

[28] Bogey, C., Bailly, C. (2005): Decrease of the effective Reynolds number with eddy-viscosity subgrid-scale modeling, AIAA J. **43**(2), 437–439

[29] Bogey, C., Bailly, C. (2006): Computation of a high Reynolds number jet and its radiated noise using large eddy simulation based on explicit filtering, Comput. Fluids **35**, 1344–1358

[30] Boris, J. P., Grinstein, F. F., Oran, E. S., Kolbe, R. L. (1992): New insights into large eddy simulation, Fluid Dyn. Res. **10**, 199–228

[31] Boussinesq, M. J. (1877): Essai sur la théorie des eaux courantes. In: Mémoires présentés par divers savants à l'Académie des Sciences, Imprimerie Nationale, Paris, France, tome XXIII, pp. 43–47

[32] Brun, C., Boiarciuc, M. P., Haberkorn, M., Comte, P. (2008): Large eddy simulation of compressible channel flow, Theor. Comput. Fluid Dyn. **22**(3–4)

[33] Cai, X. D., O'Brien, E. E., Ladeinde, F. (1997): Thermodynamic behavior in decaying, compressible turbulence with initially dominant temperature fluctuations, Phys. Fluids **9**(6), 1754–1763

[34] Cai, X. D., O'Brien, E. E., Ladeinde, F. (1998): Advection of mass fraction in forced, homogeneous, compressible turbulence, Phys. Fluids **10**(9), 2249–2259

[35] Cambon, C., Coleman, G. N., Mansour, N. N. (1993): Rapid distortion analysis and direct simulation of compressible homogeneous turbulence at finite Mach number, J. Fluid Mech. **257**, 641–665

[36] Cambon, C., Scott, J. F. (1999): Linear and nonlinear models of anisotropic turbulence, Annu. Rev. Fluid Mech. **31**, 1–53

[37] Cambon, C., Sagaut, P., private communication
[38] Canuto, C., Hussaini, M. Y., Quarteroni, A., Zang, T. A. (1988): Spectral Methods in Fluid Dynamics, Springer, Berlin
[39] Chang, P. A., Piomelli, U., Blake, W. K., (1999): Relationship between wall pressure and velocity-field sources, Phys. Fluids **11**(11), 3434–3448
[40] Chassaing, P., Antonia, R. A., Anselmet, F., Joly, L., Sarkar, S., (2002): Variable Density Turbulence, Springer, Berlin
[41] Chollet, J. P. (1984): Two-point closures as a subgrid-scale modeling tool for large-eddy simulations. In: Launder, F., Durst, B. E. (eds.), Turbulent Shear Flows IV, pp. 62–72, Springer, Heidelberg
[42] Chu, B. T., Kovasznay, L. S. G. (1957): Non-linear interactions in a viscous heat-conducting compressible gas, J. Fluid Mech. **3**, 494–514
[43] Coleman, G. N., Kim, J., Moser, R. D. (1995): A numerical study of turbulent supersonic isothermal-wall channel flow, J. Fluid Mech. **305**, 159–183
[44] Collis, S. S. (2001): Monitoring unresolved scales in multiscale turbulence modeling, Phys. Fluids **13**, 1800–1806
[45] Comte-Bellot, G., Corrsin, S. (1971): Simple Eulerian time correlation of full and narrow-band velocity signals in grid-generated isotropic turbulence, J. Fluid Mech. **48**, 273–337
[46] Comte, P., David, E., (1996): Large-eddy simulation of Görtler vortices in a curved compression ramp. In: Désidéi, J. A., Chetverushkin, B. N., Kuznetsov, Y. A., Périaux, J., Stoufflet, B. (eds.), Experimentation, Modelling and Computation in Flow, Turbulence and Combustion, vol. 1, pp. 45–61, Wiley, New York
[47] Comte, P., Lesieur, M. (1998): Large-eddy simulation of compressible turbulent flows, Lecture Series 1998-05, Advance in turbulence modeling, Von Karman Institute for Fluid Dynamics
[48] David, E. (1993): Modélisation des écoulements compressibles et hypersoniques, PhD Thesis, Grenoble University
[49] DeBonis, J. R., Scott, J. N. (2002): Large-eddy simulation of a turbulent compressible round jet, AIAA J. **40**(7), 1346–1354
[50] DeBonis, J. (2007): Progress toward large-eddy simulations for prediction of realistic nozzle systems, J. Propuls. Power **23**(5), 971–980
[51] Deck, S. (2005): Numerical simulation of transonic buffet over a supercritical airfoil, AIAA J. **43**, 1556–1566
[52] Dolling, D. S. (1998): High-speed turbulent separated flows: consistency of mathematical models and flow physics, AIAA J. **36**, 725–732
[53] Dolling, D. S., Murphy, M. T. (1983): Unsteadiness of the separation shock wave structure in a supersonic compression ramp flowfield, AIAA J. **12**, 1628–1634
[54] Domaradzki, J. A., Saiki, E. (1997): A subgrid-scale model based on the estimation of unresolved scales of turbulence, Phys. Fluids **9**, 2148–2164
[55] Domaradzki, J. A., Adams, N. A. (2002): Direct modelling of subgrid scales of turbulence in large eddy simulations, J. Turbul. **3**, 024
[56] Domaradzki, J. A., Loh, K. C., Yee, P. P. (2002): Large eddy simulations using the subgrid-scale estimation model and truncated Navier-Stokes dynamics, Theor. Comput. Fluid Dyn. **15**, 421–450
[57] Domaradzki, J. A., Xiao, Z., Smolarkiewicz, P. K. (2003): Effective eddy viscosities in implicit large eddy simulations of turbulent flows, Phys. Fluids **15**, 3890–3893

[58] Doris, L., Tenaud, C., Ta Phuoc, L. (2000): LES of spatially developing 3D compressible mixing layer, C. R. Acad. Sci., Sér. IIb **328**(7), 567–573
[59] Dubief, Y., Delcayre, F. (2000): On the coherent-vortex identification in turbulence, J. Turbul. **1**, 11
[60] Dubois, T., Temam, R. (1999): Dynamic Multilevel Methods and the Numerical Simulation of Turbulence, Cambridge University Press, Cambridge
[61] Dubois, T., Domaradzki, J. A., Honein, A. (2002): The subgrid scale estimation model applied to large eddy simulations of compressible turbulence, Phys. Fluids **14**(5), 1781–1801
[62] Ducros, F., Comte, P., Lesieur, M. (1996): Large-eddy simulation of transition to turbulence in a boundary layer developing spatially over a flat plate, J. Fluid Mech. **326**, 1–36
[63] Ducros, F., Ferrand, V., Nicoud, F., Weber, C., Darracq, D., Gacherieu, C., Poinsot, T. (1999): Large-eddy simulation of the shock/turbulence interaction, J. Comput. Phys. **152**, 517–545
[64] Dunca, A., Epshteyn, Y. (2006): On the Stolz-Adams deconvolution model for large-eddy simulation of turbulent flows, SIAM J. Math. Anal. **37**, 1890–1902
[65] Dupont, P, Haddad, C., Debieve, J. F. (2006): Space and time organization in a shock-induced separated boundary layer, J. Fluid Mech. **559**, 255–277
[66] Dussauge, J.-P., Dupont, P., Debieve, J.-F. (2006): Unsteadiness in shock wave boundary layer interactions with separation, Aerospace Sci. Technol. **10**, 85–91
[67] Dyakov, S. P. (1954): On the stability of shock waves, Zh. Eksp. Teor. Fiz. **27**, 288 [Atomic Research Agency Establishment AERE Lib./Trans. 648 (1956)]
[68] Eckhoff, K. S., Storesletten, L. (1978): A note on the stability of steady inviscid helical gas flows, J. Fluid Mech. **89**, 401
[69] Eidson, T. M. (1985): Numerical simulation of the turbulent Rayleigh-Bénard problem using subgrid modelling, J. Fluid Mech. **158**, 245–268
[70] El-Hady, N. M., Zang, T. A., Piomeli, U. (1994): Application of the dynamic subgrid-scale model to axisymmetric transitional boundary layer at high speed, Phys. Fluids **6**, 1299–1309
[71] Erengil, M. E., Dolling, D. S. (1991): Unsteady wave structure near separation in a Mach 5 compression ramp interaction, AIAA J. **29**, 728–735
[72] Erengil, M. E., Dolling, D. S. (1991): Correlation of separation shock motion with pressure fluctuations in the incoming boundary layer, AIAA J. **29**, 1868–1877
[73] Erlebacher, G., Hussaini, M. Y., Speziale, C. G., Zang, T. A. (1987): Toward the large-eddy simulation of compressible turbulent flows, NASA CR 178273, ICASE Report No. 87-20
[74] Erlebacher, G., Hussaini, M. Y., Kreiss, H. O., Sarkar, S. (1990): The analysis and simulation of compressible turbulence, Theor. Comput. Fluid Dyn. **2**, 73–95
[75] Erlebacher, G., Hussaini, M. Y., Speziale, C. G., Zang, T. A. (1982): Toward the large-eddy simulation of compressible turbulent flows, J. Fluid Mech. **238**, 155–185
[76] Erlebacher, G., Sarkar, S. (1993): Statistical analysis of the rate of strain tensor in compressible homogeneous turbulence, Phys. Fluids **5**(12), 3240–3254
[77] Fabre, D., Jacquin, L., Sesterhenn, J. (2001): Linear interaction of a cylindrical entropy spot with a shock wave, Phys. Fluids **13**(8), 2403–2422

[78] Fauchet, G. (1998): Modélisation en deux points de la turbulence isotrope compressible et validation à l'aide de simulations numériques, Thèse de Doctorat, Ecole Centrale de Lyon (in French)
[79] Fauchet, G., Bertoglio, J. P. (1999): A two-point closure for compressible turbulence, C. R. Acad. Sci. Paris, Sér. IIb **327**, 665–671 (in French)
[80] Fauchet, G., Bertoglio, J. P. (1999): Pseudo-sound and acoustic régimes compressible turbulence, C. R. Acad. Sci. Paris, Sér. IIb **327**, 665–671 (in French)
[81] Favre, A. (1971): Equations statistiques aux fluctuations turbulentes dans les écoulements compressibles : cas des vitesses et des températures, C. R. Acad. Sci. Paris **273**, 1971
[82] Feiereisen, W. J., Reynolds, W. C., Ferziger, J. H. (1981): Numerical simulation of compressible homogeneous turbulent shear flow, Report No. TF 13, Stanford University
[83] Foias, C., Manley, O., P, Temam, R. (1991): Approximate inertial manifolds and effective viscosity in turbulent flows, Phys. Fluids **A3**, 898–911
[84] Forestier, N., Jacquin, L., Geoffroy, P. (2003): The mixing layer over a deep cavity at high-subsonic speed, J. Fluid Mech. **475**, 101–145
[85] Foysi, H., Sarkar, S., Friedrich, R. (2004): Compressibility effects and turbulence scaling in supersonic channel flow, J. Fluid Mech. **509**, 207–216
[86] Fureby, C., Tabor, G. (1997): Mathematical and physical constraints on large-eddy simulations, Theor. Comput. Fluid Dyn. **9**, 85–102
[87] Fureby, C., Grinstein, F. F. (1999): Monotonically integrated large eddy simulation of free shear flows, AIAA J. **37**(5), 544–556
[88] Fureby, C., Grinstein, F. F. (2002): Large eddy simulation of high-Reynolds-number free and wall-bounded flows, J. Comput. Phys. **181**, 68–97
[89] Freund, J. (2001): Noise sources in a low-Reynolds-number turbulent jet at Mach 0.9, J. Fluid Mech. **438**, 277–305
[90] Gamet, L., Estivalezes, J. L. (1998): Application of large-eddy simulations and Kirchhoff method to jet noise prediction, AIAA J. **36**(12), 2170–2178
[91] Garnier, E., Mossi, M., Sagaut, P., Comte, P., Deville, M. (1999): On the use of shock-capturing scheme for large-eddy simulation, J. Comput. Phys. **153**, 273–311
[92] Garnier, E., Sagaut, P., Deville, M. (2001): A class of explicit ENO filters with application to unsteady flows, J. Comput. Phys. **170**, 184–204
[93] Garnier, E., Sagaut, P., Deville, M. (2002): Large eddy simulation of shock/homogeneous turbulence interaction, Comput. Fluids **31**, 245–268
[94] Garnier, E., Sagaut, P., Deville, M. (2002): Large-eddy simulation of shock/boundary-layer interaction, AIAA J. **40**, 1935–1944
[95] Germano, M., Piomelli, U., Moin, P., Cabot, W. H. (1991): A dynamic subgrid-scale eddy-viscosity model, Phys. Fluids A **3**, 1760–1765
[96] Ghosal, S., Moin, P. (1995): The basic equations for the large-eddy simulation of turbulent flows in complex geometry, J. Comput. Phys. **118**, 24–37
[97] Ghosal, S., An analysis of numerical errors in large-eddy simulations of turbulence, J. Comput. Phys. **125**, 187–206
[98] Ghosal, S., Lund, T. S., Moin, P., Akselvoll, K. (1995): A dynamic localization model for large-eddy simulation of turbulent flows, J. Fluid Mech. **286**, 229–255
[99] Ghosh, S., Sesterhenn, J., Friedrich, R. (2008): Large-eddy simulation of supersonic turbulent flow in axisymmetric nozzles and diffusers, Int. J. Heat Fluid Flow **29**, 579–590

[100] Gloerfelt, X., Bogey, C., Bailly, C. (2003): Numerical evidence of mode switching in the flow-induced oscillations by a cavity, Aeroacoustics **2**(2), 99–124
[101] Goody, M. (2004): Empirical spectral Model of surface pressure fluctuations, AIAA J. **42**(9), 1788–1793
[102] Grinstein, F. F., Fureby, C. (2002): Recent progress on MILES for high-Reynolds-number flows, Trans. ASME **124**, 848–861
[103] Grinstein, F. F., Fureby, C. (2004): From canonical to complex flows: recent progress on monotonically integrated LES, Comput. Sci. Eng. **6**, 36–49
[104] Grinstein, F. F., Fureby, C. (2006): Recent progress on flux-limiting based implicit large-eddy simulation. In: Wesseling, P., Onate, E., Periaux, J. (eds.), Proc. ECCOMAS CFD 2006
[105] Grinstein, F. F., Guirguis, R. H. (1992): Effective viscosity in the simulation of spatially evolving shear flows with monotonic FCT models, J. Comput. Phys. **101**, 165–175
[106] Grinstein, F. F., Margolin, L. G., Rider, W. J. (2007): Implicit Large Eddy Simulation—Computing Turbulent Fluid Dynamics, Cambridge University Press, Cambridge
[107] Haminh, H., Vandromme, D. (1989): Compressibility effects on turbulence in high speed flows. In: Summer Workshop, Theory and Modeling of Turbulent Flow, Ecole centrale de Lyon, Lyon
[108] Hannappel, R., Friedrich, R. (1995): Direct numerical simulation of a Mach 2 shock interacting with isotropic turbulence, Appl. Sci. Res. **54**, 501–505
[109] Harten, A., Engquist, B., Osher, S., Chakravarthy, S. R. (1987): Uniformly high order accurate essentially non-oscillatory schemes, III, J. Comput. Phys. **71**, 231–303
[110] Hayes, W. D. (1957): The vorticity jump across a gasdynamic discontinuity, J. Fluid Mech. **2**, 595–600
[111] Heisenberg, W. (1946): Zur statistischen Theorie der Turbulenz, Z. Phys. **124**, 628–657
[112] Hickel, S., Adams, N. A., Domaradzki, J. A. (2006): An adaptive local deconvolution method for implicit LES, J. Comput. Phys. **213**, 413–436
[113] Hickel, S., Adams, N. A. (2007): On implicit subgrid-scale modeling in wall-bounded flows, Phys. Fluids **19**, 105106
[114] Hickel, S., Adams, N. A., Mansour, N. N. (2007): Implicit subgrid-scale modeling for large-eddy simulation of passive-scalar mixing, Phys. Fluids **19**, 095102
[115] Hickel, S., Adams, N. A. (2008): Implicit LES applied to zero-pressure-gradient and adverse-pressure-gradient boundary-layer turbulence, Int. J. Heat Fluid Flow **29**, 626–639
[116] Hickel, S., Adams, N. A. (2008): Implicit large-eddy simulation applied to turbulent channel flow with periodic constrictions, Theor. Comput. Fluid Dyn. **22**, 227–242
[117] Hickel, S., Larsson, J. (2008): An adaptive local deconvolution model for compressible turbulence. In: Proceedings of the CTR Summer Program 2008, Center for Turbulence Research, Stanford University
[118] Hinze, J. O. (1975): Turbulence, 2nd edn., McGraw-Hill College, New York
[119] Hirt, C. W. (1968): Heuristic stability theory for finite-difference equations, J. Comput. Phys. **2**, 339–355
[120] Hirsch, C. (1988): Numerical Computation of Internal and External Flows, vol. I, Wiley, New York

[121] Honein, A., Moin, P. (2004): Higher entropy conservation and numerical stability of compressible turbulence simulations, J. Comput. Phys. **201**, 531–545
[122] Honein, A., Moin, P. (2005): Numerical aspects of compressible turbulence simulations, Report No. TF-92, Flow Physics and Computation Division, Stanford University, Stanford CA, USA
[123] Hoyas, S., Jimenez, J. (2006): Scaling of the velocity fluctuations in turbulent channels up to $Re_\tau = 2003$, Phys. Fluids **18**, 011702
[124] Howe, M. S. (1998): Acoustics of Fluid-Structure Interactions, Cambridge University Press, Cambridge
[125] Hu, Z., Morfey, C., Sandham, N. D. (2003): Sound radiation in turbulent channel flows, J. Fluid Mech. **475**, 269–302
[126] Hu, Z., Morfey, C., Sandham, N. D. (2006): Sound radiation from a turbulent boundary layer, Phys. Fluids **18**, 098101
[127] Hughes, T. J. R., Feijoo, G. R., Mazzei, L., Qunicy, J.-B. (1998): The variational multiscale method—a paradigm for computational mechanics, Comput. Methods Appl. Mech. Eng. **166**, 3–24
[128] Hughes, T. J. R., Mazzei, L., Jansen, K. E. (2000): Large eddy simulation and the variational multiscale method, Comput. Vis. Sci. **3**, 47–59
[129] Hughes, T. J. R., Mazzei, L., Oberai, A. A., Wray, A. A. (2001): The multiscale formulation of large-eddy simulation: decay of homogeneous isotropic turbulence, Phys. Fluids **13**, 505–512
[130] Hunt, D. L., Nixon, D. (1995): A very large eddy simulation of an unsteady shock wave/turbulent boundary layer interaction, AIAA paper 95-2212
[131] Jacquin, L., Cambon, C., Blin, E. (1993): Turbulence amplification by a shock wave and rapid distortion theory, Phys. Fluids A **5**(10), 2539–2550
[132] Jarrin, N., Benhamadouche, S., Laurence, D., Prosser, R. (2006): A synthetic eddy method for generating inflow conditions for large-eddy simulations, Int. J. Heat Fluid Flow **27**, 585–593
[133] von Kaenel, R., Adams, N. A., Kleiser, L., Vos, J. B. (2003): The approximate deconvolution model for large-eddy simulation of compressible flows with finite volume schemes, J. Fluids Eng. **125**, 375–381
[134] von Kaenel, R., Adams, N. A., Kleiser, L., Vos, J. B. (2003): Effects of artificial dissipation on large eddy simulation with deconvolution modeling, AIAA J. **41**(8), 1606–1609
[135] von Kaenel, R., Kleiser, L., Adams, N. A., Vos, J. B. (2004): Large-eddy simulation of shock-turbulence interaction, AIAA J. **42**, 2516–2528
[136] Kaneda, Y. (2007): Lagrangian renormalized approximation of turbulence, Fluid Dyn. Res. **39**, 526–551
[137] Kawamura, T., Kuwahara, K. (1984): Computation of high Reynolds number flow around a circular cylinder with surface roughness, AIAA-paper 84-0340
[138] Kevlahan, N., Mahesh, K., Lee, S. (1992): Evolution of the shock front and turbulence structures in the shock/turbulence interaction. In: Proceedings of the Summer Program, CTR, Stanford University
[139] Kida, S., Orszag, S. A. (1990): Energy and spectral dynamics in forced compressible turbulence, J. Sci. Comput. **5**(2), 85–125
[140] Kida, S., Orszag, S. A. (1990): Enstrophy budget in decaying compressible turbulence, J. Sci. Comput. **5**(1), 1–34
[141] Kida, S., Orszag, S. A. (1992): Energy and spectral dynamics in decaying compressible turbulence, J. Sci. Comput. **7**(1), 1–34

[142] Kim, J., Moin, P., Moser, R. (1987): Turbulence statistics in fully developed channel flow at low Reynolds number, J. Fluid Mech. **177**, 133–166

[143] Knight, D., Zhou, G., Okong'o, N., Shukla, V. (1998): Compressible large-eddy simulation using unstructured grids, AIAA-Paper 98-0535

[144] Knight, D. D., Yan, H., Panaras, A. G., Zheltovodov, A. A. (2003): Advances in CFD prediction of shock wave turbulent boundary layer interactions, Prog. Aerospace Sci. **39**, 121–184

[145] Koobus, B., Farhat, C. (2004): A variational multiscale method for the large eddy simulation of compressible turbulent flows on unstructured meshes—application to vortex shedding, Comput. Methods Appl. Mech. Eng. **193**, 1367–1383

[146] Kontorovich, V. M., (1957): To the question on stability of shock waves Sov. Phys. JETP **6**, 1179 [Atomic Research Agency Establishment AERE Lib./Trans. 648 (1956)]

[147] Kosović, B., Pullin, D. I., Samtaney, R. (2002): Subgrid-scale modeling for large eddy simulation of compressible turbulence, Phys. Fluids **14**(4), 1511–1522

[148] Kovasznay, L. S. G. (1953): Turbulence in supersonic flow, J. Aerospace Sci. **20**, 657–682

[149] Kraichnan, R. H. (1970): Diffusion by a random velocity field, J. Comput. Phys. **13**(1), 22–31

[150] Kravchenko, A. G. Moin, P., (1997): On the effect of numerical errors in large eddy simulations of turbulent flows, J. Comput. Phys. **131**, 310–322

[151] Landau, L. D., Lifshitz, E. M. (1987): Fluid Mechanics, 2nd edn., Course of Theoretical Physics, vol. 6, Butterworth-Heinemann, Stoneham

[152] Larchevêque, L., Sagaut, P., Mary, I., Labbé, O. (2003): Large-eddy simulation of a compressible flow past a deep cavity, Phys. Fluids **15**(1), 193–210

[153] Larchevêque, L., Sagaut, P., Lê, T.-H., Comte, P. (2004): Large-eddy simulation of a compressible flow in a three dimensional open cavity at high Reynolds number, J. Fluid Mech. **516**, 265–301.

[154] Larchevêque, L., Sagaut, P., Labbé, O. (2007): Large-eddy Simulation of a subsonic cavity flow including asymmetric three-dimensional effects, J. Fluid Mech. **577**, 105–126

[155] Lax, P. D., Wendroff, B. (1960): Systems of conservation laws, Commun. Pure Appl. Math. **10**, 537–566

[156] Layton, W., Neda, M. (2007): A similarity theory of approximate deconvolution models of turbulence, J. Math. Anal. Appl. **333**, 416–429

[157] Lee, S., Lele, S. K., Moin, P. (1991): Eddy shocklets in decaying compressible turbulence, Phys. Fluids A **3**, 657–664

[158] Lee, S. (1992): Large eddy simulation of shock-turbulence interaction, Center for Turbulence Research, Annual Research Briefs, Stanford University, pp. 73–84

[159] Lee, S., Lele, S. K., Moin, P. (1993): Direct numerical simulation of isotropic turbulence interacting with a weak shock wave, J. Fluid. Mech. **251**, 533–562

[160] Lee, S., Lele, S. K., Moin, P. (1997): Interaction of isotropic turbulence with shock waves: effect of shock strength, J. Fluid Mech. **340**, 225–247

[161] Lee, M. Y., Kawamura, T., Kuwahara, K. (2004): Computation of compressible fluid flow using the multi-directional method, Theor. Appl. Mech. Jpn. **53**, 223–228

[162] van Leer, B. (1979): Towards the ultimate conservative difference scheme, V: a second order sequel to Godunov's method, J. Comput. Phys. **32**, 101–136

[163] Leonard, A. (1974): Energy cascade in large eddy simulations of turbulent fluid flows, Adv. Geophys. **18**, 237–248

[164] Leonard, S., Terracol, M., Sagaut, P. (2007): Commutation error in LES with time-dependent filter width, Comput. Fluids **36**, 513–519

[165] Lele, S. K. (1992): Shock-jump relations in a turbulent flow, Phys. Fluids A **4**(12), 2900–2905

[166] Lele, S. K. (1992): Compact finite difference schemes with spectral-like resolution, J. Comput. Phys. **103**, 16–42

[167] Lele, S. (1994): Compressibility effects on turbulence, Annu. Rev. Fluid Mech. **26**, 211–254

[168] Lenormand, E., Sagaut, P., Ta Phuoc, L., Comte, P. (2000): Subgrid-scale models for large-eddy simulation of compressible wall bounded flows, AIAA J. **38**(8), 1340–1350

[169] Lenormand, E., Sagaut, P., Ta Phuoc, L. (2000): Large eddy simulation of subsonic and supersonic channel flow at moderate Reynolds number, Int. J. Numer. Methods Fluids **32**, 369–406

[170] Liu, J., Oran, E. S., Kaplan, C. R. (2005): Numerical diffusion in FCT algorithm revisited, J. Comput. Phys. **208**, 416–434

[171] Liu, X. D., Osher, S., Chan, T. (1994): Weighted essentially non-oscillatory schemes, J. Comput. Phys. **115**, 200–212

[172] Lesieur, M., Métais, O. (1996): New trends in large-eddy simulations of turbulence, Annu. Rev. Fluid Mech. 28, 45–82.

[173] Lesieur, M. (2007): Turbulence in Fluids, 4th edn., Springer, Berlin

[174] Lesieur, M., Comte, P. (2001): Favre filtering and macro-temperature in large-eddy simulations of compressible turbulence, C. R. Acad. Sci., Sér. IIb **329**, 363–368

[175] Leslie, D. C. (1973): Developments in the Theory of Turbulence, Clarendon, Oxford

[176] Liepmann, H. W. (1946): The interaction between boundary layer and shock waves in transonic flow, J. Aeronaut. Sci. **13**, 623–637

[177] Lighthill, J. (1978): Waves in Fluids, Cambridge Mathematical Library, Cambridge

[178] Lilly, D. K. (1992): A proposed modification of the Germano subgrid-scale closure method, Phys. Fluids A **4**, 633–635

[179] Liu, S., Meneveau, C., Katz, J. (1994): On the properties of similarity subgrid-scale models as deduced from measurements in a turbulent jet, J. Fluid Mech. **275**, 83–119

[180] Loginov, M. (2006): Large-eddy simulation of shock-wave/turbulent boundary layer interaction, PhD Thesis, Technische Universität München, Garching, Germany

[181] Loginov, M. S., Adams, N. A., Zheltovodov, A. A. (2006): Large-eddy simulation of shock-wave/turbulent-boundary-layer interaction, J. Fluid Mech. **565**, 135–169

[182] Lubchich, A. A., Pudovkin, M. I. (2004): Interaction of small perturbations with shock waves, Phys. Fluids **16**(12), 4489–4505

[183] Lund, T. S., Wu, X., Squires, K. D. (1998): Generation of turbulent inflow data for spatially developing boundary layer simulations, J. Comput. Phys. **140**, 233–258

[184] Lundgren, T. S. (1982): Strained spiral vortex model for turbulence fine structure, Phys. Fluids **25**, 2193–2203

[185] Mahesh, K., Lee, S., Lele, S. K., Moin, P. (1995): The interaction of an isotropic field of acoustic waves with a shock wave, J. Fluid Mech. **300**, 383–407

[186] Mahesh, K., Moin, P., Lele, S. K. (1996): The interaction of a shock wave with a turbulent shear flow, Report No. TF-69, Department of Mechanical Engineering, Stanford University

[187] Mahesh, K., Lele, S. K., Moin, P. (1997): The influence of entropy fluctuations on the interaction of turbulence with a shock wave, J. Fluid Mech. **334**, 353–379

[188] Maidi, M., Lesieur, M., Métais, O. (2006): Vortex control in large-eddy simulations of compressible round jets, J. Turbul. **7**(49)

[189] Marion, J. D., Bertoglio, J. P., Cambon, C., Mathieu, J. (1988): Spectral study of weakly compressible turbulence, part II: EDQNM, C. R. Acad. Sci. Paris, Sér. II **307**, 1601–1606

[190] Marion, J. D., Bertoglio, J. P., Cambon, C. (1988): Two-point closures for compressible turbulence. In: 11th Int. Symp. on Turb., Rolla (MO), 17–19 October 1988. Preprints (A-89-33402 13-14) University of Missouri-Rolla, B22-1–B22-8

[191] Marshall, T. A., Dolling, D. S. (1992): Computation of turbulent, separated, unswept compression ramp interactions, AIAA J. **30**, 2056–2065

[192] Martin, M. P., Piomelli, U., Candler, G. (2000): Subgrid-scale models for compressible large eddy simulations, Theor. Comput. Fluid Dyn. **13**, 361–376

[193] Mary, I., Sagaut, P. (2002): LES of a flow around an airfoil with stall, AIAA J. **40**, 1139–1145

[194] Mathew, J., Lechner, R., Foysi, H., Sesterhenn, J., Friedrich, R. (2003): An explicit filtering method for large eddy simulation of compressible flows, Phys. Fluids **15**(8), 2279–2289

[195] Mathew, J., Foysi, H., Friedrich, R. (2006): A new approach to LES based on explicit filtering, Int. J. Heat Fluid Flow **27**(4), 594–602

[196] Métais, O., Lesieur, M. (1992): Spectral large-eddy simulation of isotropic and stably stratified turbulence, J. Fluid Mech. **256**, 157–194

[197] Meyers, J., Sagaut, P. (2006): On the model coefficients for the standard and the variational multiscale Smagorinsky model, J. Fluid Mech. **569**, 287–319

[198] Meyers, J., Geurts, B. J., Sagaut, P. (2007): A computational error-assessment of central finite-volume discretizations in large-eddy simulation using a Smagorinsky model, J. Comput. Phys. **227**, 156–173

[199] Misra, A., Pullin, D. I. (1997): A vortex-based subgrid stress model for large eddy simulation, Phys. Fluids **9**, 2443–2454

[200] Miura, H., Kida, S. (1995): Acoustic energy exchange in compressible turbulence, Phys. Fluids **7**(7), 1732–1742

[201] Moin, P., Squires, K., Cabot, W., Lee, S. (1991): A dynamic subgrid-scale model for compressible turbulence and scalar transport, Phys. Fluids A **3**, 2746–2757

[202] Moin, P., Mahesh, K. (1998): Direct numerical simulation: a tool in turbulence research, Annu. Rev. Fluid Mech. **30**, 539–578

[203] Moore, F. K. (1954): Unsteady oblique interaction of a shock wave with a plane disturbance. Technical report 2879, NACA

[204] Morris, P. J., Long, L. N., Scheidegger, T. E., Boluriaan, S. (2002): Simulations of supersonic jet noise, Int. J. Aeroacoust. **1**(1), 17–41

[205] Moser, R., Kim, J., Mansour, N. N. (1999): Direct numerical simulation of turbulent channel flow up to $Re_\tau = 590$, Phys. Fluids **11**, 943–945
[206] Mossi, M., Sagaut, P. (2003): Numerical investigation of fully developed channel flow using shock-capturing schemes, Comput. Fluids **32**, 249–274
[207] Morkovin, M. V. (1961): Effects of compressibility on turbulent flows. In: Favre, A. (ed.), Mécanique de la Turbulence, 367–380, CRNS, Paris
[208] Nicoud, F., Ducros, F. (1999): Subgrid stress modeling based on the square of the velocity gradient tensor, Flow Turbul. Combust. **62**(3), 183–200
[209] Nikitin, N. (2007): Spatial periodicity of spatially evolving turbulent flow caused by inflow boundary condition, Phys. Fluids **19**, 091703
[210] Orszag, S. A. (1977). Lectures on the statistical theory of turbulence. In Balian, R., Peube, J. L. (eds.), Fluid Dynamics, pp. 235–374, Gordon and Breach, London
[211] Pamies, M., Weiss, P. E., Garnier, E., Deck, S., Sagaut, P. (2009): Generation of synthetic turbulent inflow data for large-eddy simulation of spatially-evolving wall-bounded flows, Phys. Fluids **21**, 045103
[212] Pantano, C., Sarkar, S. (2002): A study of compressibility effects in the high-speed turbulent shear layer using direct simulation, J. Fluid Mech. **451**, 329–371
[213] Pascarelli, A., Piomelli, U., Candler, G. V. (2000): Multi-block large-eddy simulations of turbulent boundary layers, J. Comput. Phys. **157**, 256–279
[214] Passot, T., Pouquet, A. (1987): Numerical simulation of compressible homogeneous flows in the turbulent regime, J. Fluid Mech. **181**, 441–466
[215] Pierce, A. D. (1989): Acoustics. An Introduction to Its Physical Principles and Applications, American Institute of Physics, New York
[216] Piomelli, U. (1999): Large-eddy simulation: achievements and challenges, Prog. Aerospace Sci. **35**, 335–362
[217] Pirozzoli, S., Grasso, F., Gatski, T. B. (2004): Direct numerical simulation and analysis of a spatially evolving supersonic boundary layer at $M = 2.25$, Phys. Fluids **16**(3), 530–545
[218] Pirozzoli, S., Grasso, F. (2004): Direct numerical simulations of isotropic compressible turbulence: influence of compressibility on dynamics and structure, Phys. Fluids **16**(12), 4386–4407
[219] Plotkin, K. J. (1975): Shock wave oscillations driven by turbulent boundary layer fluctuations, AIAA J. **13**, 1036–1040
[220] Poinsot, T., Veynante, D., (2005): Theoretical and Numerical Combustion, 2nd edn., Edwards, Ann Arbor
[221] Porter, D. H., Pouquet, A., Woodward, P. R. (1992): Three-dimensional supersonic homogeneous turbulence: a numerical study, Phys. Rev. Lett. **68**(21), 3156–3159
[222] Porter, D. H., Pouquet, A., Woodward, P. R. (1992): A numerical study of supersonic turbulence, Theor. Comput. Fluid Dyn. **4**, 13–49
[223] Porter, D. H., Pouquet, A., Woodward, P. R. (1994): Kolmogorov-like spectra in decaying three-dimensional supersonic flows, Phys. Fluids **6**(6), 2133–2142
[224] Porter, D. H., Woodward, P. R., Pouquet, A. (1998): Inertial range structures in decaying compressible turbulent flows, Phys. Fluids **10**(1), 237–245
[225] Pullin, D. I. (2000): A vortex-based model for the subgrid flux of a passive scalar, Phys. Fluids **12**, 2311–2319

[226] Ragab, S. A., Sheen, S.-C., Sreedhar, M. (1992): An investigation of finite-difference methods for large-eddy simulation of a mixing layer, AIAA paper 92-0554
[227] Razafindralandy, D., Hamdouni, A. (2008): Private communication
[228] Rai, M. M., Gatski, T. B., Erlebacher, G. (1995): Direct simulation of spatially evolving compressible turbulent boundary layer, AIAA paper 95-0583
[229] Reijasse, P., Corbel, B., Soulevant, D. (1999): Unsteadiness and asymmetry of shock induced separation in a planar two-dimensional nozzle: a flow description: AIAA Paper 99-3684
[230] Ribner, H. S. (1953): Convection of a pattern of vorticity through a shock wave. Technical report 1164, NACA
[231] Richtmyer, R. D., Morton, K. W. (1967): Difference Method for Initial-Value Problems, 2nd edn., Wiley, New York. Reprint by Krieger Publishing, 1994
[232] Rider, W. J. (2006): The relationship of MPDATA to other high-resolution methods, Int. J. Numer. Methods Fluids **50**, 1145–1158
[233] Rizetta, D. P., Visbal, M. R., Gaitonde, D. V. (2001): Large-eddy simulation of supersonic compression ramp flow by high-order method, AIAA J. **39**, 2283–2292
[234] Rizetta, D. P., Visbal, M. R. (2002): Application of large-eddy simulation to supersonic compression ramps, AIAA J. **40**, 1574–1581
[235] Rizzetta, D. P., Visbal, M. R. (2003): Large-eddy simulation of supersonic cavity flowfields including flow control, AIAA J. **41**(8), 1452–1462
[236] Rosenau, P. (1989): Extending hydrodynamics via the regularization of the Chapman-Enskog expansion, Phys. Rev. A **40**, 7193
[237] Sadiki, A., Bauer, W., Hutter, K. (2000): Thermodynamically consistent coefficient calibration in nonlinear and anisotropic closure models for turbulence, Contin. Mech. Thermodyn. **12**, 131–149
[238] Sadiki, A., Hutter, K. (2000): On thermodynamics of turbulence: development of first order closure models and critical evaluation of existing models, J. Non-Equilib. Thermodyn. **25**, 131–160
[239] Sagaut, P. (1995): Simulations numériques d'écoulement décollés avec des modèles de sous-maille, Thèse de doctorat de l'université Paris VI
[240] Sagaut, P., Grohens, R. (1999): Discrete filters for large eddy simulation, Int. J. Numer. Methods Fluids **31**, 1195–1220
[241] Sagaut, P., Comte, P., Ducros, F. (2000): Filtered subgrid-scale models, Phys. Fluids **12**(1), 233–236
[242] Sagaut, P., Garnier, E., Tromeur, E., Larchevêque, L., Labourasse, E. (2004): Turbulent inflow conditions for LES of subsonic and supersonic wall-bounded flows, AIAA J. **42**(3), 469–478
[243] Sagaut, P., Germano, M. (2005): On the filtering paradigm for LES of flows with discontinuities, J. Turbul. **6**, 23
[244] Sagaut, P. (2006): Large-Eddy Simulation for Incompressible Flows, 3rd edn., Springer, Berlin
[245] Sagaut, P., Cambon, C. (2008): Homogeneous Turbulence Dynamics, Cambridge University Press, Cambridge
[246] Salvetti, M. V., Beux, F. (1998): The effect of the numerical scheme on the subrid scale term in large-eddy simulation, Phys. Fluids **10**, 3020–3022
[247] Samtaney, R., Pullin, D. I., Kosovic, B. (2001): Direct numerical simulation of decaying compressible turbulence and shocklets statistics, Phys. Fluids **13**(5), 1415–1430

[248] Sandham, N. D., Yao, Y. F., Lawal, A. A. (2003): Large-eddy simulation of transonic turbulent flow over a bump, Int. J. Heat Fluid Flow **24**, 584–595

[249] Sarkar, S., Erlebacher, G., Hussaini, M. Y., Kreiss, H. O. (1991): The analysis and modelling of dilatational terms in compressible turbulence, J. Fluid Mech. **227**, 473–493

[250] Schlatter, P., Stolz, S., Kleiser, L. (2004): LES of transitional flows using the approximate deconvolution model, Int. J. Heat Fluid Flow **25**, 549–558

[251] Schlichting, H., Gersten, K., Boundary Layer Theory, 8th edn., Springer

[252] Schröder, W., Meinke, M., Ewert, R., El-Askary, W., (2001). LES of a turbulent flow around a sharp trailing edge. In: Direct and Large-Eddy Simulation—IV, pp. 353–363, Kluwer Academic, Dordrecht

[253] Schumann, U. (1975): Subgrid scale model for finite difference simulations of turbulent flows in plane channels and annuli, J. Comput. Phys. **18**, 376–404

[254] Scotti, A., Meneveau, C., Lilly, D. K. (1993): Generalized Smagorinsky model for anisotropic grids, Phys. Fluids A **5**, 2306–2308

[255] Mathey, F., Cokjalt, D., Bertoglio, J. P., Sergent, E. (2006): Assessment of the vortex method for the LES inlet conditions, Prog. Comput. Fluid Dyn. **6**, 58–67

[256] Settles, G. S., Fitzpatrick, T. J., Bogdonoff, S. M. (1979): Detailed study of attached and separated compression corner flowfields in high Reynolds number supersonic flow, AIAA J. **17**, 579–585

[257] Simon, F., Deck, S., Guillen, P., Sagaut, P., Merlen, A. (2007): Numerical simulation of the compressible mixing layer past an axisymmetric trailing edge, J. Fluid Mech. **591**, 215–253

[258] Simone, A., Coleman, G. N., Cambon, C. (1997): The effect of compressibility on turbulent shear flow: a RDT and DNS study, J. Fluid Mech. **330**, 307–338

[259] Singh, A., Chatterjee, A. (2007): Numerical prediction of supersonic jet screech frequency, Shock Waves **17**, 263–272

[260] Shao, L., Sarkar, S., Pantano, C. (1999): On the relationship between the mean flow and subgrid stresses in large eddy simulation of turbulent shear flows, Phys. Fluids. **11**, 1229–1248

[261] Shur, M. L., Spalart, P. R., Strelets, M. Kh. (2005): Noise prediction for increasingly complex jets, part I: methods and tests, Int. J. Aeroacoust. **4**(3&4), 213–246

[262] Shur, M. L., Spalart, P. R., Strelets, M. Kh. (2005): Noise prediction for increasingly complex jets, part II: applications, Int. J. Aeroacoust. **4**(3&4), 247–266

[263] Shu, C.-W. (1997): Essentially non-oscillatory and weighted essentially non-oscillatory schemes for hyperbolic conservation laws, Technical Report 97-65, ICASE, NASA Langley Research Center, Hampton, VA

[264] Smagorinsky, J. (1963): General circulation experiments with the primitive equations, Mon. Weather Rev. **91**, 99–164

[265] Smagorinsky, J. (1993). Some historical remarks on the use of nonlinear viscosities. In: Galperin, B., Orszag, S. A. (eds.), Large Eddy Simulation of Complex Engineering and Geophysical Flows, pp. 3–36, Cambridge University Press, Cambridge

[266] Smits, A. J., Dussauge, J. P. (2006): Turbulent Shear Layers in Supersonic Flows, 2nd edn., Springer, Berlin

[267] Smits, A. J., Muck, K. C. (1987): Experimental study of three shock wave/turbulent boundary layer interactions, J. Fluid Mech. **182**, 291–314

[268] Smolarkiewicz, P. K., (1983): A simple positive definite advection scheme with small implicit diffusion, Mon. Weather Rev. **111**, 479–486

[269] Smolarkiewicz, P. K. (1984): A fully multidimensional positive definite advection transport algorithm with small implicit diffusion, J. Comput. Phys. **54**, 325–362

[270] Smolarkiewicz, P. K. (2006): Multidimensional positive definite advection transport algorithm: an overview, Int. J. Numer. Methods Fluids **50**, 1123–1144

[271] Smolarkiewicz, P. K., Margolin, L. G. (1998): MPDATA: a finite-difference solver for geophysical flows, J. Comput. Phys. **140**, 459–480

[272] Smolarkiewicz, P. K., Szmelter, J. (2008): An MPDATA-based solver for compressible flows, Int. J. Numer. Methods Fluids **56**, 1529–1534

[273] Speziale, C. G., Erlebacher, G., Zang, T. A., Hussaini, M. Y. (1988): The subgrid-scale modeling of compressible turbulence, Phys. Fluids **31**, 940–942

[274] Spyropoulos, E. T., Blaisdell, G.A. (1996): Evaluation of the dynamic model for simulations of compressible decaying isotropic turbulence, AIAA J. **34**(5)

[275] Spyropoulos, E. T., Blaisdell, G. A. (1998): Large-eddy simulation of a spatially evolving supersonic turbulent boundary-layer flow, AIAA J. **32**, 1983–1990

[276] Stolz, S., Adams, N. A. (1999): An approximate deconvolution procedure for large-eddy simulation, Phys. Fluids **11**(7), 1699–1701

[277] Stolz, S., Adams, N. A., Kleiser, L. (2001): An approximate deconvolution model for large-eddy simulations of incompressible flows, Phys. Fluids **13**, 997–1015

[278] Stolz, S., Adams, N. A., Kleiser, L. (2001): The approximate deconvolution model for large-eddy simulations of compressible flows and its application to shock-turbulent-boundary-layer interaction, Phys. Fluids **13**, 2985–3001

[279] Stolz, S., Adams, N. A., Kleiser, L. (2001): The approximate deconvolution model for compressible flows: isotropic turbulence and shock-boundary-layer interaction. In: Friedrich, R., Rodi, W. (eds.), Proceedings of the Euromech Colloquium, vol. 412

[280] Stolz, S., Adams, N. A. (2003): Large-eddy simulation of high Reynolds number supersonic boundary layers using the approximate deconvolution model and a rescaling and recycling technique, Phys. Fluids **15**(8), 2398–2412

[281] Stolz, S., Schlatter, P., Kleiser, L. (2005): High-pass filtered eddy-viscosity models for large eddy simulations of transitional and turbulent flows, Phys. Fluids **17**, 065103

[282] Stolz, S., Schlatter, P., Kleiser, L. (2007): Large eddy simulation of subharmonic transition in a supersonic boundary layer, AIAA J. **45**(5), 1019–1027

[283] Tam, C. K. W., Webb, J. C. (1993): Dispersion-relation-preserving finite difference schemes for computational acoustics, J. Comput. Phys. **107**, 262–281

[284] Tam, C. W. K. (1995): Supersonic jet noise, Annu. Rev. Fluid Mech. **27**, 17–43

[285] Tanna, H. K. (1977): An experimental study of jet noise, part I: turbulent mixing noise, J. Sound Vib. **50**, 405

[286] Teramoto, S. (2005): Large-eddy simulation of transitional boundary layer with impinging shock wave, AIAA J. **43**, 2354–2363

[287] Terracol, M., Sagaut, P., Basdevant, C. (2001): A multilevel algorithm for large eddy simulation of turbulent compressible flows, J. Comput. Phys. **167**, 439–474

[288] Terracol, M., Sagaut, P., Basdevant, C. (2003): A time self-adaptive multilevel algorithm for large-eddy simulation, J. Comput. Phys. **184**, 339–365

[289] Thornber, B., Drikakis, D. (2008): Implicit large-eddy simulation of a deep cavity using high-resolution methods, AIAA J. **46**(10), 2634–2645

[290] Toro, E. F. (1997): Riemann Solvers and Numerical Methods for Fluid Dynamics, Springer, Berlin

[291] Tromeur, E., Garnier, E., Sagaut, P., Basdevant, C. (2003): Large eddy simulations of aero-optical effects in a turbulent boundary layer, J. Turbul. **4**(5)

[292] Tromeur, E., Garnier, E., Sagaut, P. (2006): Large-eddy simulation of aero-optical effects in a spatially developing turbulent boundary layer, J. Turbul. **7**(1)

[293] Tromeur, E., Garnier, E., Sagaut, P. (2006): Analysis of the Sutton model for aero-optical properties of compressible boundary layers, J. Fluid Eng. **128**(2), 239–246

[294] Urbin, G., Knight, D., Zheltovodov, A. A. (1995): Compressible large eddy simulation using unstructured grid—supersonic turbulent boundary layer and compression corner, AIAA paper 99-0427

[295] Urbin, G., Knight, D. (2001): Large-eddy simulation of a supersonic boundary layer using an unstructured grid, AIAA J. **39**(7), 1288–1295

[296] Uzun, A., Lyrintzis, A. S., Blaisdell, G. A. (2005): Coupling of integral acoustics methods with LES for jet noise prediction, Int. J. Aeroacoust. **4**(3), 297–346

[297] Uzun, A., Blaisdell, G. A., Lyrintzis, A. S. (2006): Impact of subgrid-scale models on jet turbulence and noise, AIAA J. **44**(6), 1365–1368.

[298] van der Bos, F., Van der Vegt, J. J. W., Geurts, B. J. (2007): A multi-scale formulation for compressible turbulent flows suitable for general variational discretization techniques, Comput. Methods Appl. Mech. Eng. **196**, 2863–2875

[299] Vasilyev, O. V., Lund, T. S., Moin, P. (1998): A general class of commutative filters for LES in complex geometries, J. Comput. Phys. **146**, 82–104

[300] van der Veen, H. (1995): A family of large eddy simulation (LES) filters with nonuniform filter widths, Phys. Fluids **7**, 1171–1172

[301] Vichnevetsky, R., Bowles, J. (1982) Fourier Analysis of Numerical Approximations of Hyperbolic Equations. SIAM Studies in Applied Mathematics, SIAM, Philadelphia

[302] Vincent, A., Meneguzzi, M. (1994): The dynamics of vorticity tubes in homogneous turbulence, J. Fluid Mech. **258**, 245–254

[303] Visbal, M. R., Rizzetta, D. P. (2002): Large-eddy simulation on curvilinear grids using compact differencing and filtering schemes, J. Fluids Eng. **124**, 836–847

[304] Visbal, M. R., Morgan, S. P., Visbal, D. P. (2003): An implicit LES approach based on high order compact differencing and filtering schemes, AIAA Paper 2003-4098

[305] Vreman, A. W., Geurts, B. J., Kuerten, J. G. M., Zandbergen, P. J. (1992): A finite volume approach to large eddy simulation of compressible, homogeneous, isotropic, decaying turbulence, Int. J. Numer. Methods Fluids **15**, 799–816

[306] Vreman, B., Geurts, B., Kuerten, H. (1995): A priori tests of large eddy simulation of the compressible plane mixing layer, J. Eng. Math. **29**, 299–327

[307] Vreman, A. W., (1995): Direct and Large Eddy Simulation of the compressible turbulent mixing layer, PhD Dissertation, University of Twente, Twente.
[308] Vreman, A. W., Geurts, B. J., Kuerten, H. (1995): Subgrid-modelling in LES of compressible flow, Appl. Sci. Res. **54**, 181–203
[309] Vreman, B., Geurts, B., Kuerten, H. (1997): Large-eddy simulation of the turbulent mixing layer, J. Fluid Mech. **339**, 357–390
[310] Wagner, C., Hüttl, T., Sagaut, P. (2007): Large-Eddy Simulation for Acoustics. Cambridge Aerospace Series
[311] Warming, R. F., Hyett, B. J. (1974): The modified equation approach to the stability and accuracy analysis of finite difference methods, J. Comput. Phys. **14**, 159–179
[312] Wasistho, B., Balachandar, S., Moser, R. D. (2004): Compressible wall-injection in laminar, transitional and turbulent regimes: numerical prediction, J. Spacecr. Rockets **41**(6), 915–924
[313] Wilcox, D. C. (1994): Turbulence Modeling for CFD, DCW Industries, Inc., La Cañada
[314] Wollblad, C., Davidson, L., Eriksson, L.-E. (2006): Large eddy simulation of transonic flow with shock wave/turbulent boundary layer interaction, AIAA J. **44**(10), 2340–2353
[315] Woodward, P. R., Colella, P. (1984): The numerical solution of two-dimensional fluid flow with strong shocks, J. Comput. Phys. **54**, 115–173
[316] Xu, S., Pino Martin, M. (2004): Assessment of inflow boundary conditions for compressible turbulent boundary layers, Phys. Fluids **16**(7), 2623–2639
[317] Yanenko, N. N., Shokin, Y. I. (1969): First differential approximation method and approximate viscosity of difference schemes, Phys. Fluids **12**, II-21–II-28
[318] Yan, H., Knight, D., Zheltovodov, A. A. (2001): Large eddy simulation of supersonic compression corner using ENO scheme. In: Liu, C., Sakell, L., Beutner, T. (eds.), DNS/LES Progress and Challenges
[319] Yan, H., Knight, D., Zeltovodov, A. A. (2002): Large-eddy simulation of supersonic flat-plate boundary layers using the monotonically integrated large-eddy simulation (MILES) technique, J. Fluids Eng. **124**, 868–875
[320] Yoshizawa, A. (1986): Statistical theory for compressible turbulent shear flows, with the application to subgrid modeling, Phys. Fluids **29**, 2152–2164
[321] Zalesak, S. T. (1979): Fully multidimensional flux-corrected transport algorithms for fluids, J. Comput. Phys. **31**, 335–362
[322] Zandonade, P. S., Langford, J. A., Moser, R. D. (2004): Finite-volume optimal large-eddy simulation of isotropic turbulence, Phys. Fluids **16**, 2255–2271
[323] Zang, T. A., Dahlburg, R. B., Dalburg, J. P. (1992): Direct and large-eddy simulations of three-dimensional compressible Navier-Stokes turbulence, Phys. Fluids A **4**(1), 127–140
[324] Zang, Y., Street, R. L., Koseff, J. R. (1993): A dynamic mixed subgrid-scale model and its application to turbulent recirculating flows, Phys. Fluids A **5**, 3186–3196
[325] Zank, G. P., Matthaeus, W. H. (1990): Nearly incompressible hydrodynamics and heat conduction, Phys. Rev. Lett. **64**(11), 1243–1245
[326] Zank, G. P., Matthaeus, W. H. (1991): The equations of nearly incompressible fluids, I: hydrodynamics, turbulence and waves, Phys. Fluids **3**(1), 69–82
[327] Zank, P., Zhou, Y., Matthaeus, W. H., Rice, W. K. M. (2002): The interaction of turbulence with shock waves: a basic model, Phys. Fluids **14**(11), 3766–3774

[328] Zel'dovich, Y. B., Raizer, Y. P. (2002): Physics of Shock Waves and High-Temperature Hydrodynamic Phenomena, Dover, New York

[329] Zhao, W., Frankel, S. H., Mongeau, L. (2001): Large eddy simulations of sound radiation from subsonic turbulent jets, AIAA J. **39**(8), 1469–1477

[330] Zheltovodov, A. A., Trofimov, M. V., Schülein, E., Yakovlev, V. N. (1990): An experimental documentation of supersonic turbulent flows in the vicinity of forward- and backward-facing ramps, Technical Report 2030, Institute of Theoretical and Applied Mechanics, USSR Academy of Sciences, Novosibirsk, Russia

Index

E. Garnier et al., *Large Eddy Simulation for Compressible Flows*,
Scientific Computation,

MIX
Papier aus verantwortungsvollen Quellen
Paper from responsible sources
FSC® C105338

If you have any concerns about our products,
you can contact us on
ProductSafety@springernature.com

In case Publisher is established outside the EU,
the EU authorized representative is:
Springer Nature Customer Service Center GmbH
Europaplatz 3, 69115 Heidelberg, Germany

Printed by Libri Plureos GmbH
in Hamburg, Germany